INHIBITION AND INACTIVATION OF VEGETATIVE MICROBES

THE SOCIETY FOR APPLIED BACTERIOLOGY
SYMPOSIUM SERIES NO. 5

INHIBITION AND INACTIVATION OF VEGETATIVE MICROBES

Edited by

F. A. SKINNER

AND

W. B. HUGO

1976

ACADEMIC PRESS

LONDON . NEW YORK . SAN FRANCISCO
A Subsidiary of Harcourt Brace Jovanovich, Publishers

ACADEMIC PRESS INC. (LONDON) LTD
24-28 OVAL ROAD
LONDON N.W.1

U.S. Edition published by
ACADEMIC PRESS INC.
111 FIFTH AVENUE
NEW YORK, NEW YORK 10003

Copyright © 1976 By the Society for Applied Bacteriology

ALL RIGHTS RESERVED

NO PART OF THIS BOOK MAY BE REPRODUCED IN ANY FORM BY PHOTOSTAT, MICROFILM, OR ANY OTHER MEANS, WITHOUT WRITTEN PERMISSION FROM THE PUBLISHERS

Library of Congress Catalog Card Number 76-22840
ISBN: 0-12-648065-6

Printed in Great Britain by
The Whitefriars Press Ltd., London and Tonbridge, England

Contributors

J. G. ANDERSON, *Department of Applied Microbiology, University of Strathclyde, George Street, Glasgow G1 1XW, Scotland*

I. J. BOUSFIELD, *National Collection of Industrial Bacteria, Torry Research Station, P.O. Box 31, Aberdeen AB9 8DG, Scotland*

CAROLE S. BURKE, *British Food Manufacturing Industries Research Association, Randalls Road, Leatherhead, Surrey KT22 7RY, England*

J. G. CARR, *Long Ashton Research Station, Long Ashton, Bristol BS18 9AF, England*

CHRISTINA M. COUSINS, *National Institute for Research in Dairying, Shinfield, Reading RG2 9AT, England*

R. DAVIES, *Department of Microbiology, National College of Food Technology, St George's Avenue, Webridge, Surrey KT13 ODE, England*

I. W. DAWES, *Department of Microbiology, University of Edinburgh, Edinburgh EH9 3JG, Scotland*

G. J. DRING, *Unilever Research, Colworth House, Sharnbrook, Bedford, England*

S. A. HAMMOND, *Long Ashton Research Station, Long Ashton, Bristol BS18 9AF, England*

W. B. HUGO, *Department of Pharmacy, The University of Nottingham, University Park, Nottingham NG7 2RD, England*

M. INGRAM, *ARC Meat Research Institute, Langford, Bristol BS18 7DY, England*

B. JARVIS, *British Food Manufacturing Industries Research Association, Randalls Road, Leatherhead, Surrey KT22 7RY, England*

L. LEISTNER, *Bundesanstalt für Fleischforschung, Institut für Bakteriologie und Histologie, 8650 Kulmbach, Blaich 4, West Germany*

A. R. MacKENZIE, *National Collection of Industrial Bacteria, Torry Research Station, P.O. Box 31, Aberdeen AB9 8DG, Scotland*

B. M. MACKEY, *ARC Meat Research Institute, Langford, Bristol BS18 7DY, England*

S. A. MALCOLM, *Unilever Research Laboratory, 455 London Road, Isleworth, Middlesex TW4 5AB, England*

Z. J. ORDAL, *Departments of Food Science & Microbiology, University of Illinois, Urbana, Illinois 61801, U.S.A.*

B. REITER, *National Institute for Research in Dairying, Shinfield, Reading RG2 9AT, England*

W. RÖDEL, *Bundesanstalt für Fleischforschung, Institut für Bakteriologie und Histologie, 8650 Kulmbach, Blaich 4, West Germany*

CONTRIBUTORS

A. D. RUSSELL, *Welsh School of Pharmacy, UWIST, King Edward VII Avenue, Cardiff CF1 3NU, Wales*

J. E. SMITH, *Department of Applied Microbiology, University of Strathclyde, George Street, Glasgow G1 1XW, Scotland*

R. I. TOMLINS, *Department of Applied Biology & Food Science, Polytechnic of the South Bank, Borough Road, London SE1 0AA, England*

M. VAN SCHOTHORST, *Rijks Instituut voor de Volksgezondheid, Antonie van Leeuwenhoeklaan 9, Postbus 1, Bilthoven, The Netherlands*

Preface

A SYMPOSIUM on the 'Inhibition and Inactivation of Vegetative Microbes' was organized by the Society for Applied Bacteriology in collaboration with the North West European Microbiological Group and held during the Summer Conference of the Society in the University of Nottingham, July 1975. The papers presented, and which comprise this volume, reviewed some of the fundamental principles of microbial inactivation and showed how they could be applied in practice, especially in the food industry.

It is surprising that after a century of non-empirical disinfection practice many problems concerning the inhibition of microbes remain to afflict industry. The solutions of these problems are costly and difficult, especially when it is necessary to inhibit or destroy micro-organisms in products to which drastic disinfection procedures cannot be applied. These difficulties reflect the remarkable ability of microbes to resist stress and to survive or grow in bizarre situations.

Fundamental inhibitory or destructive processes, heat, cold, toxic gases and other chemicals, ionizing radiation, hydrostatic pressure and decreased water activity are covered in the earlier chapters. Specialist areas of destructive and inhibitory application relate to foods in general with chapters on beverages, the dairy industry and the use of resuscitation techniques to recover bacteria from foodstuffs. Other contributions deal with the survival of bacteria in toiletries, freeze-drying and legislative aspects of the use of chemical preservatives in foodstuffs.

It is hoped that the presentation of this material may make a useful contribution to the study of microbial inactivation.

F. A. SKINNER
Rothamsted Experimental Station
Harpenden AL5 2JQ
Hertfordshire
England

W. B. HUGO
Department of Pharmacy
The University of Nottingham
Nottingham NG7 2RD
England

December, 1976

Contents

LIST OF CONTRIBUTORS v

PREFACE vii

The Inactivation of Vegetative Bacteria by Chemicals
W. B. HUGO
    Introduction 1
    The Bacterial Cell 1
    The Cell Wall as a Target 1
    The Cytoplasmic Membrane as a Target 2
    The Cytoplasm as a Target 6
    Conclusions 8
    References 9

The Inactivation of Vegetative Micro-organisms by Chemicals in the Dairying Industry
CHRISTINA M. COUSINS
    Introduction 13
    Types of Disinfectants 14
    Evaluation of Disinfectants by the Modified Lisbôa Tube Test . 16
    In-use Evaluation of Disinfectants and Methods of Application . 21
    Chlorination of Water Supplies 26
    The Microflora of Milking Equipment after Chemical Disinfection 27
    Resistance of Vegetative Micro-organisms to Dairy Disinfectants . 28
    References 28

Bacterial Inhibitors in Milk and other Biological Secretions, with Special Reference to the Complement/Antibody, Transferrin/Lactoferrin and Lactoperoxidase/Thiocyanate/Hydrogen Peroxide Systems
B. REITER
    Introduction 32
    Complement Mediated Bactericidal Activity of Specific Antibodies in Colostrum and Postcolostral Milk . . . 33

Inhibition of Bacteria by Iron Binding Proteins—Transferrin and Lactoferrin	36
The Nature of the Bactericidal Activity of Complement/Antibody and the Bacteriostatic Activity of Iron Binding Proteins	40
The Lactoperoxidase/Thiocyanate/Hydrogen Peroxide System ($LP/SCN^-/H_2O_2$)	42
Basic Proteins and other Antibacterial Factors	48
Discussion	51
References	54

Inactivation of Non-sporing Bacteria by Gases
A. D. RUSSELL

Introduction	61
Physical Properties of Gaseous Disinfectants	62
Antibacterial Activity	63
Mechanism of Action	72
Practical Uses	78
Conclusions	82
Acknowledgement	83
References	83

The Antimicrobial Activity of SO_2—with Particular Reference to Fermented and Non-fermented Fruit Juices
S. M. HAMMOND AND J. G. CARR

Introduction	89
Uses of SO_2 in Fermented and Non-fermented Fruit Beverages	90
Ionization of SO_2 in Aqueous Solutions	90
Sulphite Addition Compounds in Fruit Beverages	91
The Active Antimicrobial Principle in Sulphur Dioxide Solutions	92
Mechanism of Action of Sulphur Dioxide	94
Difficulties Encountered in Experimentation with SO_2	100
Effects of SO_2 on the Viability of Micro-organisms Present in Fermented Fruit Juices	101
Conclusions	104
Acknowledgements	105
References	106

Inactivation by Cold
M. INGRAM AND B. M. MACKEY

Introduction	111
Minimum Temperature for Growth	112

Subminimal Temperatures above Freezing 121
The Subzero Zone 128
Freezing 135
Some Practical Implications 143
Conclusion 145
References 146

Thermal Injury and Inactivation in Vegetative Bacteria
R. I. TOMLINS AND Z. J. ORDAL
Thermal Resistance 153
Thermal Injury and Recovery 168
References 184

Effect of Temperature on Filamentous Fungi
J. G. ANDERSON AND J. E. SMITH
Introduction 191
Effects of Temperature on Growth 192
Heat Resistance 194
Effects of Temperature on Cell Components and Cell Structure . 196
Effects of Temperature on Morphology 201
References 212

Inhibition of Micro-organisms in Food by Water Activity
L. LEISTNER AND W. RÖDEL
Introduction 219
Tolerance of Micro-organisms to a_w . . . 219
Measurement of a_w of Meats 223
Significance of a_w for Meats 225
References 233

The Inactivation of Vegetative Bacterial Cells by Ionizing Radiation
R. DAVIES
Introduction 239
Radiation Resistance of Wild-type Bacterial Cells . . 240
Development of Radioresistant Mutants . . . 247
Practical Implications 249
References 252

Some Aspects of the Effects of Hydrostatic Pressure on Micro-organisms
G. J. Dring

Introduction	257
The Effects of Hydrostatic Pressure on the Gross Morphology, Cell Integrity and Motility of Bacteria	258
The Interaction of Hydrostatic Pressure with Other Environmental Parameters	261
The Application of Hydrostatic Pressure Treatment to the Inactivation of Micro-organisms in Foods	272
References	275

Inactivation of Yeast
I. W. Dawes

Introduction	279
Physical Methods	280
Chemical and Biological Inhibitors	288
Selective Inactivation	290
Disruption of Yeasts	296
Summary	297
References	298

The Survival of Bacteria in Toiletries
S. A. Malcolm

Introduction	305
Contamination	306
Origins of Contamination	308
Consequences of Contamination	310
Microbiological Standards for Toiletries	311
Conclusion	314
References	314

Resuscitation of Injured Bacteria in Foods
M. van Schothorst

Introduction	317
Factors Involved in Recovery	318
Recovery of Bacteria in Foods	321
Resuscitation Treatments	323
Future Outlook	324
References	325

Inactivation of Bacteria by Freeze-drying
I. J. BOUSFIELD AND A. R. MACKENZIE

Introduction	329
The Freeze-drying Process	329
Lethal Effects of Freeze-drying	330
Protection of Bacteria During and After Freeze-drying	333
Storage	337
Resuscitation	338
Variation Between Organisms	339
References	340

Practical and Legislative Aspects of the Chemical Preservation of Food
B. JARVIS AND CAROLE S. BURKE

Introduction	345
Food Preservative Legislation in Great Britain	346
Legislation on Preservatives in Other Countries	348
Why do We Need Chemical Preservation of Foods?	354
Practical Considerations Affecting the Choice of Chemical Preservative	357
Future Trends in Chemical Preservation of Foods	362
Summary	364
Acknowledgements	365
References to Cited Food Legislation for England and Wales	365
References	367

INDEX 369

The Inactivation of Vegetative Bacteria by Chemicals

W. B. HUGO

Department of Pharmacy, The University, Nottingham, England

CONTENTS

1. Introduction 1
2. The bacterial cell 1
3. The cell wall as a target 1
4. The cytoplasmic membrane as a target 2
   (a) Leakage of small molecular weight substances and K^+ 2
   (b) Membrane enzymes 3
   (c) Attenuation of membrane electrochemical potentials 4
   (d) Vinylglycollic acid as an inhibitor of bacterial membrane transport . . . 6
5. The cytoplasm as a target 6
6. Conclusions 8
7. References 9

1. Introduction

THIS SUBJECT was last reviewed by Hugo (1967) and a further review by the same author will appear in the Proceedings of the 26th Symposium of the Society for General Microbiology, due to be published in April, 1976.

In this paper, the main targets for chemical inactivation will be considered with emphasis on recent work on membrane active substances.

2. The Bacterial Cell

The targets for chemical inactivation both by antibiotic and non-antibiotic substances have been identified as the wall, the cytoplasmic membrane and the cytoplasm.

Targets in the cytoplasmic membrane include the membrane itself and membrane associated enzymes. Within the cytoplasm targets include the ribosomes, the nucleic acids, cytoplasmic enzymes and the cytoplasm as a whole, which may be destroyed unselectively by coagulation.

3. The Cell Wall as a Target

The cell wall is a prime target for a number of important antibiotics of which the best known are the penicillins. However, some reports indicate that some chemical disinfectants cause destruction of the bacterial cell wall.

Pulvertaft & Lumb (1948) made the interesting observation that *Escherichia coli*, staphylococci and streptococci, when exposed to low concentrations of certain antiseptics, underwent lysis in media in which rapid growth was occurring. The antiseptics investigated, with appropriate lytic concentration (%), were: phenol, 0.032; merthiolate, 0.0004; formalin, 0.012; mercuric chloride, 0.0008; sodium hypochlorite, 0.005. The authors thought that the action was due to metabolic disturbance followed by uncontrolled action of lytic enzymes which function normally in cell wall synthesis during growth and cell division, and on the basis of their experiments and in the light of modern knowledge and recent researches, this seems a very reasonable explanation for their results.

Hugo (unpublished) attempted to prepare sphaeroplasts or protoplasts from susceptible cells by allowing lysis to proceed according to the above procedure but in the presence of 0.33 M sucrose. No sphaeroplasts or protoplasts could be detected and it was thought that the exposed cytoplasmic membrane was damaged by the agents used.

Other examples of lysis induced by chemical agents and possibly having the same basic mechanism are given in the work of Delphy & Champsey (1949), Bolle & Kellenberger (1958), Schaechter & Santomassino (1962) and Smith *et al.* (1975).

4. The Cytoplasmic Membrane as a Target

Three main lesions may be induced in the cytoplasmic membrane by chemical disinfectants. These are (a) induction of leakyness so that small molecular weight substances such as amino acids, purines, pyrimidines, sugars and cations, especially K^+, leave the cell, (b) the inhibition of membrane enzymes especially adenosine triphosphatase, and (c) the attenuation of the membrane electrochemical potential set up by the extrusion of H^+ during metabolism.

(a) *Leakage of small molecular weight substances and K^+*

The modern notion of cell membrane structure is of a phospholipid bi-layer in which protein molecules are embedded and is thought to be common to a variety of membranes ranging from the cytoplasmic membrane of prokaryotes to the erythrocyte and to mitochondrial membranes (Singer & Nicholson, 1972). Haemolysis, i.e. disruption of the erythrocyte membrane, is known to be induced by detergent substances (Schulman & Rideal, 1937), and by analogy Kuhn & Bielig (1940) suggested that quaternary ammonium detergents might act on the bacterial cytoplasmic membrane causing damage resulting in cell death. Domag (1935) had earlier identified the cytoplasmic membrane as a target for quaternary ammonium compounds (QAC) and Hotchkiss (1944) showed that both a QAC and hexylresorcinol induced leakage of nitrogen and phosphorus-containing compounds from *Staphylococcus aureus*. Gale & Taylor (1947) found a similar result when phenol, Aerosol OT (Cyanamid of Great Britain) and cetyltrimethylammonium bromide acted on *Staph. aureus* and *Streptococcus faecalis* and declared that this lesion explained the disinfectant action of these two compounds. Salton (1950, 1951) related leakage to 99.99% kill in *Staph. aureus, Bacillus pumilis, E. coli, Strep. faecalis* and *Pseudomonas fluorescens*.

Moving away from the traditional detergents and phenols Hugo & Longworth (1964) demonstrated leakage in *E. coli* and *Staph. aureus* by the diguanidide, chlorhexidine. However, they concluded from parallel studies on cell viability that leakage was associated with bacteriostasis. Woodroffe & Wilkinson (1966) also linked leakage with bacteriostasis in *Staph. aureus* when treated with tetrachlorosalicylanilide (TCS). TCS is also an uncoupling agent; see Section 4(c).

Many other drugs have been shown to promote leakage from bacterial cells and it is very important to relate bacteriostasis or death with leakage using comparable drug concentrations and cell numbers.

Lambert & Hammond (1973) using potassium-sensitive electrodes showed that K^+ efflux is probably the first sign of membrane damage.

Pullman & Reynolds (1965) clearly demonstrated repair to bacterial membranes, rendered leaky by disinfectants, if their action was arrested in time.

It is clear that no hard and fast rules can be postulated about leakage and disinfectant mechanisms. Cells in a non-metabolizing environment and in the presence of sufficient membrane-active substance may be damaged beyond repair and a bactericidal effect will be induced. In other cases, leakage will cause an upset in the delicately balanced environment resulting in bacteriostasis but this process, under appropriate circumstances, may be reversed.

(b) *Membrane enzymes*

Membrane-bound adenosine triphosphatase has been implicated in the utilization of the trans-membrane pH and proton gradient to power energy dependent transport in microbial cells as postulated in Mitchell's chemiosmotic theory; see Section 4(c). Harold *et al.* (1969) showed that chlorhexidine inhibited this enzyme in *Strep. faecalis*. Chlorhexidine was also shown to inhibit the same enzyme in the anaerobic *Clostridium perfringens* (Daltrey & Hugo, 1974).

The membrane-bound electron transport chain of *B. megaterium* was shown to be inhibited by hexachlorophene (Frederick *et al.*, 1974). These workers showed that the inhibition could be reversed by the artificial electron carrier menadione.

Mercuric ion at low concentrations, 10^{-6} M inhibited membrane enzymes—containing thiol groups; this inhibition was reversed by thioglycollate or cysteine (Passow *et al.*, 1961).

(c) *Attenuation of membrane electrochemical potentials*

Mitchell proposed a mechanism whereby the activities of the respiratory chain or oxidative processes at the substrate level, i.e. anaerobic metabolism, generated a gradient of pH (ΔpH) and electrical potential ($\Delta\Psi$) across the bacterial, and also the mitochondrial and chloroplast, membrane. The total electrochemical potential or protonmotive force. (Δp) thus envisaged was expressed thus:

$$\Delta p = \Delta\Psi - Z\Delta pH$$

where $Z = 2.303RT/F$ and is a factor converting ΔpH to electrical units, i.e. the same units as Δp and $\Delta\Psi$. At 37°, $Z = 62$ mV.

This protonmotive force, which expresses itself with the cell interior negative and alkaline with respect to the external milieu, in turn provides energy for the transport of certain substances, e.g. sugars and amino acids into the cell and, working in conjunction with a membrane adenosine triphosphatase (ATPase), the generation of adenosine triphosphate (ATP). For a recent discussion of this hypothesis see Grenville (1969), Mitchell (1972), Simoni & Postma (1975), and for the origin of the membrane potential and its role in transport, Harold & Papineau (1972). Kashket & Wilson (1973) have made an elegant demonstration of the accumulation of thiomethyl-β-D-galactopyranoside by the anaerobe *Strep. lactis* (despite its deprivation of a metabolic energy source) to a concentration 20 times the concentration outside the cell, by exposing the organisms to an external pH environment of 6.0, i.e. artifically creating a pH gradient.

Niven & Hamilton (1974) by ingenious manipulation of the external environment of *Staph. aureus,* have been able to distinguish the components of the total electrochemical potential (protonmotive force) responsible for the uptake of certain amino acids by this organism. Thus, lysine was transported by the membrane potential, $\Delta\Psi$, isoleucine by the total protonmotive force $\Delta p = \Delta\Psi - Z\Delta pH$, while glutamic acid owed its uptake to the pH gradient, $-Z\Delta pH$.

Several workers have shown that a group of chemicals will attenuate the membrane electrochemical potential generated by metabolism and, as a consequence, inhibit anabolic reactions dependent on it. These are the traditional and newer uncoupling agents, some of which are also used as disinfectants.

Classically, 2,4-dinitrophenol falls in this category and its action as an uncoupler and an inhibitor of substrate uptake has been known for at least four decades.

More recently TCS has been shown to inhibit energy-dependent uptake of amino acids and phosphate and the energy-dependent incorporation of lysine and glucose into cellular material. The apparent leakage of amino acids was thought to be due not to a change in permeability but to the inhibition of the driving force for the maintenance of the intracellular pool (Hamilton, 1968).

A direct action on the pH gradient as demonstrated by a facilitated proton uptake resulting in the dissipation of this gradient, has been shown to occur when *Strep. faecalis,* in the appropriate experimental situation, was treated with TCS (Harold & Baarda, 1968). These workers showed that at the same time the energy-dependent uptake of rubidium, phosphate and amino acids was also inhibited.

Hugo & Bloomfield (1971*b*) demonstrated a similar inhibition of energy-dependent substrate uptake with 2,2'-dihydroxy-5,5'-dichloro-diphenyl sulphide (Fentichlor) and *E. coli* and *Staph. aureus.* Later Bloomfield (1974) showed that Fentichlor was able to dissipate the membrane pH gradient in these two organisms.

The first demonstration of a pH gradient in a strict anaerobe, *Cl. perfringens,* was made by Daltrey & Hugo (1974). This gradient was totally dissipated by 2,4-dinitrophenol and partially by 4-ethylphenol and TCS. Phenol and chlorhexidine had no effect.

The experiments described above all refer to the dissipation of a pH gradient, however, it has recently been shown that the other component of the protonmotive force, the membrane potential, $\Delta\Psi$, may be evaluated and its dissipation shown to occur in the presence of uncoupling agents.

It is possible to use the dibenzyldimethylammonium ion (DDA^+) to measure $\Delta\Psi$ in bacteria.

DDA^+ is allowed to equilibrate across the bacterial cytoplasmic membrane and $\Delta\Psi$ calculated from the Nernst equation:

$$\Delta\Psi = \frac{RT}{F} \ln \frac{\text{concentration of } DDA^+ \text{ outside the cell}}{\text{concentration of } DDA^+ \text{ inside the cell}}$$

where R is the gas constant, $8.314 \text{ JK}^{-1} \text{ mol}^{-1}$, T the absolute temperature and F the Faraday constant, $9.65 \times 10^4 \text{ C mol}^{-1}$. $\Delta\Psi$ is the potential difference across the membrane in volts.

Harold & Papineau (1972) have shown that the active accumulation of DDA^+ by metabolizing *Strep. faecalis* ceases upon adding TCS and the accumulated DDA^+ leaves the cells.

It is apparent from the brief account given above that this aspect of membrane biochemistry and the possible involvement of disinfectants in this metabolic function form an area in which our understanding of the mode of action of chemical disinfectants has been most advanced during the past decade. Studies on the mode of action of disinfectants found to be membrane active by the more traditional methods of seeking for leakage of cellular contents must be refined by looking for their possible effects on energy-dependent substrate accumulation, the activity of membrane ATPase and the effect on the membrane pH and electrical gradient, which together constitute the protonmotive force.

For a very full account of aspects of this area of investigation the review of Harold (1970) should be consulted.

Although in the foregoing account, the action of uncoupling agents and disinfectants belonging to this class have been explained in terms of the Mitchell chemiosmotic hypothesis, there are some who hold that the hypothesis is not the final and unifying theory of membrane function (see Skulachev, 1970). Be that as it may, it is very widely held that a trans-membrane pH gradient is generated across membranes as a consequence of metabolic activity especially aerobic and anaerobic glycolysis, that this gradient is responsible for some membrane functions and that the gradient is eliminated by uncoupling agents by making the membrane permeable to protons.

(d) *Vinylglycollic acid as an inhibitor of bacterial membrane transport*

A transport mechanism, believed to be limited to bacteria, was described by Kundig *et al.* (1964). In essence it is as follows:

$$\text{phosphoenolpyruvic acid} + \text{sugar} \rightarrow \text{sugar phosphate} + \text{pyruvic acid};$$

three enzymes and various factors are involved in the process and the sugar is transported across the membrane as a 'sugar' phosphate. Amongst the 'sugars' known to be transported in this way are glucose, fructose, mannose, mannitol and sorbitol.

Vinylglycollic acid, 2-hydroxy-3-butenoic acid, is a specific inhibitor of this process (Walsh & Kaback, 1973).

The fact that this transport mechanism is unique to bacteria poses the interesting question as to whether vinylglycollate has a place in the therapy of human and animal infections, but it may fail if alternative uninhibited energy sources, i.e. lactose or amino acids, are available to the organism as they would be in the animal body.

5. The Cytoplasm as a Target

The cytoplasm, i.e. all that part of the cell lying within the cytoplasmic membrane, is a viscous liquid containing as operative substructures the ribosomes, DNA, RNA and an array of enzymes.

This system may be coagulated irreversibly by high concentrations of certain chemical disinfectants which include the phenols, mercuric salts and chlorhexidine.

Cytoplasmic coagulation was noted early in investigations on the mechanism of disinfectants, i.e. the UV microscope studies of Bancroft & Richter (1931) and indeed many of the earlier disinfectants were classified, and often dismissed, as general protoplasmic poisons.

Even today, many adhere to this inexact generalization despite a wealth of evidence to suggest more subtle actions. However, if the concentrations of disinfectants used are sufficiently high, irreversible coagulation is the expressed and final lesion. It is possible to demonstrate the effect of disinfectants on bacterial cytoplasm by methods other than direct microscopic observation.

Cells may be disrupted and the agent under examination added in graded doses to the cell juice. Any precipitate formed may be collected and analysed for its protein and nucleic acid content. By this method Hugo & Longworth (1966) studied the effect of chlorhexidine on the precipitation of protein and nucleic acid from the cell free cytoplasm of *E. coli* and *Staph. aureus*.

If bacterial cells are converted to protoplasts by degradation of their cell walls in a medium of high osmotic pressure, equivalent to 0.33 M sucrose, dilution of the system leads to lysis. If the cells are pretreated with disinfectants at concentrations causing cytoplasmic coagulation then on dilution as above the protoplasts remain intact as spheres of coagulated cytoplasm (Tomcsik, 1955; Hugo & Longworth, 1964; Daltrey & Hugo, 1974). In contrast, Hugo & Bloomfield (1971a) showed the failure of Fentichlor to coagulate protoplasts of *B. megaterium*.

A further interesting aspect of coagulation may be seen on the effect of dose of drug on leakage of cytoplasmic constituents. Leakage increases steadily with dose and then one of two events may happen. With some antibacterial agents the plot of leakage against dose continues parallel to the concentration axis. Fentichlor, acting upon *E. coli*, shows this pattern of behaviour (Hugo & Bloomfield, 1971a). With others, the curve, after reaching its maximum, dips again such that at higher concentrations less or no leakage is induced. Chlorhexidine acting on *E. coli* and *Staph. aureus* (Hugo & Longworth, 1964) or on *Cl. perfringens* (Daltrey & Hugo, 1974) show this behaviour. The downward sweep of the diphasic curve is due to the progressive coagulation of the cytoplasm with trapping of the small molecular weight material.

An overall assessment of the action of disinfectants on cytoplasmic enzymes comes from the vast array of work on the action of these agents on metabolic oxido-reductions as measured by dye reduction or oxygen uptake. This field was reviewed by Hugo (1967) and will not be dealt with in detail here. Dye reduction of oxygen uptake often stops at very low drug concentrations but it is not likely to be a cause of death but a contributor to bacteriostasis. Loss of co-enzymes and inhibition of transport mechanisms are more likely to be associated with metabolic inhibition in whole cells. Cell free systems, provided that adequate substrate and co-factors are present, are often more resistant to inhibition than whole cells. It is the author's opinion that studies of this nature have been the most unrewarding in endeavours to find out how chemical disinfectants inhibit or kill microbial cells.

However, a group of chemical disinfectants which have as their target the thiol group will attack all such groups in the cell. If a thiol enzyme is part of a metabolic chain then metabolic inhibition will result. Heavy metal ions, notably Hg^{2+}, have been known to act in this manner at low concentrations, higher concentrations are cytoplasmic coagulants. More recently, 2-bromo-2-nitro-propan-1-3-diol, bronopol, has been shown to react with cellular thiol groups (Bowman & Stretton, 1972; Stretton & Manson, 1973). The —SH group is also likely to be a target for oxidizing agents such as hypochlorites and iodine; the amino group is thought to be the site of action of formaldehyde.

The ribosome, a target for many antibiotics, is not a specific target for chemical disinfectants.

Some antibacterial dyestuffs belonging to the acridine series have been found to have as their target the nucleic acid, which in prokaryotes resides naked in the protoplasm, unsurrounded by a nuclear membrane as in eukaryotes.

Dyes such as proflavine fit, or intercalate, into the double-stranded DNA helix and thereby prevent its functioning. For a recent account of this specific lesion see Franklin & Snow (1975).

6. Conclusions

Chemical disinfectants have been, and sometimes still are, classified as general protoplasmic poisons but it is clear from the large field of published work on their mode of action that such a gross generalization is no longer tenable.

The ability of a group of chemical disinfectants to act as uncoupling agents and to prevent active transport, probably because they are able to modify the ability of cells to maintain a proton gradient across the cytoplasmic membrane, represents an example.

General changes in the permeability of the cytoplasmic membrane to metabolites other than protons is another characteristic lesion of many chemical disinfectants.

The specific interaction of acridine dyes with DNA represents another chemically defined lesion defined with precision.

Finally, it should be remembered that many chemical disinfectants have differing modes of action depending on the concentration at which they act and may exert a bacteriostatic effect at one concentration and a bactericidal effect at another.

Only at high concentrations may the epithet, general protoplasmic poison, be applied meaningfully.

7. References

BANCROFT, W. D. & RICHTER, G. H. (1931). The chemistry of disinfection. *Journal of Physical Chemistry* **35**, 511-530.
BLOOMFIELD, S. F. (1974). The effect of the phenolic antibacterial agent Fentichlor on energy coupling in *Staphylococcus aureus. Journal of Applied Bacteriology* **37**, 117-131.
BOLLE, A. & KELLENBERGER, E. (1958). The action of sodium lauryl sulphate on *E. coli. Schweizerische Zeitschrift für allgemeine Pathologie und Bakteriologie* **21**, 714-740.
BOWMAN, W. R. & STRETTON, R. J. (1972). Antimicrobial activity of a series of halo-nitro compounds. *Antimicrobial Agents and Chemotherapy* **2**, 504-505.
DALTREY, D. C. & HUGO, W. B. (1974). Studies on the mode of action of the antibacterial agent chlorhexidine on *Clostridium perfringens*. 2. Effect of chlorhexidine on metabolism and on the cell membrane. *Microbios* **11**, 113-146.
DELPHY, P. L. & CHAMPSEY, H. M. (1949). Sur la stabilisation des suspensions sporulées de *B. anthracis* par l'action de certains antiseptiques. *Compte rendu hebdomadaire des séances de l'Académie des sciences, Paris* **225**, 1071-1073.
DOMAG, K. (1935). Chemotherapy of bacterial infections. *Deutsche medizinische Wochenschrift* **61**, 250-253.
FRANKLIN, T. J. & SNOW, T. J. (1975). *Biochemistry of Antimicrobial Action*, 2nd edn. London: Chapman & Hall.
FREDERICK, J. J., CORNER, T. R. & GERHARDT, P. (1974). Antimicrobial actions of hexachlorophene: Inhibition of respiration in *Bacillus megaterium. Antimicrobial Agents and Chemotherapy* **6**, 712-721.
GALE, E. F. & TAYLOR, E. S. (1947). The action of tyrocidin and some detergent substances in releasing amino acids from the internal environment of *Streptococcus faecalis. Journal of General Microbiology* **1**, 77-84.
GRENVILLE, G. D. (1969). A scrutiny of Mitchell's chemiosmotic hypothesis of respiratory chain and photosynthetic phosphorylation. *Current Topics in Bioenergetics* **3**, 1-78.
HAMILTON, W. A. (1968). The mechanism of the bacteriostatic action of tetrachlorosalicylanilide. *Journal of General Microbiology* **50**, 441-458.
HAROLD, F. H. (1970). Antimicrobial agents and membrane function. *Advances in Microbial Physiology* **4**, 45-104.
HAROLD, F. M. & BAARDA, J. R. (1968). Inhibition of membrane transport in *Streptococcus faecalis* by uncouplers of oxidative phosphorylation and its relation to proton conduction. *Journal of Bacteriology* **96**, 2025-2034.
HAROLD, F. M. & PAPINEAU, D. (1972). Cationic transport and electrogenesis in *Streptococcus faecalis*. I. The membrane potential. *Journal of Membrane Biology* **8**, 27-44.
HAROLD, F. M., BAARDA, J. R., BARON, C. & ABRAMS, A. (1969). Dio 9 and chlorhexidine: inhibitors of membrane bound ATPase and of cation transport in *Streptococcus faecalis. Biochimica et biophysica acta* **183**, 129-136.
HOTCHKISS, R. D. (1944). Gramicidin, tyrocidin and tyrothricin. *Advances in Enzymology* **4**, 153-199.
HUGO, W. B. (1967). The mode of action of antibacterial agents. *Journal of Applied Bacteriology* **30**, 17-50.
HUGO, W. B. & BLOOMFIELD, S. F. (1971a). Studies in the mode of action of the phenolic antibacterial agent Fentichlor against *Staphylococcus aureus* and *Escherichia coli*. II. The effects of Fentichlor on the bacterial membrane and the cytoplasmic constituents of the cell. *Journal of Applied Bacteriology* **34**, 569-578.
HUGO, W. B. & BLOOMFIELD, S. F. (1971b). Studies in the mode of action of the phenolic antibacterial agent Fentichlor against *Staphylococcus aureus* and *Escherichia coli*. III. The effect of Fentichlor on the metabolic activities of

Staphylococcus aureus and *Escherichia coli. Journal of Applied Bacteriology* **34**, 579-591.
HUGO, W. B. & LONGWORTH, A. R. (1964). Some aspects of the mode of action of chlorhexidine. *Journal of Pharmacy and Pharmacology* **16**, 655-662.
HUGO, W. B. & LONGWORTH, A. R. (1966). The effect of chlorhexidine on the electrophoretic mobility, cytoplasmic contents, dehydrogenase activity and cell walls of *Escherichia coli* and *Staphylococcus aureus. Journal of Pharmacy and Pharmacology* **18**, 569-578.
KASHKET, E. R. & WILSON, T. H. (1973). Proton-coupled accumulation of galactoside in *Streptococcus lactis* 7962. *Proceedings of the National Academy of Sciences of the United States of America* **70**, 2866-2869.
KUHN, R. & BIELIG, H. J. (1940). Uber Invertseifen. I. Die Einwirkung von Invertseifen auf Eiwess-Stoffe. *Berichte der Deutschen chemischen Gesellschaft* **73**, 1080-1091.
KUNDIG, W., GHOSH, S. & ROSEMAN, S. (1964). Phosphate bound to histidine in a protein as an intermediate in a novel phospho-transferase system. *Proceedings of the National Academy of Sciences of the United States of America* **52**, 1067-1074.
LAMBERT, P. A. & HAMMOND, S. M. (1973). Potassium fluxes. First indications of membrane damage in micro-organisms. *Biochemical and Biophysical Research Communications* **54**, 796-799.
MITCHELL, P. (1972). Chemiosmotic coupling in energy transduction: a logical development of biochemical knowledge. *Journal of Bioenergetics* **3**, 5-24.
NIVEN, D. F. & HAMILTON, W. A. (1974). Mechanisms of energy coupling to the transport of amino acids in *Staphylococcus aureus. European Journal of Biochemistry* **37**, 244-248.
PASSOW, H., ROTHSTEIN, A. & CLARKSON, T. W. (1961). The general pharmacology of the heavy metals. *Pharmacological Reviews* **13**, 185-224.
PULLMAN, J. E. & REYNOLDS, B. L. (1965). Some observations on the mode of action of phenol on *Escherichia coli. Australian Journal of Pharmacy* **46**, S80-S84.
PULVERTAFT, R. J. V. & LUMB, G. D. (1948). Bacterial lysis and antiseptics. *Journal of Hygiene, Cambridge* **46**, 62-64.
SALTON, M. R. J. (1950). The bactericidal properties of certain cationic detergents. *Australian Journal of Scientific Research* **B3**, 45-60.
SALTON, M. R. J. (1951). The adsorption of cetyltrimethylammonium bromide by bacteria, its action in releasing cellular constituents and its bactericidal effect. *Journal of General Microbiology* **5**, 391-404.
SCHAECHTER, M. & SANTOMASSINO, K. A. (1962). The lysis of *Esch. coli* by sulphydryl binding agents. *Journal of Bacteriology* **84**, 318-325.
SCHULMAN, J. H. & RIDEAL, E. K. (1937). Molecular interactions in monolayers. I. Complexes between large molecules. *Proceedings of the Royal Society* **B122**, 29-45.
SIMONI, R. D. & POSTMA, P. W. (1975). The energetics of bacterial active transport. *Annual Review of Biochemistry* **44**, 523-554.
SINGER, S. J. & NICHOLSON, G. L. (1972). The fluid mosaic model of the structure of cell membranes. *Science, New York* **175**, 720-731.
SKULACHEV, V. P. (1970). Energy transformation in the respiratory chain. *Current Topics in Bioenergetics* **4**, 127-190.
SMITH, A. R. W., LAMBERT, P. A., HAMMOND, S. M. & JESSUP, C. (1975). The differing effects of cetyltrimethylammonium bromide and cetrimide B.P. upon growing cultures of *Escherichia coli* NCIB 8277. *Journal of Applied Bacteriology* **38**, 143-149.
STRETTON, R. J. & MANSON, T. W. (1973). Some aspects of the mode of action of the antibacterial compound bronopol (2-bromo-2-nitropropan-1,3-diol). *Journal of Applied Bacteriology* **36**, 61-76.
TOMCSIK, J. (1955). Effects of disinfectants and of surface active agents on bacterial protoplasts. *Proceedings of the Society of Experimental Biology and Medicine* **89**, 459-463.

WALSH, C. T. & KABACK, H. R. (1973). Vinylglycolic acid. An inactivator of the phosphoenolpyruvate–phosphate transferase system in *Escherichia coli*. *Journal of Biological Chemistry* **248**, 5456-5462.

WOODROFFE, R. C. S. & WILKINSON, B. E. (1966). The antibacterial activity of tetrachlorsalicylanilide. *Journal of General Microbiology* **44**, 343-352.

The Inactivation of Vegetative Micro-organisms by Chemicals in the Dairying Industry

CHRISTINA M. COUSINS

National Institute for Research in Dairying, Shinfield, Reading, England

CONTENTS

1. Introduction 13
2. Types of disinfectants 14
    (a) Approved chemical agents in the U.K. 14
    (b) Other disinfectants 15
3. Evaluation of disinfectants by the modified Lisbôa Tube Test 16
4. In-use evaluation of disinfectants and methods of application 21
    (a) Circulation or in-place cleaning and disinfection of pipeline milking machines 21
    (b) Milk processing plant 22
    (c) Cold cleaning and disinfection of farm bulk tanks and transport tankers . . 23
    (d) Washing of cows' udders and teats 24
    (e) Post-milking disinfectant teat dips 25
5. Chlorination of water supplies 26
6. The microflora of milking equipment after chemical disinfection 27
7. Resistance of vegetative micro-organisms to dairy disinfectants 28
8. References 28

1. Introduction

IN THE UNITED KINGDOM only those brands of 'chemical agents' (disinfectants or combined detergent–disinfectants) approved by the authorities concerned in accordance with the appropriate milk and dairy legislation—e.g. Milk & Dairies (General) Regulations, 1959—are permitted as alternatives to scalding with steam or boiling water for treatment of utensils and equipment used for milk and cream; equipment for milk products is not covered by this legislation. In the trade these chemical agents are sometimes described as sanitizers or more frequently, but incorrectly, as chemical sterilizing agents and detergent-sterilizers (*Anon.*, 1976). There are no requirements concerning the effect of these disinfectants on human pathogens and they are not expected to inactivate bacterial spores.

The main purpose of using approved chemical agents, in addition to complying with statutory requirements, is to prevent the introduction of spoilage and other undesirable micro-organisms into raw and processed milk and milk products. Treatment of the equipment probably constitutes the most widespread use of disinfectants.

The perishable nature of milk and cream at ambient temperatures has undoubtedly accounted for the attention paid to the disinfection of dairy equipment for the last 50 or more years. However, the widespread use of refrigeration throughout the industry and in the home has reduced the rapid spoilage effects of inadequate disinfection. On the other hand, buyers and consumers expect milk and milk products to have a much longer 'life' under refrigeration and, therefore, at the present time effective disinfection to prevent contamination, particularly from psychotrophic bacteria, is still important.

Cleaning and disinfection in-place, which may be controlled automatically, for both milk producing equipment and processing plant has removed some of the hazards associated with manual application of chemical disinfectants but problems of inactivating micro-organisms in certain situations still remain, and will be discussed.

The effects of the presence of some commonly used dairy disinfectants in milk on its microflora and on starter cultures, reviewed by Swartling (1959), and the preservation of milk by addition of hydrogen peroxide (Lück, 1962; Tentoni *et al.*, 1968) are beyond the scope of the present review.

Methods of application of disinfectants have been adequately covered elsewhere, for example, Clegg (1967, 1970), Cousins (1967), Society of Dairy Technology (*Anon.*, 1968, 1972*b*, 1975*b*) and British Standards Institution (*Anon.*, 1975*a*) and, except to illustrate principles, methods will not be discussed.

2. Types of Disinfectants

Many countries have legislation or other requirements governing the types of disinfectants permitted for use in the dairy industry, but there is little unanimity in tests applied or control exercised. The present paper relates mainly to types of disinfectants and detergent–disinfectants used and their evaluation in the U.K.

(a) *Approved chemical agents in the U.K.*

A list of brands of approved chemical agents (sterilizing agents and detergent-sterilizers), classified according to type of active agent and giving approved use-concentrations, is published periodically by H.M.S.O. for the Ministry of Agriculture, Fisheries and Food. This list is also applicable to Northern Ireland. Information on products approved for use in Scotland is available from the Scottish Home & Health Department.

It is likely that chemicals based on available chlorine are the most extensively used. These and the other chemical agents approved in the U.K. may be either disinfectants alone or combined detergent–disinfectants as indicated in Table 1.

Sodium hypochlorite may be added to solutions of suitable detergents for combined cleaning and disinfection. Iodophors and acid-wetting agents.

although classed as detergent–disinfectants in the U.K., are also used for disinfecting previously cleaned surfaces of equipment.

A review of the chemical and bactericidal properties of disinfectants based on available chlorine and iodine and some of their practical applications has been

Table 1
Types of disinfectant and combined detergent–disinfectants included in the U.K. lists of approved chemical agents

Types of disinfectant	Active ingredient of	
	Disinfectant alone	Combined detergent–disinfectant
Available chlorine		
sodium hypochlorite	+	+
chlorinated tri-sodium phosphate	+	+
dichloro-*iso*-cyanurate	–	+
dichlorodimethyl hydantoin	–	+
Iodine (iodophor)	–	+
QAC	+	+
Acid-wetting agent	–	+
Amphoteric surfactant*	+	+

+ Active ingredient present.
QAC = quaternary ammonium compounds.
* Only one example of this type is included in the lists.

made by Trueman (1971). Twomey (1968, 1969) has also reviewed iodophors with particular reference to their use in the dairy industry. The general properties of quaternary ammonium compounds (QAC) and their applications as disinfectants have been reviewed by Lawrence (1968), and the synergistic bactericidal effects of mixtures of acids and anionic surfactants, the acid-wetting agents, are described and discussed by Dychdala (1968).

(b) *Other disinfectants*

Disinfectants other than approved chemical agents are also used in the industry. For example, caustic soda, the main ingredient of bottle-washing detergents, at the temperatures (63-70°) and concentrations (0.75-1%) used in mechanical bottle washers, disinfects the bottles effectively (Hobbs & Wilson, 1943).

Gaseous chlorine, and to a lesser extent, chlorine dioxide (Palin, 1974) are used for disinfecting water used for cooling canned products after sterilization, for washing butter and cottage cheese curd (*Anon.*, 1972*b*) and for final rinsing of washed milk bottles (*Anon.*, 1968).

Formalin may be used to disinfect base-exchange beds of water softeners which, when they become contaminated with bacteria, give rise to high bacterial counts in the softened water (Davis, 1959; Anon., 1968). Generation of formaldehyde from formalin is recommended for reducing contamination with yeasts and moulds in cheese storerooms. In Denmark, nitric acid (0.06 N) is approved for disinfecting milking machines (Pedersen & Møller-Madsen, 1959).

Iodophors with a relatively low acid content are widely recommended and used for the post-milking disinfection of cows' teats; this application will be discussed in Section 4(d).

3. Evaluation of Disinfectants by the Modified Lisbôa Tube Test

For most practical purposes involving plant and equipment the selection of a suitable disinfectant or detergent–disinfectant normally depends on the type of equipment, the method of application, the time available for disinfection, risks of corrosion and other factors such as convenience in handling and cost, with the overriding requirement that the products from which the selection is to be made are all highly effective bactericides. Because of the wide variety of types and species of spoilage bacteria which might gain access to milk and dairy products during production, processing and manufacture, dairy disinfectants should be non-selective bactericides. Sodium hypochlorite, which is well recognized as active against a wide spectrum of vegetative bacteria (Trueman, 1971), is very suitable in this respect.

Tests of the suspension, capacity and carrier type for disinfectants and detergent–disinfectants used in the dairy and food industries (e.g. *Anon., 1962a,b*; Shaffer & Stuart, 1968) normally employ pure cultures of *Escherichia coli* I and a Gram positive organism such as *Staphylococcus aureus* or *Micrococcus caseolyticus*. However, at the National Institute for Research in Dairying, bulked raw milk aged at $22°$ has been used for many years as a soiling medium for studies on chemical disinfection, and in the U.K. for the purpose of approval under the Milk and Dairies (General) Regulations, 1959, the test for bactericidal efficiency involves a surface soiled with aged raw milk and a comparison with sodium hypochlorite. Until recently 5-gallon milk cans were used as the test surfaces for examination of both detergent–disinfectants and disinfectants alone in the Hoy Can Test (Cousins *et al.,* 1960). However, using stainless steel tubes (lengths of milking machine pipeline, 0.325 m long × 25.4 mm diameter) in place of milk cans, Lisbôa (1959) devised a scaled-down version of the Can Test. Following modifications and extensive collaborative trials this Tube Test has replaced the Can Test as the official approval test.

The apparatus and test procedure for detergent–disinfectants are described in detail elsewhere (*Anon.,* 1967). Briefly, the tubes, 8-12 for each day's test, are soiled with aged bulk raw milk having a methylene blue reduction time at $37°$ of 15-30 min and total colony count (Yeastrel milk agar, 3 days at $30°$) of 10^6-10^7/ml. Incubation of the tubes for 18 h at 22-$24°$ causes the residual milk

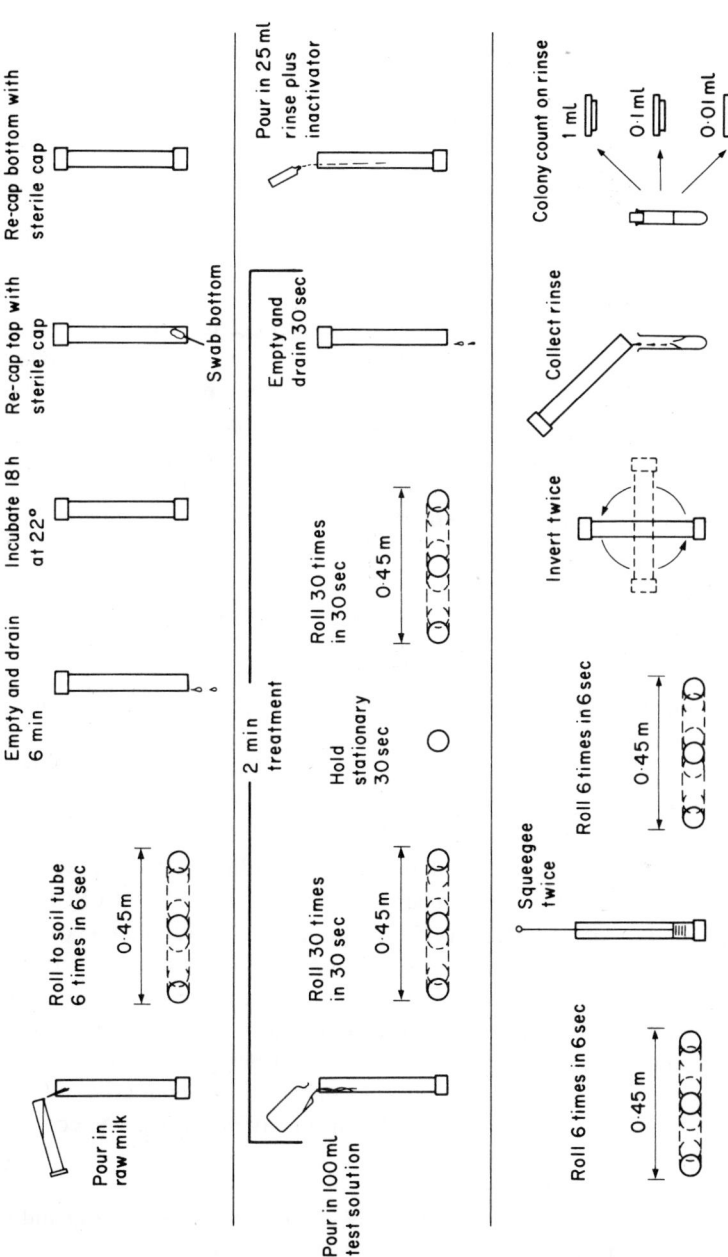

Fig. 1. The Lisbôa Tube Test.

to dry in the form of a thin film in which bacterial multiplication has occurred leaving the inner surface of each tube, the test surface, coated with a film of milk solids containing 10^7-10^8 bacteria. Its predominant microflora consists of micrococci, streptococci, Gram negative rods (including coliforms and pseudomonads) and, less abundantly, Gram positive non-sporing coryneform rods; the proportions of each type vary from time to time. Solutions of detergent–disinfectants under test are prepared in a standard hard water containing some alkaline hardness (Ca and Mg bicarbonates) which has a neutralizing effect on acidic materials such as iodophors and tends to reduce their bactericidal efficiency.

A diagrammatic representation of the whole test procedure for one tube is shown in Fig. 1. One of the test solutions, 100 ml at 44°, is poured into a soiled tube and after a contact time of 2 min during which time the tube is rotated on a horizontal surface for two ½-min periods and finally drained, the solution is replaced with a rinse consisting of quarter-strength Ringer solution containing suitable disinfectant inactivators (Cousins, 1963). The tube is 'squeegeed' to remove any residual milk film and bacteria into the rinse which is then plated for colony count.

If a disinfectant alone is to be tested the tube is treated first with a detergent solution, 0.25% (w/v) Na_2CO_3, and then with the test disinfectant, each treatment lasting 2 min.

Each test material at the proposed use-concentration is compared with a standard detergent–hypochlorite solution, 0.25% (w/v) Na_2CO_3 containing 300 mg/l of available chlorine; in addition each test material, diluted to $\frac{2}{3}$ of the use-concentration, is compared with the standard solution similarly diluted. Tubes treated with the standard solutions are included with each set of tubes treated with the test materials.

Normally five test materials and the detergent–hypochlorite standard, each at the two concentrations, 12 treatments in all, are allocated at random to 12 tubes and the procedure is repeated after six days. An analysis of variance is carried out on the logarithmic transformations, $y = \log (X + 1)$, of the original colony count/ml of rinse, X, of each treated tube. Table 2 shows a typical set of results with treatment means and significance tests (allowance being made for one missing value). Materials A and B have failed the test, A being significantly worse than the standard at both the proposed use-concentration and at $\frac{2}{3}$ of this concentration, whereas B was significantly worse only at the latter concentration.

The detergent–hypochlorite solution in 2 min at 44°, with no pre-rinse, reduces the bacterial contamination in a tube by c. 5 log cycles (99.999%) and is therefore remarkably effective.

With such a high proportion of the bacteria inactivated and removed, the chemicals at approved use-concentrations are unlikely to be selective for any major components of the microflora and for materials based on available chlorine, iodine, QAC and acid-wetting agents this has been confirmed by isolation and examination of colonies from platings of rinses of disinfectant treated tubes; furthermore a few tests using milk inoculated with pure cultures of bacteria have shown that similar numbers of *E. coli* (NCDO 555) and *Staph. aureus* (NCDO 771) survived treatment with a detergent-QAC formulation

Table 2

Results of a 6-day Tube Test on five test materials, A to E, and the standard detergent–hypochlorite, H

Test material	Transformed colony counts (X)/ml of rinse: log $(X + 1)$		Detransformed colony count/ml, geometric mean
	Geometric mean	Difference from H	
Use-concentration			
A	2.436	1.100*	272.2
B	2.179	0.842	150.0
C	1.096	− 0.241	11.5
D	1.752	0.415	55.5
E	1.351	0.014	21.4
H (standard)	1.337	—	20.7
2/3 use-concentration			
A	3.322	1.651†	2097.9
B	3.101	1.430†	1261.4
C	2.336	0.665	215.8
D	2.180	0.509	150.3
E	2.345	0.674	220.2
H (standard)	1.671	—	45.9
S.E. of a difference (54df)	—	0.4814	—
Least significant difference (5%)	0.966	0.966	9.24-fold

* ($P < 0.05$).
† ($P < 0.01$).

(Cousins, unpublished data). Soprey & Maxcy (1968) observed that cells of *E. coli* and *Pseudomonas fluorescens* adapted to become tolerant to QAC by growing in nutrient broth plus QAC, were no more resistant than unadapted cells when exposed to normal use-concentrations of QAC in a standard disinfectant test.

As normally used the Tube Test measures the combined effects of detergency and disinfection, and experience of testing a wide variety of formulations has indicated that the detergent ingredients of a formulation may markedly affect in various ways its performance in the test, such as efficiency in removing milk film and bacteria from the test surface, interaction with the disinfectant to improve or reduce bactericidal efficiency, or because the detergent has inherent bactericidal properties. For example, it is now well recognized that the addition of hypochlorite to some types of detergents improves their soil removing efficiency as demonstrated by Clegg & Cousins (1970). Formulations of QAC containing non-ionic detergents of low alkalinity require a much higher content of QAC to pass the Tube Test than do those containing a similar QAC with suitable inorganic alkaline detergents, not only because QAC are better disinfectants at higher pH values (Soike et al., 1952) but also because non-ionic detergents are generally less effective than alkaline detergents in removing milk film (Cousins & McKinnon, 1962) and, furthermore, some non-ionics may have an inactivating effect on some QAC (Dvorkovitz et al., 1951).

Table 3 shows the concentrations of active disinfectant ingredients in various types of approved chemical agents which have passed the Tube or Can Tests; the difference between the two types of detergent–QAC formulations is striking.

Table 3

Contents of active ingredients in approved use-concentrations of disinfectants and detergent–disinfectants determined by comparison with a standard detergent–hypochlorite solution (300 mg/l of available chlorine)

Type of active ingredient	Content (mg/l) in approved use-concentrations
Available chlorine	150-300
QAC (disinfectant alone)	200
QAC, combined with detergent	
inorganic alkaline	150-250
non-ionic liquid	400-500
Iodine (iodophor)	35-70
	(pH ≤ 2.3)
Acid-wetting agent	pH ≤ 2.1*
Amphoteric	500

* Content of acid to give pH ≤ 2.1 depends on type of acid used.

All laboratory disinfectant tests have limitations and the Tube Test does not take into account the effects of repeated use of the material, different methods and temperatures of application, surfaces other than stainless steel, the incidence of joints, gaskets, crevices, etc. in dairying equipment and the presence of milk stone. However, the assessment of a test material is based on a comparison with hypochlorite and factors reducing the effectiveness of this disinfectant are likely to apply to others. Some hazards in practice, e.g. the presence of milk stone and corroded metal or perished rubber surfaces, which provide protection for micro-organisms or are difficult to clean, thereby reducing the efficacy of any disinfectant, can usually be recognized by the user and action can be taken to remove or overcome them. Other problems may only be revealed by practical trials.

4. In-use Evaluation of Disinfectants and Methods of Application

Field trials are well recognized as a means of determining the efficacy of different types of chemical disinfectants and of new methods, particularly for milking equipment (Clegg, 1955), and sometimes circumstances are revealed where, in spite of correct application, chemicals fail to inactivate micro-organisms. The failure being attributed to an invisible film protecting the bacteria on equipment surfaces, reaction of disinfectants with residual organic matter and consequent decline in activity, the presence of resistant micro-organisms or failure of chemicals to penetrate milk stone and into crevices; this latter is the most likely and frequently occurring problem.

(a) *Circulation or in-place cleaning and disinfection of pipeline milking machines*

Continuous unjointed sections of milking equipment can easily be disinfected. For example, an in-line cooler consisting of an 18 m length of stainless steel tubing formed into two concentric coils, was cleaned and disinfected twice daily by circulating a warm (45°) detergent–hypochlorite solution through it and the pipeline milking machine with which it was used. For a period of seven months, bacteriological rinses of the in-line cooler recovered on average <500 bacteria/930 cm^2 whereas similar rinses of the milking machine recovered > 10^6/930 cm^2 (Slade, 1974).

A combined cleaning–disinfectant treatment involving circulation of a suitable detergent–hypochlorite solution at 42-45° for 5-10 min after every milking (Cuthbert, 1960), effective in keeping surfaces of glass jars, pipelines, rubber tubing, etc. visibly clean for as long as six months, nevertheless will fail to keep complex pipeline milking installations in a satisfactory bacteriological condition. When such machines, known to have received the correct treatment, are examined by means of bacteriological rinses within a few hours of circulating the detergent–disinfectant solution large numbers of bacteria (10^6-10^7/930 cm^2 of surface area) can be recovered by the rinses (Cousins, 1963). This is because during milking, traces of milk and bacteria penetrate beneath gaskets and into tap seatings and joints formed with rubber and metal or glass (Basić et al., 1968) without which it would be impossible to assemble the necessary components for milking cows. It has become apparent that during circulation of solution neither the detergent nor the disinfectant can penetrate into the joints sufficiently to remove all residues of milk and, protected from the disinfectant, the bacteria present survive and multiply. Bacteria which are recovered by bacteriological rinses and which have survived the detergent–hypochlorite treatment are not resistant to the disinfectant because if exposed to 1-10 mg/l of free available chlorine they are inactivated within minutes (Cousins, unpublished data).

This phenomenon of bacteria in joints resisting prolonged exposure to disinfectants has also been observed in cases where disinfection of water mains, after laying or repair work, has failed. After treatment of the mains by repeated flushing with solutions containing as much as 100 mg/l of available chlorine, coliforms within one of the joints emerged freely into unchlorinated water subsequently flowing through the main (Hutchinson, 1971).

The failure of warm solutions to disinfect pipeline milking installations has led to the recommendation that detergent–disinfectant solutions should be circulated at 60-70°, in effect 'pasteurizing' the machine; however, where some types of joint insulate bacteria from the heat, reduction in bacterial numbers may be less than expected.

In view of the failure of warm (45°) detergent–disinfectant solutions to prevent a progressive build-up of bacteria in this type of equipment in spite of maintaining satisfactory visual cleanliness of glass jars, pipelines, etc., the prospects of achieving an acceptable level of disinfection with cold solutions would not appear to be good. However, because of the increasing cost of heating water, efforts to improve the effectiveness of cold or warm chemical disinfectants and minimizing the use of hot water are justified. It is possible that satisfactory results may be obtained by allowing the disinfectant a much longer contact time with plant surfaces to facilitate more effective penetration into joints, etc. Twice daily cold cleaning using a caustic detergent followed by separate disinfection with solutions of iodophor or acid-wetting agent disinfectants and allowing the plant surfaces to remain wet with the solution for the period between milkings was reported to give satisfactory plant rinse counts (Palmer & O'Shea, 1973; Palmer, 1974), but the results using iodophors have not been confirmed (McKinnon *et al.*, 1975). Furthermore, the procedure is time consuming and would require automatic control of the cleaning cycle. Practical trials to evaluate the use of different chemicals and methods in the cold within limits of cost and labour acceptable to farmers are in progress.

(b) *Milk processing plant*

This type of plant does not present the same problems as milking machines. Not only are there fewer rubber connections but cleaning, normally done prior to disinfection, is effected by circulating solutions at 70-75° for at least 20 min. Such a treatment is bactericidal, and thereafter if chemicals are used for disinfection, circulation of a cold solution of hypochlorite (100 mg/l of available chlorine) or QAC (200 mg/l) immediately before processing gives satisfactory results.

(c) *Cold cleaning and disinfection of farm bulk tanks ana transport tankers*

Most refrigerated farm tanks in the U.K. operate by means of an ice bank which would be unnecessarily dissipated if hot cleaning solutions were introduced and these would of course be rapidly cooled. In-place cleaning should therefore preferably be accomplished using cold solutions. To determine the suitability of chemicals and automatic methods for cleaning and disinfection of these farm tanks, in-use evaluation by means of bacteriological rinse tests and inspection for adequate cleaning was essential (Cousins *et al.,* 1966; McKinnon & Cousins, 1969). Of the available approved detergent–disinfectants iodophors were found most suitable. It should be pointed out that, unlike milking machines, the tanks present large areas of smooth surface free from joints and crevices and should therefore be more easily disinfected. However, in extremely cold and, therefore, difficult conditions, to achieve satisfactory disinfection (i.e. rinse counts of $< 5 \times 10^4/930$ cm^2 of tank surface) the iodophor solution has to be allowed to remain in contact with the tank surfaces for at least 10 min before the final clean water rinse. Manual cleaning at least once a month has been found necessary to prevent build up of milk residues.

The plugs of bulk tanks provide another example of a crevice condition where the shaft of the plug passes into the rubber bung causing failure of disinfection. Perhaps because of deformation of the rubber each time the plug is inserted into and withdrawn from the tank outlet, milk and bacteria can eventually accumulate undetected between the metal and the rubber. Even with prolonged immersion in cleaning and disinfecting solution, because the rubber is not under suitable stress, the solution cannot penetrate. Consequently if the plug is compressed when subjected to a bacteriological rinse micro-organisms are forced out of the crevice into the rinse solution (McKinnon & Cousins, 1969). Using this rinse method we have obtained colony counts of the order of 10^8-10^9/plug on occasions, whereas swabs of the plug show relatively low counts, frequently $< 10^4$/plug.

It has been reported that transport tankers may be successfully cleaned and disinfected by spraying with cold detergent–disinfectant solutions (Reeves & Tilley, 1972; Palmer, 1975). Again, in-use evaluation by means of rinses and swabs and visual examination was necessary for determining the effectiveness of disinfecting procedures. Both reports point out the need for periodic special treatments to remove deposits which may have built up. The failure of cold solutions to keep surfaces clean for more than a few weeks is the most probable cause of high bacterial counts on the surfaces rather than the effect of the low temperature on the disinfection process.

The Tube Test may prove useful for screening formulations unsuitable for use in the cold by detecting those which are relatively inactive at 5-10° because of impairment of detergent or disinfectant activity. Table 4 shows a comparison between a detergent–disinfectant, X, and the standard detergent–hypochlorite in a test where the tubes were treated with solutions at both 5° and 44°. Material X, comparable at 44° with hypochlorite, was significantly inferior at 5° (Longman & Cousins, unpublished data).

Table 4

Extracted results of a Tube Test comparing the effects of solution temperature on the efficiency of use-concentrations of a material, X, and the standard detergent–hypochlorite solution, H

Test material	Solution temperature (°)	Geometric mean colony count/ml of tube rinses
X	44	16
H	44	7
X	5	11,000
H	5	131

(d) *Washing of cows' udders and teats*

Soiled teats and udders are the main sources of 'sediment' (visible extraneous matter consisting of particles of dirt, dung and bedding materials such as straw, sawdust and sand) in milk; they are also sources of bacterial contaminants.

Johns (1962) reported colony counts of 10^5/ml of milk from heifers with heavily soiled unwashed teats but the effect of washing teats with water or disinfectant solution in reducing bacterial numbers in milk has never been clearly established, mainly because of the variation in numbers of bacteria excreted from apparently healthy udders and also because cows' teats are not uniformly dirty under farm conditions. It has, however, been demonstrated that washing teats with hypochlorite or iodophor solutions is not completely effective in eliminating mastitis pathogens from the skin (Neave et al., 1969). If cows' teats are heavily soiled the dirt should be removed prior to milking to prevent its entry into the milk. In milking parlours, teats are most conveniently washed using a jet of water from a hose and the gloved hand to assist in removing the dirt. A disinfectant, most commonly hypochlorite or an iodophor, may be entrained into the water used to feed the hose and, if the teats are dried, paper towels are often used. To study the effects of this method of washing teats, the addition of hypochlorite (600 mg/l of available chlorine) to the washing water, drying the teats or leaving them wet, ten cows with healthy udders giving milk having bacterial counts of < 10/ml were selected. Representative samples of each cow's total yield of milk were taken by means of sterilized teat cup clusters and in-line samplers (McKinnon et al., 1973) which ensured that any bacteria present were derived only from the teat surfaces. For a period of six weeks the cows were kept in a yard where the bedding, either straw or sawdust, was allowed to become heavily soiled or, during one 14-day period, was kept clean by daily addition of fresh straw. The different treatments applied to the teats before milking, and allocated to pairs of cows for 14-day periods according to a predetermined experimental design, are listed in Table 5 which shows a summary of the mean colony count/ml of milk for samples taken after each treatment (McKinnon *et al.,* unpublished data). Although the use of hypochlorite solution followed by drying the teats with a paper towel brought about a significant and worthwhile reduction in the mean bacterial count of the milk as compared with the other treatments it is evident that drying was also an important factor. These results confirm that, in spite of using a relatively strong solution of hypochlorite (600 mg/l of available chlorine) teat skin is not easily disinfected in the short time available. It should be borne in mind that to give a bacterial count of 10^3/ml of milk the surface of each teat of any one cow must have contributed c.

Table 5

Contamination of milk with micro-organisms from surfaces of cows' teats after different hose washing treatments

Treatment of teats	Geometric mean colony count/ml of milk
Unwashed	7490
Washed with water and left wet	7947
Washed with water and dried	4153
Washed with NaOCl* and left wet	4140
Washed with NaOCl and dried	1519

* Solution containing 600 mg/l of available chlorine.

10^6 bacteria to that cow's total milk yield. The mean results conceal the large variation in bacterial counts of milk from individual cows, from time to time, and for any one treatment.

(e) *Post-milking disinfectant teat dips*

Dipping cow's teats in a suitable disinfectant immediately after removing the teat cup cluster has been shown to inactivate staphylococcal and streptococcal mastitis pathogens transferred to or near the teat orifice during milking. Disinfectants have been assessed for this purpose using teats both artificially and naturally exposed to contamination with mastitis pathogens. Swabbing one hour after dipping the teats in disinfectant is a measure of efficacy. Typical results are shown in Table 6 (Dodd & Neave, 1970). All tests indicate that 4% hypochlorite is very nearly completely effective in destroying pathogens left on healthy teats after milking. However, the real value of a disinfectant teat dip is the extent by which it reduces the incidence of new mastitis infections. In field trials, both closely controlled small scale experiments and with commercial herds, hypochlorite (40,000 mg/l of available chlorine) and iodophor (5000 mg/l of free iodine) teat dips reduced the incidence of new infections by about half in 2-3 years. Where there are teat lesions, which may be colonized by mastitis pathogens, results may not be as good.

Table 6
The skin-disinfecting properties of some teat dips*

Teat dip	Concentration	Other additive	pH	Geom. mean count of *Staph. aureus* from teat swabs†
Iodophor	0.5% Iodine	Dichlorophenol and 5% glycerol	4.4	538
Hypochlorite (Na)	0.1% Chlorine		8.8	416
Hexachlorophane	0.5%	Alcohol	3.9	234
Iodophor	0.5% Iodine	33% Glycerol	4.6	206
PVP iodine	0.5% Iodine	33% Glycerol	5.2	139
Iodophor	0.5% Iodine	15% Glycerol	4.7	107
Chlorhexidine	1.0%	Pyrollidine	6.2	40
Iodophor	0.5% Iodine	2.5% Lanolin	2.2	23
Iodophor	0.5% Iodine		4.9	17
Hypochlorite (Na)	1.0% Chlorine		10.3	14
Hypochlorite (Na)	4.0% Chlorine		10.9	6

* Each dip was tested on 18-24 teats using 55 cows free of teat lesions, udder infections and teat contamination. An hour before using the teat dips the teats were immersed once in milk containing 5×10^7 *Staph. aureus*/ml. Swabs of the teats were taken one hour after using the disinfectant dips.
† Figures differing by *c.* 3.3-fold are significant at 5% level.
Dodd & Neave (1970).

Laboratory evaluation of the bactericidal efficiency of disinfectants for teat dipping has been attempted (O'Shea *et al.*, 1975) but the authors conclude that their data show it is not possible to extrapolate from *in vitro* or simulated *in vivo* studies on teat dips to their effectiveness in preventing new infection.

5. Chlorination of Water Supplies

In the dairy industry supplies of water free from pathogenic micro-organisms are essential to prevent contamination of milk and dairy products where cold water is used for the final rinsing of plant and equipment prior to use. In addition the absence of spoilage micro-organisms is a necessity for economic reasons. The use of mains water may not be sufficient safeguard against spoilage organisms because such water can become contaminated from storage tanks, hoses and water softeners. The types of micro-organisms present in water supplies, microbiological control and practical application of disinfectants have been covered in detail by Davis (1959), and chemical aspects of water chlorination have been reviewed by Palin (1974). It is sufficient to say here that the level of chlorination required or employed depends more on the nature and chemical quality of the water supply than on the micro-organisms present because these factors influence the content of available chlorine. Where operational control of chlorination is based on frequent or continuous monitoring of residual chlorine, with bacteriological tests at suitable intervals, dosages giving 1-5 mg/l of available chlorine are often adequate.

On the other hand in the absence of chemical control, for example in the treatment of some farm dairy water supplies to enable them to meet the requirement of the Milk and Dairies (General) Regulations, 1959 concerning suitability for dairying purposes, dosage by hand with approved hypochlorite at a level giving *c.* 50 mg/l of available chlorine is required to ensure an adequate safety margin. This level is also recommended for the treatment of the final rinse water to avoid contamination from the wash trough during cleaning and disinfection by hand of milking equipment (*Anon.*, 1972*a*).

Polluted waters containing lipolytic psychotrophs including *Ps. putrifaciens* were chlorinated with doses ranging from 5-200 mg/l of available chlorine and used for washing butter during butter making (Lewis *et al.*, 1958). Chlorination not only rendered the water practically sterile but also gave butter of superior quality to that washed with untreated water. No chlorine taint was detected in the butter at any level of chlorination in the wash water.

6. The Microflora of Milking Equipment after Chemical Disinfection

There is little published information on the influence of types of disinfectants on the microflora of milk processing and manufacturing equipment, but that of milking machines and farm bulk tanks has been studied. Thomas *et al.* (1964) observed that Gram negative rods occurred more frequently on equipment disinfected with QAC as compared with chlorine-based chemicals and caustic soda, particularly in rinses of equipment with low bacterial counts ($< 10^4/\text{ft}^2$). Predominance of Gram negative rods in the microflora of milking machines disinfected by QAC was also observed by Jackson & Clegg (1965). However, in subsequent studies (Druce & Thomas, 1972) involving pipeline milking machines and refrigerated bulk tanks for which QAC are seldom if ever used, the diversity of the microfloras of the milking machines and the predominance of Gram negative and psychrotrophic bacteria in farm bulk tanks suggest that factors other than the type of disinfectant used also influence the composition of the microflora.

7. Resistance of Vegetative Micro-organisms to Dairy Disinfectants

Hypochlorite, other chlorine-based disinfectants and iodine, are relatively non-selective inactivators of vegetative bacterial cells and yet the concentrations used in practice cover a very wide range as illustrated in this review. Teat dip disinfectants have to be used at high concentration, 5000 mg/l of iodine or 40,000 mg/l of available chlorine, whereas water supplies can be effectively disinfected by treatment giving 1 to 10 mg/l of available chlorine. However, the micro-organisms involved, mastitis pathogens and undesirable types present in water supplies, do not differ markedly in their resistance to chlorine-based disinfectants.

It is evident that for non-selective disinfectants the situation of the micro-organisms to be inactivated is probably the most important factor determining their apparent resistance to the disinfectant or treatment.

8. References

ANON. (1962a). *Standard Suspension Test for the Evaluation of the Disinfectant Activity of Dairy Disinfectants.* IDF 18. Brussels: International Dairy Federation.

ANON. (1962b). *Standard Capacity Test for the Evaluation of the Disinfectant Activity of Dairy Disinfectants.* IDF 19. Brussels: International Dairy Federation.

ANON. (1967). *Tube Test for the Evaluation of Detergent–Disinfectants for Dairy Equipment.* IDF 44. Brussels: International Dairy Federation.

ANON. (1968). *Bottle Washing.* London: Society of Dairy Technology.

ANON. (1972a). *Hand Cleaning and Sterilization of Farm Dairy Equipment Using Chemicals.* Ministry of Agriculture, Fisheries & Food Advisory Leaflet No. 422. London: H.M.S.O.

ANON. (1972b). *Quality Control of Milk Products.* London: Society of Dairy Technology.

ANON. (1975a). *Recommendations for Cleaning and Sterilization of Pipeline Milking Machine Installations.* BS 5226. London: British Standards Institution.

ANON. (1975b). *Cream Processing Manual.* London: Society of Dairy Technology.

ANON. (1976). *Glossary of Terms Relating to Disinfectants.* BS 5283. London: British Standards Institution.

BACIC, B., COUSINS, C. M. & CLEGG, L. F. L. (1968). Studies on the laboratory soiling of milking equipment. *Journal of Dairy Research* 35, 247-256.

CLEGG, L. F. L. (1955). Laboratory and field evaluation of chemical sterilizers for dairy farms. *Journal of Applied Bacteriology* 18, 358-373.

CLEGG, L. F. L. (1967). Disinfectants in the dairy industry. *Journal of Applied Bacteriology* 30, 117-140.

CLEGG, L. F. L. (1970). Disinfection in the dairy industry. In *Disinfection,* ed. Benarde, M. New York: Marcel Dekker.

CLEGG, L. F. L. & COUSINS, C. M. (1970). A technique for studying the build-up and prevention of milk film on hard surfaces. *Journal of Dairy Research* 37, 61-76.

COUSINS, C. M. (1963). Methods for the detection of survivors on milk handling equipment with reference to the use of disinfectant inhibitors. *Journal of Applied Bacteriology* 26, 376-386.

COUSINS, C. M. (1967). The cleaning and disinfection of milk plant on the farm. *Journal of the Society of Dairy Technology* 20, 198-203.

COUSINS, C. M. & McKINNON, C. H. (1962). The evaluation of some chemical agents alone and in combination with detergents for the disinfection of farm dairy utensils. *Proceedings of the XVI International Dairy Congress, Copenhagen, A,* 479-485.

COUSINS, C. M., HOY, W. A. & CLEGG, L. F. L. (1960). The evaluation of surface active disinfectants for use in milk production. *Journal of Applied Bacteriology* **23**, 359-371.
COUSINS, C. M., DAWKINS, J. & HOYLE, J. B. (1966). Automatic spray cleaning of bulk milk tanks. *Farm Mechanization* **18** (204), 16-19.
CUTHBERT, W. A. (1960). Bacteriological problems arising from circulation cleaning. *Journal of the Society of Dairy Technology* **13**, 142-146.
DAVIS, J. G. (1959). The microbiological control of water in dairies and food factories. *Proceedings of the Society for Water Treatment and Examination* **8**, 31-54.
DODD, F. H. & NEAVE, F. K. (1970). Mastitis Control. *Biennial Review, National Institute for Research in Dairying*, 1970, pp. 21-60.
DRUCE, R. G. & THOMAS, S. B. (1972). Bacteriological studies on bulk milk collection: pipeline milking plants and bulk milk tanks as sources of bacterial contamination of milk—a review. *Journal of Applied Bacteriology* **35**, 253-270.
DVORKOVITZ, V., CROCKER, C. K. & GALLOWAY, S. (1951). Studies with sanitizers based on quaternary ammonium salts. *Journal of Milk and Food Technology* **14**, 18-22.
DYCHDALA, G. R. (1968). Acid-anionic surfactant sanitizers. In *Disinfection, Sterilization and Preservation*, eds Lawrence, C. A. & Block, S. S. Philadelphia: Lea & Febiger.
HOBBS, B. C. & WILSON, G. S. (1943). The cleaning and sterilization of milk bottles. *Journal of Hygiene, Cambridge* **43**, 96-120.
HUTCHINSON, M. (1971). The disinfection of new water mains. *Chemistry and Industry*, No. 5, 30 January 1971, 139-142.
JACKSON, H. & CLEGG, L. F. L. (1965). Effect of preliminary incubation on microflora of raw bulk tank milk, with some observations on microflora of milking equipment. *Journal of Dairy Science* **48**, 407-409.
JOHNS, C. K. (1962). The coliform count of raw milk as an index of udder cleanliness. *Proceedings of the XVI International Dairy Congress, Copenhagen, C.* 365-371.
LAWRENCE, C. A. (1968). Quaternary ammonium surface active disinfectants. In *Disinfection, Sterilization and Preservation*, eds Lawrence, C. A. & Block, S. S. Philadelphia: Lea & Febiger.
LEWIS, J., ALLISON, C., DRUCE, R. G., GEORGE, G. & THOMAS, S. B. (1958). The effect of untreated and chlorinated water on the keeping quality of butter. *Journal of the Society of Dairy Technology* **11**, 186-193.
LISBÔA, N. P. (1959). A tube test for evaluating agents possessing both detergent and sterilizing properties. *Proceedings of the XV International Dairy Congress, London* **3**, 1816-1822.
LUCK, H. (1962). The use of hydrogen peroxide in milk and dairy products. In *Milk Hygiene.* World Health Organization Monograph Series No. 48, 423. Geneva: World Health Organization.
McKINNON, C. H. & COUSINS, C. M. (1969). Automatic spray cleaning of bulk tanks: the results of a field trial. *Journal of the Society of Dairy Technology* **22**, 227-232.
McKINNON, C. H., COUSINS, C. M & FULFORD, R. J. (1973). An in-line milk sampler for determining the numbers of bacteria derived from teat surfaces and udder infections of cows milked in recorder machines. *Journal of Dairy Research* **40**, 47-52.
McKINNON, C. H., LONGMAN, A. G. & COUSINS, C. M. (1975). Cold cleaning of milking machines. *Report 1973-74. National Institute for Research in Dairying, Shinfield, Reading*, p. 100.
NEAVE, F. K., DODD, F. H., KINGWILL, R. G. & WESTGARTH, D. R. (1969). Control of mastitis in the dairy herd by hygiene and management. *Journal of Dairy Science* **52**, 696-707.
O'SHEA, J., MEANEY, W. J., LANGLEY, O. H. & PALMER, J. (1975). Comparisons of the effectiveness of iodophor and hypochlorite disinfectant teat dips in reducing new intramammary infections in dairy cows. *Irish Journal of Agricultural Research* **14**, 99-105.
PALIN, A. T. (1974). Chemistry of modern water chlorination. *Water Services* **78** (No. 935) 4-7, (No. 936) 8-11.

PALMER, J. (1974). Evaluation of a nitric acid based detergent-sterilizer as replacement for iodophor in cold circulation cleaning of milking machines. *Irish Journal of Agricultural Research* **13**, 231-234.

PALMER, J. (1975). Low temperature cleaning of milk tankers. *Dairy Industries* **40**, 180-182.

PALMER, J. & O'SHEA, J. (1973). Cold-circulation cleaning of milking machines. *Irish Journal of Agricultural Research* **12**, 175-185.

PEDERSEN, A. H. & MØLLER-MADSEN, A. (1959). Experiments regarding the disinfection of milk cans and milking machines. *Beretning fra Statens Forsøgsmejeri* **121**.

REEVES, E. W. & TILLEY, N. (1972). An improved iodophor—its use for the cold cleaning of milk collection tankers: Part 1. Initial development of C.I.P. system. Part 2. Extended field trials of C.I.P. system. *Dairy Industries* **37**, 363-370, 433-437.

SHAFFER, C. H. & STUART, L. S. (1968). Methods of testing sanitizers and bacteriostatic substances. In *Disinfection, Sterilization and Preservation,* eds Lawrence, C. A. & Block, S. S. Philadelphia: Lea & Febiger.

SLADE, J. R. (1974). The development of an in-line refrigerated milk cooler. *Journal of the Society of Dairy Technology* **27**, 98-103.

SOIKE, K. F., MILLER, D. D. & ELLIKER, P. R. (1952). Effect of pH of solution on germicidal activity of quaternary ammonium compounds. *Journal of Dairy Science* **35**, 764-771.

SOPREY, P. R. & MAXCY, R. B. (1968). Tolerance of bacteria for quaternary ammonium compounds. *Journal of Food Science* **33**, 536-540.

SWARTLING, P. (1959). The influence of the use of detergents and sanitizers on the farm with regard to the quality of milk and milk products. *Dairy Science Abstracts* **21**, 1-10.

TENTONI, R., PASTORE, M. & OTTOGALLI, G. (1968). Hydrogen peroxide for milk collection under difficult conditions. *Annali di Microbiologia ed Enzimologia* **18**, 85-123.

THOMAS, S. B., HOBSON, P. M. & ELSON, K. (1964). The microflora of milking equipment cleansed by chemical methods. *Journal of Applied Bacteriology* **27**, 15-26.

TRUEMAN, J. R. (1971). The halogens. In *Inhibition and Destruction of the Microbial Cell,* ed. Hugo, W. B. London: Academic Press, pp. 137-183.

TWOMEY, A. (1968). Iodophors: their physical, chemical and bactericidal properties, and use in the dairy industry—a review. Part I. *Australian Journal of Dairy Technology* **23**, 162-165.

TWOMEY, A. (1969). Iodophors: their physical, chemical and bactericidal properties, and use in the dairy industry—a review. Part II. *Australian Journal of Dairy Technology* **24**, 29-32.

Bacterial Inhibitors in Milk and Other Biological Secretions, with Special Reference to the Complement/Antibody, Transferrin/Lactoferrin and Lactoperoxidase/Thiocyanate/Hydrogen Peroxide Systems

B. REITER

National Institute for Research in Dairying, Shinfield, Reading, England

CONTENTS

1. Introduction 32
2. Complement mediated bactericidal activity of specific antibodies in colostrum and postcolostral milk 33
   (a) Occurrence of complement and specific antibodies to coliforms 33
   (b) The occurrence of specific antibodies against *Escherichia coli* of human serotypes in bovine colostrum 35
   (c) The Neisser-Wechsberg or prozone effect in colostrum 35
   (d) The role of lysozyme 35
3. Inhibition of bacteria by iron binding proteins—transferrin and lactoferrin . . . 36
   (a) Iron requirement of bacteria 36
   (b) Transferrin (TF) 36
   (c) Lactoferrin (LF) 37
   (d) Inhibition of *Staphylococcus epidermis* by lactoferrin 37
   (e) The 'apparent' inhibition of spore germination of *Bacillus stearothermophilus* by lactoferrin 37
   (f) Inhibition of *E. coli* by lactoferrin in bovine colostrum (the role of citrate and bicarbonate) 39
4. The nature of the bactericidal activity of complement/antibody and the bacteriostatic activity of iron binding proteins 40
5. The lactoperoxidase/thiocyanate/hydrogen peroxide system (LP/SCN$^-$/H$_2$O$_2$) . 42
   (a) Lactoperoxidase 42
   (b) Hydrogen peroxide 43
   (c) Thiocyanate 43
   (d) The mode of inhibition of lactic acid bacteria by the LP system . . . 44
   (e) The nature of the inhibitory oxidation product of SCN$^-$ 45
   (f) The bactericidal activity of the LP system against *E. coli, Ps. fluorescens* and other Gram negative organisms 45
   (g) The nature of the bactericidal activity of the LP system 46
   (h) Preservation of milk by the LP system 47
6. Basic proteins and other antibacterial factors 48
   (a) β-lysin 48
   (b) Agglutinins 49
   (c) Properdin 50
   (d) Conglutinin 51
   (e) Vitamin B$_{12}$ and folate protein binders 51

7. Discussion 51
   (a) The function of the antibacterial system of colostrum and milk 51
   (b) Mammary gland 51
   (c) Milk inhibitors and defence mechanisms against infection in the neonate . . 53
8. References 54

1. Introduction

THIS REVIEW may appear discursive but it would not do justice to the subject if it merely referred to previous reviews and then added the more recent findings. Each review was biased towards some particular aspect or dealt only with certain bacterial species. It also seems to me rewarding to relate our present knowledge to early work in which phenomena were correctly interpreted long before sophisticated techniques of purification and identification became available. The bactericidal and bacteriostatic properties of milk were first described (Hesse, 1894) at about the same time as those of blood but naturally have not subsequently attracted the same attention. However, veterinary and medical workers have been interested in the inhibition of streptococci and salmonellae in relation to the resistance of the bovine udder to infection and the spreading of disease through the consumption of (raw) milk, (e.g. Hanssen, 1924; Jones & Little, 1927; Jones, 1928; Jones & Simms, 1930; Wilson & Rosenblum, 1952*a, b*). Dairy workers became concerned with the inhibition of lactic acid streptococci which are essential for the manufacture of cheese and other dairy products (Auclair & Hirsch, 1953; Auclair, 1954; Wright & Tramer, 1957, 1958; Portmann & Auclair, 1959; Jago & Morrison, 1962; Reiter *et al.,* 1963; Reiter & Møller-Madsen, 1963).

The milk inhibitor(s) affecting streptococci became known and are still referred to as 'lactenin(s)' or 'Inhibine' in text books. These terms must now be regarded as obsolete because many of these inhibitors have been identified and shown to occur in secretions other than milk; they are also to be found in leucocytes and constitute part of their bacteriostatic and bactericidal systems. Several reviews of milk inhibitors have appeared in the past (Berridge, 1955; Reiter & Møller-Madsen, 1963; Auclair, 1964; Reiter & Oram, 1967; Hanson & Winberg, 1972; Goldman & Smith, 1973; Reiter, 1973). This review deals also with recent work concerning Gram negative organisms including psychotrophs, the complement-mediated bactericidal effect of specific antibodies in bovine colostrum and postcolostral milk, the non-specific bacteriostatic activity of transferrin and lactoferrin, and the non-specific bactericidal activity of the lactoperoxidase/thiocyanate/hydrogen peroxide system. An attempt will be made to arrive at a unified concept of how some of these systems affect the bacterial cell. An outline of how the lactoperoxidase system could be employed to preserve milk will be discussed.

2. Complement Mediated Bactericidal Activity of Specific Antibodies in Colostrum and Postcolostral Milk

(a) *Occurrence of complement and specific antibodies to coliforms*

During an investigation of the relationship between numbers of bacteria and dye reduction time (as used for the assessment of the hygienic quality of milk) it became evident that a strain of *Klebsiella aerogenes* 1 failed to reduce resazurin because it died out in the milk during incubation at 37° (Reiter *et al.*, 1965). The organisms were agglutinated in milk, indicating the presence of specific antibodies (Ab), and were not killed in milk heated at 56° for 30 min, the accepted time-temperature combination which inactivates some components of complement (C). An attempt was also made to detect the presence of C by haemolysis of erythrocytes but we failed to lyse sheep erythrocytes which are commonly used. A search of the literature revealed that the C in bovine milk does not lyse sheep erythrocytes but readily lyses guinea-pig erythrocytes (Pfaudler & Moro, 1907).

It is curious that Morris & Edwards (1950) characterized the same phenomenon in some detail (destruction of activity by heating and restoration by the addition of blood serum, reduction of activity after storage at 4°, absorption by homologous organisms) but did not identify or even mention the

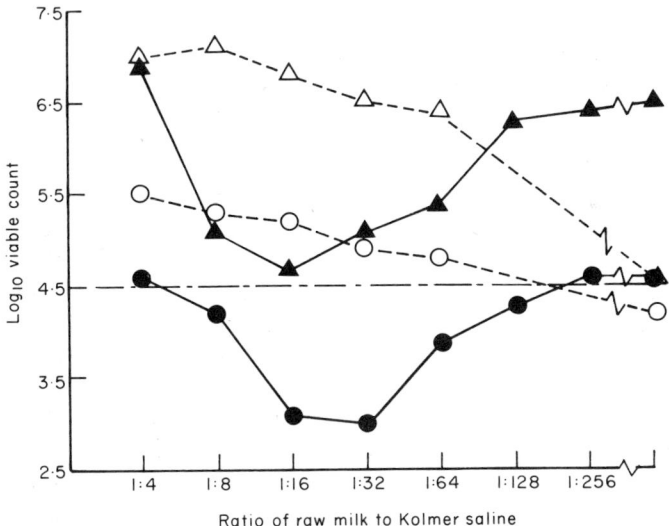

Fig. 1. The effect on the growth of *E. coli* 9703 of colostral whey diluted in raw milk or in Kolmer saline (KS). Dilutions in raw milk: ●——●, viable count at 3 h; ▲——▲, at 6 h. Dilutions in KS: ○— — —○, viable count at 3 h; △- - - -△, at 6 h. —·—, inoculum. From Reiter & Brock (1975).

C/Ab system. Using guinea-pig erythrocytes (the reason why sheep erythrocytes are not lysed by C in milk is still unknown) we were able to detect C in colostrum, postcolostral milk and in the secretion of the dry udder. C was also detected by the conglutinin agglutination technique which only detects the C components up to C_4, while haemolysis and bactericidal activity requires also the later acting C components ('cascade' effect).

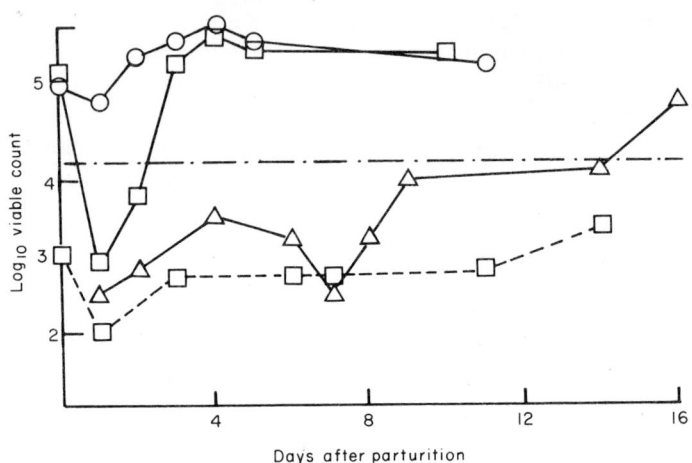

Fig. 2. Viable counts of *E. coli* 9703 after 6 h growth in colostrum and postcolostral milk from three cows. △——△, counts for cow 618; ○——○, counts for cow 423, with and without addition of 5% precolostral calf serum (PCCS) to fluids tested; □——□, counts for cow 488 (without addition of 5% PCCS); □– – –□, counts for cow 488 (with addition of 5% PCCS). – · –, inoculum. From Reiter & Brock (1975).

Since C can be detected in colostrum it is surprising that colostrum has been repeatedly reported to be non-bactericidal for Gram negative organisms (Caroll & Jain, 1969. Bullen *et al.*, 1972; Wilson, 1972). We confirmed these results but found the colostrum became bactericidal against a strain of *Escherichia coli* 0111 when diluted in milk containing C, or in Kolmer saline containing precolostral calf serum as a source of antibody-free C—Fig. 1 (Reiter & Brock, 1975). Inactivation of C by heating or adding C3-inhibitor *N*-acetyltyrosine ethyl ester (Basch, 1965) reversed the bactericidal activity.

It follows that the gradual secretory change from colostrum to postcolostrum milk is analogous to dilution of colostrum with milk. Figure 2 shows indeed that the milk of cow 618 was bactericidal for about one week and bacteriostatic for another week. The milk of cow 488 became bactericidal only after addition of C (precolostral calf serum) showing that the milk contained specific antibodies but not C. The milk of cow 423 failed to show any bactericidal activity even after addition of C, and therefore also lacked specific antibodies.

(b) *The occurrence of specific antibodies against* Escherichia coli *of human serotypes in bovine colostrum*

It is of interest that bovine colostrum contains specific antibodies against a human enteropathogenic strain of *E. coli* which was used throughout these experiments—NCTC 9703, serotype 0111 K_{58} (B_4) H-(non-motile). We have since found that a sample of diluted bovine colostrum was bactericidal against 24 of 50 enteropathogenic and 10 of 50 non-pathogenic isolates of various serotypes, from baby faeces, kindly supplied by Dr P. Rowe of the Salmonella and Shigella Reference Laboratory, Colindale Avenue, London (unpublished).

It is not understood whether the appearance of antibodies in cow serum and colostrum is due to direct experience of the animal with *E. coli* of human serotypes or to cross antigenicity between completely unrelated bacterial species. For instance, it has been found that strictly anaerobic rumen bacteria (strains of *Butyrivibrio*) possessed common antigens with a serotype of *Salmonella typhimurium* and also cross-reacted with *Mycobacterium tuberculosis* of human but not of bovine origin (Sharpe *et al.*, 1969, 1975; Minden *et al.*, 1972; Sharpe & Reiter, 1972).

(c) *The Neisser-Wechsberg or prozone effect in colostrum*

It remained to be explained why undiluted colostrum remained non-bactericidal even after addition of a surplus of C (Reiter & Brock, 1975). This appeared to be due to the high concentration of IgG_1, the predominant immunoglobulin class in bovine colostrum. It has long been known (Neisser & Wechsberg, 1901) that high titre antisera decrease or prevent the C-mediated bactericidal effect. This phenomenon is also referred to as prozone effect caused by high concentration of IgG in the antiserum (Muschel *et al.*, 1969; Normann, 1972).

(d) *The role of lysozyme*

While it is well known that IgM antibodies are 100- to 1000-fold more efficient than IgG for bactericidal activity, sIgA is generally considered not to be bactericidal because the original report of its bactericidal activity in the presence of lysozyme (Adinolfi *et al.*, 1966) was not confirmed by numerous workers. However, recently Hill & Porter (1974) supported the original finding but it has now been disputed again by Heddle *et al.* (1975). It is interesting in this context that human (and porcine?) milk is rich in both lysozymes and sIgA while bovine milk is poor in both. Bovine milk contains on average 13 μg of lysozyme/100 ml compared to 39 mg/100 ml in human milk (Candran *et al.*, 1964). Only when the bovine udder is infected does the lysozyme increase appreciably in the milk and this appears to correlate well with the leucocyte content which increases dramatically in mastitic udders (Korhonen, 1973). Since bovine leucocytes do not contain lysozyme (Padgett & Hirsch, 1967) the increased lysozyme content appears to be derived from the blood of the cows (see also Section 3(c)).

3. Inhibition of Bacteria by Iron Binding Proteins—Transferrin and Lactoferrin

(a) *Iron requirements of bacteria*

Bacteria have an absolute requirement for Fe but the quantitative requirements differ greatly amongst bacterial species. For instance *E. coli* requires 0.02–0.33 p/m of Fe in the medium. *Pseudomonas aeruginosa* with a higher cytochrome content requires appreciably more Fe (Waring & Werkman, 1942) while lactic acid streptococci require as little as 0.0002–0.01 p/m of Fe, according to strain (Reiter & Oram, 1968). The differences between bacterial species is not surprising in view of their different contents of the enzymes in which Fe plays an essential role: such as haem enzymes (e.g. cytochromes, catalase, peroxidase), the oxygenases and the metaloflavoproteins (e.g. xanthine dehydrogenase).

(b) *Transferrin (TF)*

Schade & Caroline (1944) were the first to show that egg white, or rather conalbumin, inhibited growth of *E. coli* and other bacteria and that this inhibition could be reversed by the addition of iron. Later (1946) the same authors showed that serum inhibited the growth of many bacterial species and that the effect was reversed by iron; this inhibition was attributed to siderophilin, now known as transferrin. Schade and his collaborators also demonstrated that a strain of *Staphylococcus aureus* propagated in an iron-limited medium ('low Fe' cells) had an increased dependence upon glycolysis and decreased oxidation compared with 'high Fe' cells; also, although this strain normally produced coagulase, hyaluronidase, haemolysins and staphylokinase, it failed to elaborate these pathogenesis-linked factors when grown in serum (Schade, 1960; Schade, 1963; Theodore & Schade, 1965*a, b*; Schade *et al.*, 1968*a, b*).

Bullen and collaborators published a series of papers on the bactericidal and bacteriostatic effects of serum, and its abolition by iron compounds, on *Clostridium welchii* (Rogers, 1967; Bullen *et al.*, 1967), *Pasteurella septica*, and *E. coli* (Bullen & Rogers, 1969; Bullen *et al.*, 1971). Iron in excess of the binding capacity of TF was shown to permit growth of *E. coli* even in the presence of complement and antibody, thus abolishing the bactericidal activity of the latter system (Bullen & Rogers, 1969; Fletcher, 1971). The mode of action—selective inhibition of macromolecular synthesis—will be discussed in a later section.

(c) *Lactoferrin (LF)*

The iron-binding protein lactoferrin (formerly known as the red protein of milk, lactosiderophiline, lactotransferrine or ekkrinosiderophiline) was isolated and identified by several workers concurrently (Johansson, 1960; Groves, 1960; Montreuil & Mullet, 1960). Although TF and LF show similar biological activity in binding iron, they are immunologically different proteins which do not show any cross reaction (for a monograph on LF see Masson, 1970). Their distribution also differs. While LF does not occur in blood and was first isolated from milk, it also occurs in other secretions, tears, saliva, nasal, bronchial and gastro-intestinal secretions, seminal fluid, cervical mucus, and in polymorphonuclear leucocytes. The distribution is in fact very similar to that of lactoperoxidase and secretory IgA.

The concentration of TF and LF vary greatly according to the animal species; human milk contains >2 mg of LF/ml and <50 μg of TF/ml, whilst bovine and porcine milk contains only 20-200 μg of LF/ml and TF/ml (Masson, 1970). However, when we recently determined by immunodiffusion the LF content of the colostrum and milk of three sows, the colostrum contained >1 mg/ml and the level in milk declined to 0.5 mg/ml within two weeks. Bovine colostra contain up to 5 mg of LF/ml (unpublished).

(d) *Inhibition of* Staphylococcus epidermidis *by lactoferrin*

The inhibitory activity of serum against several bacterial species was established before the isolation of TF, while the inhibitory activity of LF was only discovered some years after the protein had been isolated. Masson *et al.* (1966) reported the inhibition of *Staph. albus* by apolactoferrin (free of bound iron) isolated from human milk. However, Gladstone found that high concentrations of apolactoferrin were not only bacteriostatic, but were bactericidal for *Staph. epidermidis* (Gladstone & Walton, 1970, 1971; Gladstone, 1973). Moreover, the bactericidal activity could not be reversed by iron and in view of the high concentration of LF necessary the authors attributed the bactericidal effect to an impurity in the LF preparation.

(e) *The 'apparent' inhibition of spore germination of* Bacillus stearothermophilus *by lactoferrin*

Wolin & Kosikowski (1958) first demonstrated that raw or dried milk produced zones of inhibition on agar seeded with spores of *B. stearothermophilus*. A similar but weaker effect was observed when spores were suspended in ultra-high-temperature sterilized milk (momentarily heated at 135°) (Franklin *et al.*, 1958). Twenty-three out of 60 strains of thermophilic sporeformers tested were found to be inhibited and the factor proved to be stable to heat (80° for 60 min) and low pH, but lost its activity in the presence of $FeSO_4$ (to a lesser extent in the presence of $MgSO_4$) (Cheeseman & Jayne-Williams, 1964). We considered that the inhibitor might be lactoferrin (Oram & Reiter, 1966a, 1968) and indeed a preparation of lactoferrin obtained from bovine milk inhibited both the germination of the spores and the growth of *B. stearothermophilus*; both effects were reversed by iron. An attempt was made to prepare lactoferrin from dry secretion (see p. 39), which appeared to be extremely inhibitory for *B. stearothermophilus* spores (Auclair, 1964); the inhibitory effect in dry secretion was associated with lactoferrin but was less reversible by Fe than was 'pure' lactoferrin and this was attributed to contamination with β-lysin or lysozyme (Reiter & Oram, 1968). These results are frequently quoted in the literature on lactoferrin. However, more recent work has confirmed that our preparation of LF was impure because it contained traces of lysozymes, which could be detected after chromatography on Sephadex 200 and assay with *Micrococcus lysodeikticus* (unpublished).

Furthermore, we have now found that egg white lysozyme also inhibits the germination of spores of *B. stearothermophilus*, showing zones of inhibition on agar seeded with spores. Egg white lysozyme is believed not to be affected by Fe but lysozyme activity measured as zones of inhibition against *B. stearothermophilus* or lysis of *M. lysodeikticus* (OD) can be reversed by iron in phosphate buffer but not citrate buffer (unpublished).

Table 1

Effect of citrate and bicarbonate on the bacteriostatic activity of dialysed colostrum

HCO_3^- (μM/ml)	Citrate (μM/ml)			
	0	0.1	1	10
0	0.7*	1.9	2.6	2.4
0.7	0.7	0.9	2.5	2.5
7.0	1.1	1.1	1.1	2.4
70.0	1.0	0.7	0.7	0.8

* Increase in $\log_{10}$ viable count after 6 h incubation.
From Reiter *et al.* (1975).

The suggestion of Gladstone & Walton (1971) that an impurity in the preparation of LF may have been responsible for the inhibition of *Staph. epidermidis* seems to be indirectly supported by our results. The particular preparation of their LF was obtained from human milk, which is known to be very rich in lysozymes; the likelihood of contamination with this enzyme is therefore even greater than with our preparation of LF obtained from bovine milk, which is extremely low in lysozymes. The inhibition of the growth of vegetative organisms of *B. stearothermophilus* (Oram & Reiter, 1968) is still likely to be due to lactoferrin but the experimental work will have to be repeated. I am glad to have this opportunity of throwing doubt on the results obtained in the past—a good example of the difficulties in judging the purity of a protein—in this instance by immunological methods (immunoelectrophoresis, immunodiffusion), determination of molecular weight by gel filtration and sedimentation coefficient on the ultracentrifuge; biological activity, in this case lysis of *M. lysodeikticus,* proved capable of detecting contamination which was not shown by any of the other tests.

(f) *Inhibition of* E. coli *by lactoferrin in bovine colostrum (the role of citrate and bicarbonate)*

Colostrum has been reported to have no bactericidal or bacteriostatic activity (Carroll & Jain, 1969; Wilson, 1972). However, Bullen *et al.* (1972), demonstrated that adjustment of the pH of bovine colostrum to 7.2 made it bacteriostatic against *E. coli* 0111. This was not a pH effect because adjusting the pH with phosphate or tris instead of bicarbonate failed to make colostrum bacteriostatic whereas adjustment with bicarbonate achieved the bacteriostatic effect (Reiter *et al.,* 1975). This was not surprising because LF (and TF) bind Fe and bicarbonate mole for mole (Schade *et al.,* 1949; Masson & Heremans, 1968). It was also found that colostrum became bacteriostatic after dilution in Kolmer saline or dialysis. It appeared, therefore, that some low mol. wt component prevented the inhibition of *E. coli* by LF and this was found to be citrate which is known to exchange Fe from TF molecules (Aisen & Leibman, 1968) and ferric citrate can be actively taken up by and promote growth of *E. coli* in low Fe media (Wang & Newton, 1969). The effect of citrate and bicarbonate is illustrated in Table 1 which shows that the more citrate is present the more bicarbonate is required for inhibition. Milk contains only traces of bicarbonate in contrast to blood while the citrate is high in milk but low in blood, hence precolostral calf serum, as a source of antibody-free TF, was found to be more effective in inhibiting growth of *E. coli* in diluted colostrum (diminishing the citrate content) than an equivalent amount of LF (Reiter *et al.,* 1975).

More recently it was found that the secretion of non-lactating udders ('dry secretion') strongly inhibited strains of *E. coli* which cause severe mastitis of the lactating udder and which can lead to the death of the animal unless treated (Reiter & Bramley, 1975). During drying off, the LF content increases appreciably as assayed by immunodiffusion (up to 60 mg/ml). Milk components such as casein, fat and citrate are absorbed into the blood, while blood components, including immunoglobulins and bicarbonate move into the secretion, thus creating ideal conditions for inhibition of *E. coli.* As soon as colostrum is produced towards the time of parturition the bicarbonate and citrate levels are reversed and the colostrum consequently becomes non-inhibitory. This course of events may be responsible for the fact that the lactating udder but not the dry udder is easily infected by *E. coli*. Infections which are dormant during the dry state of the udder flare up during the colostral stage and cows are known to go down with severe coliform mastitis after parturition.

4. The Nature of the Bactericidal Activity of Complement/Antibody and the Bacteriostatic Activity of Iron Binding Proteins

Although bacteria are more complex than erythrocytes in their metabolism, the bactericidal effect of the C/Ab system was generally thought to be similar to the haemolytic effect. This appeared to be confirmed by the demonstration of 'holes' or ultrastructural lesions on the surfaces of both bacteria and erythrocytes. These lesions have a characteristic shape and diameter and were at first thought to cause lysis, however, it now appears that the relevance of the ultrastructural lesion to the functional lesion which leads to lysis is not yet fully explained (see reviews of Humphrey & Dourmashkin, 1969; Dourmashkin *et al.*, 1972; Polley *et al.*, 1972). However, evidence is accumulating of an active detergent effect of complement (C_5-b) in releasing lipids and proteins from the outer membrane of *E. coli* (Kolb *et al.*, 1972; Wilson & Glynn, 1975). In the case of bacteria the metabolic changes caused by the C/Ab system have attracted increased attention.

Amano *et al.* (1955) reported a rapid loss in ability of bacteria exposed to the C/Ab system to synthesize adaptive enzymes. Melching & Vas (1971) and Griffiths (1971*a, b*) have now shown independently that the C/Ab system causes a decrease in RNA accumulation and subsequent inhibition of macromolecular synthesis; DNA, protein and lipids, hence the findings of the release of cell wall associated enzymes, leakage and inhibition of active transport (Heppel, 1967; Davis *et al.*, 1969). Degradation of RNA and ribosome take place at the same time and the death of the cell appears, at least in the case of *P. septica,* to depend on the bacterial oxidative energy metabolism because inhibitors of oxidative phosphorylation protect the bacteria against the bactericidal action of the C/Ab systems (Griffiths, 1974 *a, b*).

This sequence of events fails, however, to explain how the bactericidal activity of the C/Ab system in serum is reversed by the addition of iron (Bullen & Rogers, 1969; Fletcher, 1971), as shown in Fig. 3. In the case of *P. septica,* although iron compounds clearly abolish the bactericidal effect of antiserum, the antibacterial mechanism cannot be explained simply by bacterial iron deficiency. The process is more complex and involves other factors (Griffiths, 1975). This is in contrast to the situation in *E. coli* where the bacteriostatic action of sera can be explained by an interference with bacterial iron supply.

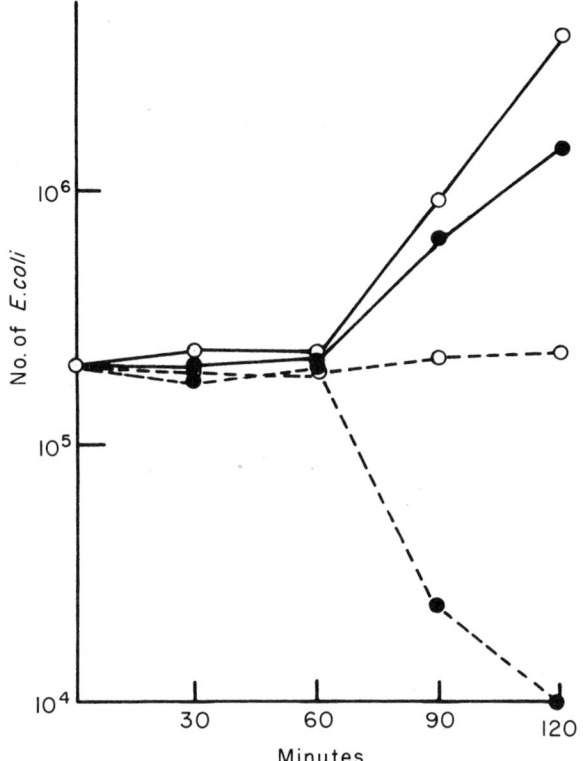

Fig. 3. The effect of iron, 2×10^{-4} M, on the growth or killing of *E. coli* in fresh (●) and complement inactivated (○) serum. ———, with iron; – – –, without iron. From Fletcher (1971).

When *E. coli* is cultured in media containing $< 10^{-7}$M Fe^{3+}, abnormal species of phenylalanyl-tRNA appear (Wettstein & Stent, 1968). Griffiths (1972) discovered that *E. coli* inhibited by serum also contained abnormal phenylalanyl-tRNA, suggesting that the bacteria were deficient in iron. More recently, Griffiths (pers. comm.) chromatographed phenylalanyl-, tyrosyl- and tryptophanyl-tRNA from *E. coli* 0111 inhibited by bovine colostrum adjusted to pH 7.4 with HCO_3^-, and showed that c. 90% of the tRNA was abnormal but was rapidly replaced by the normal form after addition of Fe. Addition of Fe also resulted in the resumption of bacterial growth. It is suggested that the low level of normal tRNA may not be able to support protein synthesis thus causing bacteriostasis, an event which apparently takes place in the presence of purified LF. In the presence of antibodies the inhibition by LF is appreciably increased (Bullen *et al.*, 1972). The latter observation needs some qualification, because we have recently tested in bovine colostrum many enteropathogenic and also non-pathogenic strains of *E. coli* of bovine, human and porcine serotypes. All of these strains were inhibited by bovine colostrum and it seems to be unlikely that specific antibodies occur in bovine colostrum against all of these diverse serotypes (to be published).

5. The Lactoperoxidase/Thiocyanate/Hydrogen Peroxide System (LP/SCN$^-$/H$_2$O$_2$)

(a) *Lactoperoxidase*

As early as 1924 Hanssen observed that freshly drawn milk was bactericidal against '*B. typhosa*' and '*B. paratyphosa*'. He associated this effect with the presence of oxidizing enzymes because the heat destruction of the bactericidal activity (75° for 15 min) correlated with the loss of the oxidizing activity of the milk, measured by the *p*-phenylenediamine colour reaction after the addition of H$_2$O$_2$. Mistakenly, he attributed the observed increased bactericidal activity of summer milk to the 'transfer' of peroxidative enzymes from plants to milk when cows were on pasture. Wright & Tramer (1958) associated the inhibition of some strains of *Streptococcus cremoris* (group N streptococci) in milk with the presence of lactoperoxidase, based also on the inactivation of the enzyme by heat and also by its sensitivity to acid. The lactoperoxidase was, according to these authors, identical with the so-called lactenin 2, first described by Auclair (1954) and thought to be responsible for the inhibition of lactic streptococci and later for lactobacilli also (Auclair & Portmann, 1959; Portmann *et al.*, 1962).

By then a simple method of LP purification had been developed (Morrison *et al.*, 1957) and Portmann & Auclair (1959) confirmed that a preparation of LP reactivated the inhibitory activity of heated milk against lactic streptococci and also restored the bactericidal activity against *Strep. pyogenes*. Previous to this work, unsuccessful attempts had been made to purify and identify the bactericidal agent in milk active against 'scarlet fever streptococci' and the inhibitor in milk which acted temporarily against a streptococcus isolated from a mastitic udder (Jones & Little, 1927; Jones, 1928; Jones & Simms, 1930). Wilson & Rosenblum (1952*a*, *b*) confirmed the findings of Jones and showed that Group A streptococci were always killed in milk but only when < 1000 colony forming units (cfu) were suspended. Also, strains of other serological groups F, G, H, K and L varied in their sensitivity; some were strongly inhibited or killed and others were resistant. By contrast, strains of Groups B (mastitis streptococci), C, D and G could be temporarily inhibited but ultimately achieved full growth in raw milk as in heated milk. The same authors discounted the hypothesis of Hanssen that an oxidizing enzyme was involved because they failed to reverse the inhibition by cyanide which would inactivate a haem enzyme such as lactoperoxidase. The effect of cyanide remains difficult to understand because Jago & Morrison (1962) who used concentrations of cyanide which decreased the oxidation of guaiacol by LP in milk failed to reverse the inhibition of a strain of lactic acid streptococcus. Higher concentrations of cyanide directly inhibited the organisms.

(b) *Hydrogen peroxide*

Wright & Tramer (1958) suggested that the inhibitory activity of LP in the presence of H_2O_2 produced by the metabolism of the streptococci was due to the formation of an oxidation product of a quinonoid present in milk. This was supported by the observation of Wilson & Rosenblum (1952a, b) that inhibition occurred only under aerobic conditions and was reversed by a number of reducing agents. However, the same authors had discarded the hypothesis of a peroxidative reaction because they failed to reverse the inhibition by catalase. Later, Jago & Morrison (1962) showed that H_2O_2 was necessary for the reaction and that catalase reversed the inhibition only if present in high concentration, since LP competes effectively for H_2O_2 with the catalase. Under some circumstances, particularly in synthetic media (Oram & Reiter, 1966b) the inhibition by LP was often unsatisfactory because the organisms did not produce sufficient H_2O_2 to activate the LP system. To avoid such failures H_2O_2 can be produced enzymically by the addition of glucose oxidase and glucose; it has also been observed that ascorbic acid (which is present in varying concentrations in milk of different species) promotes H_2O_2 production.

(c) *Thiocyanate*

The hypothesis of Wright & Tanner (1958) that the inhibitor for lactic acid streptococci might be produced by oxidation was investigated later. Reiter *et al.* (1963) established that a third factor (besides LP and H_2O_2) was involved because dialysed milk or synthetic medium containing LP was non-inhibitory unless c. 20% of non-dialysed milk was added. It was possible to discard the suggestion of Wright & Tramer that a quinonoid-like substance would be oxidized by LP/H_2O_2, because quinones were found to be inhibitory but only non-selectively and in milk only some strains of streptococci were inhibited. Selectivity was characteristic for the natural milk inhibitor associated with LP.

Next, the effects of naturally occurring oxidizable substances in milk such as iodine and indoxyl sulphuric acid (indican) were investigated (Reiter *et al.*, 1964). These substances were strongly inhibitory in the presence of LP but only at concentrations higher than those occurring in milk, (non-physiological) and they also lacked the specificity of the natural milk inhibitory system.

Eventually, large scale chromatography with a strong anionic exchange resin yielded sufficient material for it to be identified by Dr J. D. S. Goulden (NIRD) by IR spectroscopy as thiocyanate (SCN^-). Since removal of SCN^- from milk by chromatography with strong ion-exchange resins made milk non-inhibitory, it was proven that SCN^- was the third factor, next to H_2O_2, which was required to activate the LP system.

(d) *The mode of inhibition of lactic acid bacteria by the LP system*

In synthetic media LP, H_2O_2 and SCN^- inhibited the growth and acid production of the same strains of group N streptococci that are inhibited in milk. This system inhibited O_2 uptake and oxidized reduced nicotinamide adenine dinucleotide (NADH) in proportion to the amount of H_2O_2 added. The latter oxidation accounted for the observed inhibition of triose phosphate dehydrogenase. At the same time as SCN^- disappeared an intermediate oxidation product (absorbing at 235 nm: '235 compound') was detected (Reiter *et al.*, 1964). Of the glycolytic enzymes hexokinase was most strongly inhibited, aldolase and 6-phosphogluconate dehydrogenase were only partially inhibited but several other enzymes remained unaffected. Strains which were resistant to the LP system did not fail to produce H_2O_2 or dissociate it, inactivate LP or use an alternative pathway of carbohydrate metabolism as suggested by Stadhouders & Veringa (1962), but they possessed a 'reversal factor' which catalysed the oxidation of $NADH_2$ in the presence of an intermediate oxidation product of SCN^- ($NADH_2$-oxidizing enzyme) and also reversed in purified form or as cell-free extract the inhibiton of glycolysis of the LP sensitive strains. This resistance of strains of the same species (in this case *Strep. cremoris*) should be noted because later workers generally investigated only one strain of a bacterial species, although Wilson & Rosenblum (1952*a*, *b*) had already shown that strains of the same sera group of streptococci could be sensitive or resistant to 'lactenin'. Mickelson (1966) concluded that those strains of *Strep. agalactiae* (Group B) which were unable to utilize an oxidative pathway for their energy supply (cyanide sensitive respiration) were the most inhibited.

In becoming independent of the fermentative pathway, these organisms were no longer sensitive to the $LP/SCN^-/H_2O_2$ system and started to multiply. The same author gave the only adequate interpretation up to now of the fact that *Strep. pyogenes* (Group A) was not inhibited temporarily but was killed by the LP system: he concluded that these organisms have to obtain their energy by the fermentative pathway because they have no oxidative capacity and hence were therefore most affected by the LP system and killed. Since the glyceraldehyde phosphate dehydrogenase was inhibited by the LP system and reversed by cysteine and glutathione, he postulated the peroxidative conversion of essential enzymic sulphydryl groups thus interfering with the energy metabolism of *Strep. pyogenes*.

(e) *The nature of the inhibitory oxidation product of SCN$^-$*

The nature of the intermediate oxidation product of SCN$^-$ has not yet been established. The end products of the oxidation of SCN$^-$ catalysed by LP/H$_2$O$_2$ are sulphate, carbon dioxide and ammonia, none of which is inhibitory. The intermediate products are cyanate, sulphite and the '235 compound'. These intermediates do not inhibit either growth or glycolysis. However, sulphur dicyanide which can be produced by the perchloric acid-catalysed oxidation of SCN$^-$ by H$_2$O$_2$ (Wilson & Harris, 1961) does inhibit glycolysis. The relationship between the 235 compound (E$_{235}$) and sulphur dicyanide remains obscure.

Because the inhibitor was too unstable to be identified by direct chemical means, Hogg & Jago (1970*a, b*) attempted to identify it by its properties in aqueous solutions. The inhibitor displayed a polarographic reduction wave of which the half-wave potential was pH-dependent. Studies of the variation of the polarographic half-wave potential and of the UV extinction with pH indicated that it had a pK$_a$ 5.1 ± 0.1. The inhibitor decomposed by a mechanism involving H$^+$ and thiocyanate, the kinetics varying according to whether the inhibitor was in the acidic or basic form. From these studies it was concluded that the inhibitor was either cyanosulphurous acid or cyanosulphuric acid. Hogg & Jago (1970*a, b*) also found, in contrast to Reiter *et al.* (1964) and Oram & Reiter (1966*a, b, c*), that the inhibitor was responsible for the increase in E$_{235}$ and was not sulphur dicyanide as the latter compound is unstable in aqueous solutions at pH 7.0. However, as sulphur dicyanide displayed a polarographic wave with a half-wave potential close to that of the inhibitor at pH 7.0, it was thought that sulphur dicyanide might, upon decomposition, give rise to the inhibitor.

(f) *The bactericidal activity of the LP system against* E. coli, Ps. fluorescens *and other Gram negative organisms*

Although lactic acid bacteria become temporarily self-inhibited in the presence of LP/SCN$^-$ and metabolically-produced H$_2$O$_2$ they are not killed. Hydrogen peroxide can also be supplied exogenously at sublethal levels by enzyme reactions such as glucose oxidase/glucose which assure reliable and reproducible results; for optimal LP-catalysed oxidation of SCN$^-$ a proportion of 3-3.5 mol H$_2$O$_2$/mol SCN$^-$ was required (Oram & Reiter, 1966*b*). Gram negative organisms, however, cannot be self-inhibited by the LP system; no free or available H_2O_2 is produced because of their catalase activity. If *E. coli*, however, is suspended in milk (or whey) which contains LP and sufficient SCN^- (~ 10 μg/ml) the catalase of the milk or whey and the catalase of the organisms can be overcome by producing enzymically sufficient H_2O_2 to activate the LP system. Since it is difficult to measure the H_2O_2 level in milk or whey, the events are best explained by following the reaction in synthetic media to which LP, SCN^-, glucose oxidase and glucose are added. As long as H_2O_2 production is only sufficient to oxidize the SCN^- present, little or no free H_2O_2 can be detected, even though the organisms are being rapidly killed. *Escherichia coli* cells are therefore not only inhibited like the lactic acid bacteria but killed by the LP system (Fig. 4, Reiter *et al.*, 1976).

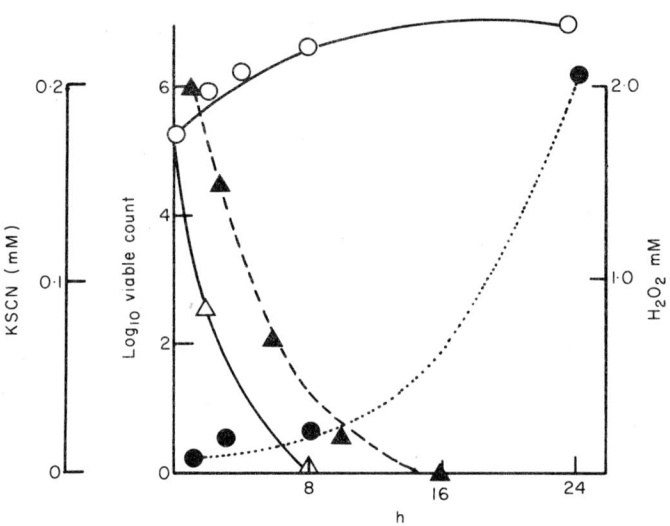

Fig. 4. Bactericidal effect against *E. coli* 0111 of lactoperoxidase (1.5 μg/ml), SCN^- (0.225 mM) and glucose oxidase (0.1 μg/ml) in a synthetic medium. o———o, control, growth in medium only; △———△, viable count in presence of bactericidal system; ▲———▲, oxidation of SCN^-; ●– – – –●, H_2O_2 level. From Reiter *et al.* (1976).

(g) *The nature of the bactericidal activity of the LP system*

Preliminary work (Marshall & Reiter, unpublished) indicates that synthesis of TCA-precipitable protein and of RNA ceases rapidly after exposure of *E. coli* to the system in a synthetic medium. Since the incorporation of glucose and amino acids into the amino acid pool is also inhibited, it appears that the active transport across the membrane is affected.

In contrast to these rapid metabolic changes, the morphological changes are much delayed, but after 1½-2 h exposure to the LP system, death of the organism occurs and cytoplasm is extruded but always in a polar position. This pole may be the last formed cell envelope dividing the two daughter cells (Plate 1).

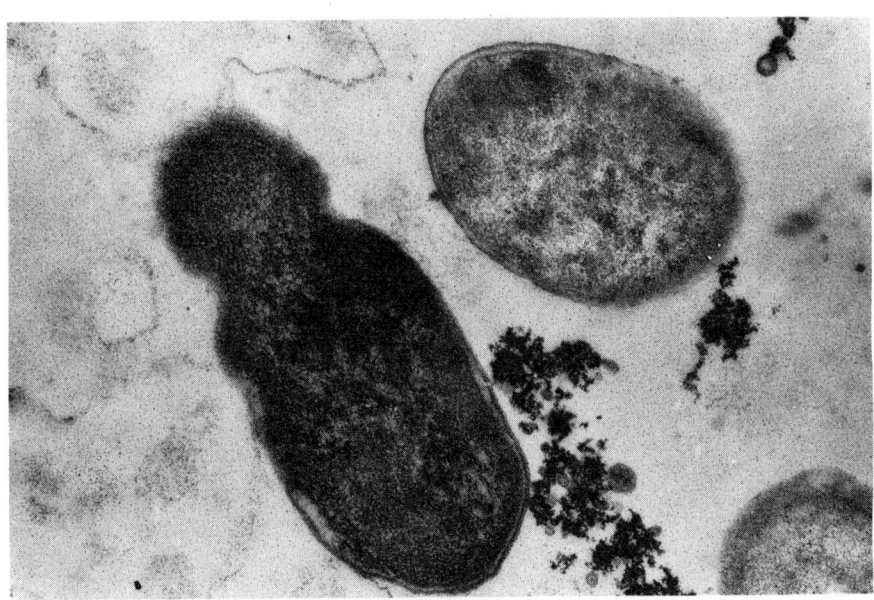

Plate I. Cells were collected on a Millipore filter (0.22 μm pore size), covered with a thin layer of agar and fixed for 1 h in 0.2 M cacodylate–HCl buffered 3% gluteraldehyde. The filter was washed in buffer, fixed for 1 h in 1% osmium tetroxide and *en bloc* stained with 1% uranyl acetate for ½ h. After dehydration in a graded series of alcohol–water mixtures, it was embedded in Araldite. (By courtesy of B. E. Brooker and D. E. Hobbs, NIRD.)

(h) *Preservation of milk by the LP system*

After prolonged storage even at refrigeration temperatures, milk is spoiled because of the multiplication of psychrotrophs. To investigate whether such organisms are affected by the LP system (Björck *et al.*, 1975), *Ps. fluorescens* EF 1998 was suspended in raw milk after the addition of glucose oxidase and glucose. The enzymically produced H_2O_2 had only a slight bactericidal effect but after addition of SCN⁻ the bactericidal effect was quite dramatic in the first 6 h or so and the continuous production of H_2O_2 was sufficient to maintain bacteriostasis for 24 h (Fig. 5). Eight isolates of Gram negative rods, capable of

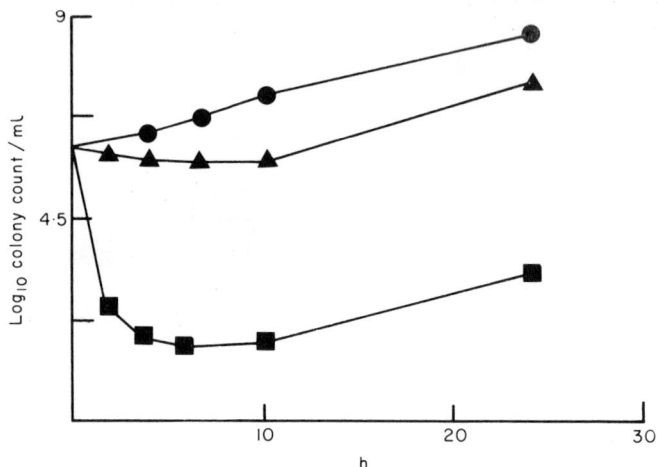

Fig. 5. *Ps. fluorescens* EF 1998 in raw milk containing 0.085 mM SCN⁻. Supplements: ●, none; ▲, 0.1 U of glucose oxidase/ml and 0.3% glucose; ■, 0.1 U of glucose oxidase/ml, 0.3% glucose and 0.085 mM SCN⁻. From Björck *et al.* (1975).

multiplying at 4°, were also tested and 90% of the inoculum of each isolate were found to be killed after 4 h under these conditions. Under pilot plant conditions milk could be maintained palatable for ∼ 3 weeks. Similar results were obtained with an immobilized enzyme system. This process of milk preservation has been patented by Alfa Laval.

6. Basic Proteins and Other Antibacterial Factors

(a) *β-lysin*

The bactericidal property of blood serum against Gram positive bacteria was first described by Fodor (1887) and the component became known as β-lysin—a basic protein (like lysozyme, LF and LP) which is released from blood platelets during the formation of the blood clot. It has a similar effect on bacteria as that of the basic protein isolated from polymorphonuclear leucocytes (Zeya & Spitznagel, 1966*a, b*) affecting cell permeability and thus causing leakage of nucleotide components and affecting the oxidative metabolism. Iron reverses the bactericidal activity of β-lysin as also that of the cationic protein(s) isolated from polymorphonuclear leucocytes (Gladstone & Walton, 1971; Gladstone, 1973). The concentration of β-lysin varies greatly according to the species of animals but is present only in very low concentrations in the blood plasma. Since it was also detected in saliva (Jensen *et al.*, 1967) it could occur in milk but so far has not been investigated although the technique of isolation is simple (Donaldson *et al.*, 1964).

Basic proteins isolated from the keratin of the teat canal are active against *Strep. agalactiae* and *Staph. aureus* (Hibbitt & Cole, 1968; Hibbitt *et al.*, 1969), but recent work has shown that such proteins can be rapidly inactivated in milk or casein (Russell & Reiter, 1975; Russell *et al.*, 1976). In this context it is interesting to note that Hirsch (1958) and Hirsch & Cohn (1960) found the bactericidal activity of phagocytin (the basic protein extracted from leucocytes, and probably identical with the basic protein isolated from the same source by Spitznagel and his collaborators) to be abolished by casein. The ease with which basic proteins can be inactivated is overcome during the intracellular killing of bacteria by leucocytes. After ingestion of bacteria during phagocytosis a vacuole is formed and the lysosomes containing various antibacterial substances, including basic proteins, migrate towards the vacuole; the membranes of the vacuole and lysosome fuse and the content (e.g. basic protein) is discharged into the vacuole, thus ensuring that the basic protein hits the target.

Recently, Whittlestone (1975) cited the discovery by Shannon, Twomey & Molan of a small basic polypeptide which is not only bactericidal to a wide range of bacteria but retains its activity even when adsorbed to proteins. It has been named 'ubiquitin' and occurs in bovine seminal plasma, walls of leucocytes, saliva, pancreas, and in parts of the digestive tract. It is secreted in large quantities by the Fürstenberg rosette at the top of the teat canal and may be important for the defence of the udder against bacterial infection.

(b) *Agglutinins*

Wright & Tramer (1957) suggested that lactenin 1 (L_1) (Auclair & Hirsch, 1953) was associated with the agglutinins which agglutinate fat globules and during the creaming of milk carry some strains of *Strep. cremoris* to the surface. McPhillips (1958) and Portmann *et al.* (1960) established that milk can contain specific agglutination antibodies of low titres (1:4 to 1:8) against such strains, and used these antibodies to serotype a number of strains of *Strep. cremoris*. Reiter, *et al.* (1964) showed later that infusion of streptococci into the dry udder produced specific antibodies of high titre in the colostrum which could be used to serotype Group N streptococci into groups which were found to agree with the grouping of these organisms by their susceptibility to lytic phages. The agglutinins indirectly cause inhibition of the streptococci by being swept up with the cream in whole milk, and in skim milk the streptococci are deposited and do not acidify the milk uniformly, as in the manufacture of cottage cheese (Emmons *et al.*, 1963).

The question remains as to how the streptococci attach to the fat globules to be swept up with the cream. This phenomenon is not unique because the well known *Brucella abortus* 'ring test' is based on the rise of the organisms to the top of the milk in the presence of brucella antibodies. Dowden *et al.* (1967) demonstrated that antisera against bovine fat globule membranes agglutinate bovine erythrocytes. This is not surprising because the fat globule membrane is a true animal cell membrane. Since serum contains 'cold agglutinins' for erythrocytes, and milk creams best at low temperatures (8-10°), it was argued that the agglutinins of the fat globules could be derived from the blood. This was confirmed by suspending erythrocytes in milk at low temperatures which greatly diminished the creaming of the milk and by the restoration of the creaming properties of the milk when the erythrocytes were resuspended in milk at elevated temperatures (37°) at which cold agglutinins are desorbed (Reiter, 1967). The presence of cryoglobulins in milk was confirmed by Stadhouders & Hup (1970) who postulated that the attachment required yet a third type of agglutinin of unknown origin. In the light of present knowledge of attachment of bacteria this seems to be unnecessary. Brooker (1976) has recently shown by electron microscopy that some streptococci have a surface carbohydrate which appears to be filamentous and thus capable of attaching to the curd matrix of cheese. Also, in milk autoclaved to destroy antibodies, streptococci were seen to attach in a similar manner to fat globules and to casein micelles; thus the clumps of agglutinated bacteria may also attach to fat globules removing a great number of bacteria and reducing appreciably the population of streptococci in raw milk. Most recent work indicates that cells of *E. coli* K_{88}, a pathogen for piglets (Sojka, 1971), are also capable of attaching to fat globules and fat globule membranes at 37°. In contrast to the non-specific attachment of these organisms to erythrocytes which occurs only at 4° (Jones & Rutter, 1974) this attachment appears to be quite specific (Reiter & Brown, unpublished). It is well known that milk also contains agglutinating antibodies against streptococci, staphylococci and *E. coli*; although these antibodies are capable of serving as opsonins for phagocytosis (Russell & Reiter, 1975; Russell *et al.*, 1976) and bactericidal activity if sufficient complement is present (Reiter & Brock, 1975) they are unable to prevent mastitis (Reiter *et al.*, 1970).

(c) *Properdin*

Serum contains a high mol. wt protein which, together with C and magnesium is reputed to be antibacterial and antiviral. It was originally isolated by Pillemer *et al.* (1954) but its precise mode of action and relation to other serum components remains to be determined. Blanc (1964) detected low titres of properdin in milk.

(d) *Conglutinin*

Sera of ruminants contain a relatively heat stable protein which greatly enhances agglutination of cells in the presence of specific antibody and C. So far, conglutinin has only been shown to occur in colostrum (Ingram & Mitchell, 1970), and was found to be absent from milk (Korhonen, 1973).

(e) *Vitamin B_{12} and folate protein binders*

The milk of many species contains proteins which bind strongly these vitamins; the concentration and relative proportion of saturated and unsaturated binder proteins varies widely between animal species. Bacteria that take up free vitamins rapidly are unable to take up protein-bound vitamins thus presumably making the vitamins available to the young animal. Since many of the bacterial species tested colonize the stomach and intestines, the vitamin binders may strongly influence the ecology of the intestinal flora of the neonate and should be included in the antibacterial factors occurring in milk (e.g. Ford, 1974; Ford *et al.*, 1975).

7. Discussion

(a) *The function of the antibacterial system of colostrum and milk*

The introduction indicated the interest of medical, veterinary and dairy scientists in the antibacterial activity of milk. The effect of the milk inhibitors on cheese making and preservation of milk are manufacturing problems; their possible physiological significance can only relate to the defence of the udder and/or the neonate against infection. Immunity is generally defined as resistance to infection through the presence of protecting antibody systems. However, non-antibody factors of the milk, such as the LP and TF/LF systems, may play an important role in immunity, not only augmenting antibody action but also giving protection before specific antibody responses become effective.

(b) *Mammary gland*

The antibody defence of the bovine mammary gland (for a recent review see Reiter & Bramley, 1975), including phagocytosis, is very inefficient. There is no lack of opsonins in milk for the ingestion of staphylococci for instance, but the main reason for the extreme susceptibility to bacterial infection is that phagocytosis is inefficient. Gram negative organisms can be eliminated not only by phagocytosis but also by direct bactericidal action of C/Ab. Nevertheless, invasion of the lactating udder with Gram negative organisms causes an extremely rapid development of mastitis and the transfer of C/Ab from the blood is apparently too slow to influence the disease (Reiter & Bramley, 1975). In any case the majority of mastitic coliforms are serum resistant (A. J. Bramley, NIRD, pers. comm.). The C/Ab system appears not to constitute an effective defence mechanism, at least in the cow. The TF/LF system is inefficient in the lactating udder because the level of these proteins is too low and the citrate content too high, but in the dry or non-lactating udder it was shown to be efficient.

So far, no *in vivo* experiments have been performed to prove or disprove whether the LP system can prevent streptococcal mastitis. LP is always present in milk in sufficient concentration but the SCN^- level depends on the feeding regime. SCN^- is derived through the rhodanase-catalysed reaction in the liver and kidney, from the detoxification of cyanide (present in such feed as clover) in the rumen, and from glucosides (present in *Brassica* and *Raphanas* spp., Virtanen, 1961). The third component of the system, peroxide, is present in freshly cannulated milk only in very low concentrations, 2-4 µg/ml (Cousins & Longman, NIRD, pers. comm.), which may be insufficient to activate the system unless the invading organisms (e.g. streptococci) can produce additional H_2O_2 *in vivo*.

Hydrogen peroxide may be derived from the metabolism of the tissue or the oxidation of an unknown substrate by xanthine oxidase which is attached to the fat globule membrane. There is, however, some indirect evidence that the LP system may be active in the cow. *Strep. uberis* was monitored in the milk and on the skin of the udder of a herd throughout a year and frequently isolated except during the summer months (Forbes, 1970). This coincides with the time when the SCN^- content of the milk is at its highest level (Boulangé, 1959) and *Strep. uberis* like *Strep. agalactiae* is susceptible to the LP system (unpublished).

The second indirect evidence is the virtual absence of mastitis caused by Group A streptococci (*Strep. pyogenes*), the most common streptococcus infecting man but which is killed in raw milk.

By the ease with which the udder can be infected by introducing bacteria through the teat canal into the teat cistern, it appears that the main defence is the teat canal itself. Although this aspect does not concern this review, it is interesting to speculate whether besides the physical barrier of the teat canal and its lining the newly discovered bactericidal basic protein ubiquitin plays an important part in the defence against penetration of bacteria (for a recent review see Reiter & Bramley, 1975).

The situation in the (dry) udder is very different. LF and TF definitely prevent infection with *E. coli* and it is possible that phagocytosis is more efficient in the dry than in the lactating udder. However, this still needs to be investigated. Altogether the study of the susceptibility and resistance of the dry udder to bacterial infection may well be the most rewarding facet to investigate in relation to antibody and non-antibody factors.

(c) *Milk inhibitors and defence against infection in the neonate*

It is well known that neonates are more resistant to enteric infections when suckled than when fed heated milk or formulated feeds. This is irrespective of whether they are born with humoral immunoglobulins (e.g. human infant), or without immunoglobulins and relying on transfer in the gut from the colostrum into the blood (e.g. calves and piglets). The role of the immunoglobulins in the intestine is now increasingly appreciated; it involves prevention of attachment of enteropathogenic strains of *E. coli* to the intestinal epithelium, bactericidal activity, opsonization and neutralization of viruses (for recent reviews see Shearman *et al.*, 1972; Walker & Haig, 1973). The immunoglobulins of milk can be considered to bridge the immunological gap until the newborn animal begins to synthesize its own antibodies.

It is now becoming evident that non-antibody factors contained in milk may contribute towards the establishment of a balanced microflora in the intestinal tract and therefore contribute towards the defence of the newborn against microbial infections.

Bullen *et al.* (1972) showed that in suckled guinea-pigs dosed with *E. coli* 0111 the counts of this organism in the intestine were appreciably lower and lactobacilli counts higher compared with those in artificially fed guinea-pigs, in which *E. coli* was predominant. These authors produced evidence that the milk actually depressed *E. coli* levels. Dosing the suckling guinea-pigs with haematin greatly increased the coli counts, probably because it reversed the bacteriostatic activity of the LF of the milk *in vivo* as *in vitro*. In this context it is interesting that the predominant faecal flora of suckled babies consists of lactic acid bacteria in contrast to artificially fed babies which have a predominant coliform flora.

Guinea-pig milk, like human milk, contains a high concentration of LF compared with bovine milk; the latter, however, contains high concentrations of LP. Recently we have shown (unpublished) that the bactericidal system of LP is active *in vivo* and reduces the *E. coli* count in the stomachs and small intestines of the calf and piglet. This is encouraging because pig LP is particularly active against streptococci (Morrison & Steele, 1968) and the calf contains only 10-20% of the adult level of LP in the first post-natal days (Morrison & Steele, 1968); it may be significant that the concentration of LP in bovine milk is highest between 2 and 14 days after parturition (Kiermeier & Kayser, 1960) when the neonate is most susceptible to infection.

It seems likely that the non-antibody factors discussed in this review play as vital a part as the immunoglobulins in the defence of the neonate against infection until the animal develops its own defence mechanisms.

8. References

ADINOLFI, M., GLYNN, A. A., LINDSAY, M. & MILNE, C. M. (1966). Serological properties of A antibodies to *Escherichia coli* present in human colostrum. *Immunology* **10**, 517.

AISEN, P. & LEIBMAN, A. (1968). The stability constants of the Fe^{3+} conalbumin complexes. *Biochemical & Biophysical Research Communications* **30**, 407-413.

AMANO, T., KINJO, K., NISHIMOTO, M., INOUE, K. & YACHIKU, H. (1955). Studies on the immune bacteriolysis. VI. Cause of the death of bacteria by immune bacteriolysis. *Medical Journal of Osaka University* **6**, 57. Cited by Melching, L. & Vas, S. I. Effects of serum components on Gram-negative bacteria during bactericidal reactions. *Infection & Immunity* **3**, 107-115.

AUCLAIR, J. E. (1954). The inhibition of microorganisms by raw milk. III. Distribution and properties of two inhibitory substances, lactenin 1 and lactenin 2. *Journal of Dairy Research* **21**, 323-336.

AUCLAIR, J. E. (1964). Les substances antibactériennes du lait cru et leur rôle en technologie laitière. In *Microbial Inhibitors in Food*, ed. Molin, N. Stockholm: Almquist & Wiksell.

AUCLAIR, J. E. & HIRSCH, A. (1953). The inhibition of microorganisms by raw milk. I. The occurrence of inhibitory and stimulatory phenomena. Methods of estimation. *Journal of Dairy Research* **20**, 45-59.

AUCLAIR, J. E. & PORTMANN, A. (1959). Action inhibitrice des lacténines sur les streptocoques lactiques des levains de fromagerie. *Proceedings of the 15th International Dairy Congress, London* **2**, 580-586.

BASCH, R. S. (1965). Inhibition of the complement system by derivatives of aromatic amino acids. *Journal of Immunology* **94**, 629-640.

BERRIDGE, N. J. (1955). Inhibitory substances of bacterial and other organisms in milk and milk products. *Journal of the Science of Food and Agriculture* **2**, 65-72.

BJÖRCK, L., ROSEN, C. G., MARSHALL, V. & REITER, B. (1975). The antibacterial activity of the lactoperoxidase system in milk against pseudomonads and other Gram-negative bacteria. *Applied Microbiology* **3**, 199-204.

BLANC, B. (1964). Les protéines du lactosérum. Les relations avec l'immunité et la métabolisme du fer. *Thèse, Médecine et Hygiene, Genèvé*.

BOULANGÉ, M. (1959). Fluctuation saisonière du taux des thiocyanates dans le lait frais de vache. *Compte rendu des séances de la Société de biologie, Paris* **153**, 2019-2020.

BROCK, J. H., STEEL, E. D. & REITER, B. (1975). The effect of intramuscular and intramammary vaccination of cows on antibody levels and resistance to intramammary infection by *Staphylococcus aureus*. *Research in Veterinary Science* **19**, 152-158.

BROOKER, B. E. (1976). Cytochemical observations of the extracellular carbohydrate produced by *Streptococcus cremoris*. *Journal of Dairy Research* **43**, 283-290.

BULLEN, J. J. & ROGERS, H. J. (1969). Bacterial iron metabolism and immunity to *Pasteurella septica* and *Escherichia coli*. *Nature, London* **224**, 380-382.

BULLEN, J. J., CUSHNIE, G. H. & ROGERS, H. J. (1967). The abolition of the protective effect of *Clostridium welchii* Type A antiserum by ferric iron. *Immunology* **12**, 303-312.

BULLEN, J. J., ROGERS, H. J. & LEWIN, J. E. (1971). The bacteriostatic effect of serum on *Pasteurella septica* and its abolition by iron compounds. *Immunology* **20**, 391-406.

BULLEN, J. J., ROGERS, H. J. & LEIGH, L. (1972). Iron-binding proteins in milk and resistance to *Escherichia coli* infection in infants. *British Medical Journal* **1**, 69-75.

CAROLL, E. J. & JAIN, N. C. (1969). Bactericidal activity of normal milk, mastitic milk and colostrum against *Aerobacter aerogenes*. *American Journal of Veterinary Research* 30, 1123-1132.

CHANDRAN, R. C., SHAHANI, K. M. & HOLLY, R. G. (1964). Lysozyme content of human milk. *Nature, London* 204, 76-77.

CHEESEMAN, G. C. & JAYNE-WILLIAMS, D. J. (1964). An inhibitory substance present in milk. *Nature, London* 204, 688-689.

DAVIS, S. D., BOATMAN, E. S., GEMSA, D., IANETTA, A. & WEDGWOOD, R. J. (1969). Biochemical and fine structural changes induced in *Escherichia coli* by human serum. *Microbiology* 1, 69-86.

DAVIS, S. D., IANETTA, A. & WEDGWOOD, R. J. (1972). Bactericidal reactions of serum. In *Biological Activities of Complement*, ed. Ingram, D. G. Basel: Karger.

DONALDSON, D. M., ELLSWORTH, B. & MATHESON, A. (1964). Separation and purification of β lysin from normal serum. *Journal of Immunology* 92, 896-901.

DOURMASHKIN, R. R., HESKETH, R., HUMPHREY, J. H., MEDHURST, F. & PAYNE, N. (1972). Electron microscopic studies of the lesions in cell membranes caused by complement. In *Biological Activities of Complement*, ed. Ingram, D. G. Basel: Karger.

DOWDEN, R. M., BRUNNER, J. R. & PHILPOTT, D. E. (1967). Studies on milk fat globule membranes. *Biochimica et biophysica acta* 135, 1-10.

EMMONS, D. B., ELLIOTT, J. H. & BECKETT, D. C. (1963). Agglutination of starter bacteria, sludge formation, and slow acid development in Cottage cheese manufacture. *Journal of Dairy Science* 46, 600.

FLETCHER, J. (1971). The effect of iron and transferrin on the killing of *Escherichia coli* in fresh serum. *Immunology* 20, 493-500.

FODOR, J. (1887). Die Fähigkeit des Blutes Bakterien zu vernichten. *Deutsche medizinische Wochenschrift* 13, 745. Cited in Skarnes, R. C. & Watson, D. W. (1957). Antimicrobial factors of normal tissues and fluids. *Bacteriological Reviews* 21, 273-294.

FORBES, D. (1970). The survival of Micrococcaceae in bovine teat canal keratin. *British Veterinary Journal* 126, 268-274.

FORD, J. E. (1974). Some observations on the possible nutritional significance of vitamin B_{12}- and folate-binding proteins in milk. *British Journal of Nutrition* 31, 243-257.

FORD, J. E., SCOTT, K. J., SANSOM, B. F. & TAYLOR, P. J. (1975). Some observations on the possible nutritional significance of vitamin B_{12}- and folate-binding proteins in milk. Absorption of (^{58}Co) cyanocobalamin by suckling piglets. *British Journal of Nutrition* 34, 469-492.

FRANKLIN, J. G., WILLIAMS, D. J., CHAPMAN, H. R. & CLEGG, L. F. L. (1958). Methods of assessing the sporicidal efficiency of an ultra-high-temperature milk sterilizing plant. II. Experiments with suspensions of spores in milk. *Journal of Applied Bacteriology* 21, 47-50.

GLADSTONE, G. P. (1973). *Staphylococci and Staphylococcal infections*, ed. Jelzaszewicz, J. *Proceedings of 2nd International Symposium, Warszawa*. Polish Medical Publishers.

GLADSTONE, G. P. & WALTON, E. (1970). Effect of iron on the bactericidal proteins from rabbit polymorphonuclear leucocytes. *Nature, London* 227, 849-851.

GLADSTONE, G. P. & WALTON, E. (1971). The effect of iron and haematin on the killing of staphylococci by rabbit polymorphs. *British Journal of Experimental Pathology* 52, 452-464.

GOLDMAN, A. S. & SMITH, C. W. (1973). Heat resistant factors in human milk. *Journal of Pediatrics* 82, 1082-1090.

GRIFFITHS, E. (1971*a*). Selective inhibition of macromolecular synthesis in *Pasteurella septica* by antiserum and its reversal by iron. *Nature, London* 232, 89-90.

GRIFFITHS, E. (1971*b*). Mechanism of action of specific antiserum on *Pasteurella septica*. Selective inhibition of net macromolecular synthesis and its reversal by iron compound. *European Journal of Biochemistry* 23, 69-76.

GRIFFITHS, E. (1972). Abnormal phenylalanyl-tRNA found in serum inhibited *Escherichia coli* strain 0111. *Febs Letters* 25, 159-164.

GRIFFITHS, E. (1974a). Rapid degradation of ribosomal RNA in *Pasteurella septica*. *Biochimica et biophysica acta* **340**, 400-412.
GRIFFITHS, E. (1974b). Metabolically controlled killing of *Pasteurella septica* by antibody and complement. *Biochimica et biophysica acta* **340**, 598-602.
GRIFFITHS, E. (1975). Effect of pH and haem compounds on the killing of *Pasteurella septica* by specific antiserum. *Journal of General Microbiology* **88**, 345-354.
GROVES, M. L. (1960). The isolation of red protein from milk. *Journal of the American Chemical Society* **82**, 3345-3350.
HANSON, L. Á. & WINBERG, J. (1972). Breast milk and defence against infection in the newborn. *Archives of Disease in Childhood* **47**, 845-848.
HANSSEN, F. W. (1924). The bactericidal property of milk. *British Journal of Experimental Pathology* **5**, 271-280.
HEDDLE, R. J., KNOP, J., STEELE, E. J. & ROWLEY, D. (1975). The effect of lysozyme on the complement action of different antibody classes. *Immunology* **28**, 1061-1066.
HEPPEL, L. A. (1967). Selective release of enzymes from bacteria. *Science, New York* **156**, 1451-1455.
HESSE, W. (1894). Über die Beziehungen zwischen Kuhmilch und Cholerabacillen. *Zeitschrift für Hygiene und Infektionskrankheiten* **17**, 238-271.
HIBBITT, K. G. & COLE, C. B. (1968). The antibacterial activity of teat-canal cationic proteins. *Biochemical Journal* **106**, 39P.
HIBBITT, K. G., COLE, C. B. & REITER, B. (1969). Antibacterial proteins isolated from the teat canal of the cow. *Journal of General Microbiology* **56**, 365-371.
HILL, I. R. & PORTER, P. (1974). Studies of bactericidal activity to *Escherichia coli* of porcine serum and colostral immunoglobulins and the role of lysozyme with secretory IgA. *Immunology* **26**, 1239-1250.
HIRSCH, J. G. (1958). Bactericidal action of histone. *Journal of Experimental Medicine* **108**, 925-944.
HIRSCH, J. G. & COHN, Z. A. (1960). Comparative bactericidal activities of blood serum and plasma serum. *Journal of Experimental Medicine* **112**, 15-22.
HOGG, D. McC. & JAGO, G. R. (1970a). The antibacterial action of lactoperoxidase. The nature of the bacterial inhibitor. *Biochemical Journal* **117**, 779-790.
HOGG, D. McC. & JAGO, G. R. (1970b). The oxidation of reduced nicotinamide nucleotides by hydrogen peroxide in the presence of lactoperoxidase and thiocyanate, iodide or bromide. *Biochemical Journal* **117**, 791-797.
HUMPHREY, H. & DOURMASHKIN, R. R. (1969). The lesions of cell membranes caused by complement. *Advances in Immunology* **11**, 75-115.
INGRAM, D. G. & MITCHELL, W. R. (1970). Conglutinin levels in dairy cattle. Changes associated with parturition. *American Journal of Veterinary Research* **31**, 487-492.
JAGO, G. R. & MORRISON, M. (1962). Antistreptococcal activity of lactoperoxidase. *Proceedings of the Society for Experimental Biology and Medicine* **111**, 585-588.
JENSEN, R. S., TEW, J. G. & DONALDSON, D. M. (1967). Extracellular β lysin and muramidase in body fluids and inflammatory exudates. *Proceedings of the Society for Experimental Biology and Medicine* **124**, 545-548.
JOHANSSON, B. (1960). Isolation of an iron-containing red protein from human milk. *Acta chemica scandinavica* **14**, 510.
JONES, F. S. (1928). The properties of the bactericidal substance in milk. *Journal of Experimental Medicine* **47**, 877-888.
JONES, F. S. & LITTLE, R. B. (1927). The bactericidal properties of cow's milk. *Journal of Experimental Medicine* **45**, 319-335.
JONES, F. S. & SIMMS, H. F. (1930). The bacterial growth inhibitor (lactenin) of milk. I. The preparation in concentrated form. *Journal of Experimental Medicine* **51**, 327-339.
JONES, G. W. & RUTTER, M. (1974). The association of K_{88} antigen with haemagglutinin activity in porcine strains of *Escherichia coli*. *Journal of General Microbiology* **84**, 135-144.
KIERMEIER, F. & KAYSER, C. (1960). Zur Kenntnis der Lactoperoxidase. 1. Mitt.

Verteilung der Lactoperoxidase-Aktivität in Kuhmilch und Abhängigkeit von biologischen Einflüssen. *Zeitschrift für Lebensmitteluntersuchung und -forschung* 112, 481-498.

KOLB, W. P., HAXBY, J. A., ARROYAVE, C. M. & MULLER-EBERHARD (1972). Molecular analysis of the membrane attack mechanism of complement. *Journal of Experimental Medicine* 135, 549-566.

KORHONEN, H. (1973). Ph.D. Thesis Institut für Hygiene der Bundesanstalt für Milchforschung in Kiel. Untersuchungen zur Bakterizidie der Milch und Immunisierung der bovinen Milchdrüse.

MASSON, P. (1970). *La Lactoferrine.* Editions Arscia S.A. Bruxelles. Librairie Maloine SA. Paris VI.

MASSON P. L., HEREMANS, J. F., PRIGNOT, J. & WAUTERS, G. (1966). Immunohistochemical localization and bacteriostatic properties of an iron binding protein from bronchial mucus. *Thorax* 21, 538-544.

MASSON, P. L. & HEREMANS, J. F. (1968). Metal binding properties of human lactoferrin (red milk protein). I. The involvement of bicarbonate in the reaction. *European Journal of Biochemistry* 6, 579-584.

McPHILLIPS, J. (1958). Specificity of Agglutinins in Milk. *Nature, London* 182, 869.

MELCHING, L. & VAS, S. I. (1971). Effects of serum components on Gram-negative bacteria during bactericidal reactions. *Infection and Immunity* 3, 107-115.

MICKELSON, M. N. (1966). Effect of lactoperoxidase and thiocyanate on the growth of *Streptococcus pyogenes* and *Streptococcus agalactiae* in chemically defined culture medium. *Journal of General Microbiology* 43, 31-43.

MINDEN, P., McCLATCHY, J. K. & FARR, R. S. (1972). Shared antigens between heterologous bacterial species. *Infection and Immunity* 6, 574-582.

MONTREUIL, J. & MULLET, S. (1960). Isolement de la lactosiderophiline du lait humain. *Compte rendu hebdomadaire des séances de l'Académie des sciences* 250, 1736-1737.

MORRIS, C. S. & EDWARDS, M. A. (1950). Further investigations on the presence in raw milk of a bactericidal substance specific for certain strains of coliform organisms. *Journal of Dairy Research* 17, 253-260.

MORRISON, M. & STEELE, W. F. (1968). Lactoperoxidase, the peroxidase in the salivary gland. In *Biology of the Mouth,* ed. Parson, P. H. American Association for the Advancement of Science.

MORRISON, M. H., HAMILTON, B. & STOTZ, E. (1957). The isolation and purification of lactoperoxidase by ion exchange chromatography. *Journal of Biological Chemistry* 228, 767-776.

MUSCHEL, L. H., GUSTAFSON, L. & LARSEN, L. J. (1969). Re-examination of the Neisser-Wechsberg (antibody prozone) phenomenon. *Immunology* 17, 525-538.

NEISSER, M. & WECHSBERG, F. (1901). Über die Wirkungsart bakterieider Sera. *Münchener medizinische Wochenschrift* 48, 697-673.

NORMANN, B. E. (1972). Inhibition by rabbit anti-bacterial IgG of the bactericidal effect on *Salmonella typhimurium* 395 MR by normal cattle serum. *Acta pathologica et microbiologica scandinavica* Section B, 80, 140-148.

ORAM, J. D. & REITER, B. (1966a). Inhibitory substances present in milk and secretion of the dry udder. Report of National Institute for Research in Dairying, p. 93.

ORAM, J. D. & REITER, B. (1966b). The inhibition of streptococci by lactoperoxidase, thiocyanate and hydrogen peroxide. The effect of the inhibitory system on susceptible and resistant strains of Group N streptococci. *Biochemical Journal* 100, 373-381.

ORAM, J. D. & REITER, B. (1966c). The inhibition of streptococci by lactoperoxidase, thiocyanate and hydrogen peroxide. The oxidation of thiocyanate and the nature of the inhibitory compound. *Biochemical Journal* 100, 382-388.

ORAM, J. D. & REITER, B. (1968). Inhibition of bacteria by lactoferrin and other iron-chelating agents. *Biochimica et biophysica acta* 170, 351-365.

PADGETT, G. A. & HIRSCH, J. G. (1967). Lysozyme: its absence in tears and leucocytes of cattle. *Australian Journal of Experimental Biology and Medical Science* 45, 569.

PFAUDLER, M. & MORO, E. (1907). Über hämolytische Substanzen der Milch. *Zeitschrift für experimentelle Pathologie und Therapie* **4**, 451-458.
PILLEMER, L., BLUM, L., LEPOW, I. H., ROSS, O. A., TODD, E. W. & WARDLAW, A. C. (1954). The properdin system and immunity. I. Demonstration and isolation of a new serum protein, properdin, and its role in immune phenomena. *Science, New York* **120**, 279-285.
POLLEY, M. J., MULLER-EBERHARD, H. J. & FELDMAN, J. D. (1972). Production of ultrastructural membrane lesions by the fifth component of complement. In *Biological Activities of Complement*, ed. Ingram, D. G. Basel: Karger.
PORTMANN, A. & AUCLAIR, J. E. (1959). Relation entre la lacténine L_2 et la lactoperoxidase. *Lait* **39**, 147-158.
PORTMANN, A., PLOMMET, M. & AUCLAIR, J. (1960). Le caractère immunologique des lacténines du lait de vache. *Annales de l'Institut Pasteur, Paris* **103**, 141-158
PORTMANN, A., GATÉ, Y. & AUCLAIR, J. (1962). Influence de la lactoperoxidase et des agglutinines du lait sur l'activité des bactéries des levains thermophiles. *XVIth International Dairy Congress, Kφbenhavn. B*, 729.
REITER, B. (1967). Relationship of the agglutinins of the milk fat globule to the cold agglutinins of the erythrocytes. Report of the National Institute for Research in Dairying, p. 89.
REITER, B. (1973). Some thought on cheese starters. *Journal of the Society of Dairy Technology* **26**, 3-15.
REITER, B. & BRAMLEY, A. J. (1975). Defence mechanisms of the udder and their relevance to mastitis control. *Proceedings of a Seminar on Mastitis Control, International Dairy Federation*, eds Dodd, F. H., Griffin, T. K. & Kingwill, R. G., pp. 210-222.
REITER, B. & BROCK, J. H. (1975). Inhibition of *Escherichia coli* by bovine colostrum and postcolostral milk. I. Complement-mediated bactericidal activity of antibodies to a serum susceptible strain of *E. coli* of the serotype 0111. *Immunology* **28**, 71-82.
REITER, B. & MØLLER-MADSEN, A. (1963). Reviews of the progress of dairy science Section B. Cheese and butter starters. *Journal of Dairy Research* **30**, 419-456.
REITER, B. & ORAM, J. D. (1967). Bacterial inhibitors in milk and other biological fluids. *Nature, London* **216**, 328-330.
REITER, B. & ORAM, J. D. (1968). Iron and vanadium requirements of lactic acid streptococci. *Journal of Dairy Research* **35**, 67-69.
REITER, B., PICKERING, A., ORAM, J. D. & POPE, G. S. (1963). Peroxidase-thiocyanate inhibition of streptococci in raw milk. *Journal of General Microbiology* **33**, xii.
REITER, B., Di BIASE, C. & NEWBOULD, F. H. S. (1964*a*). A note on the serological typing of some strains of *Streptococcus cremoris*. *Journal of Dairy Research* **31**, 125-129.
REITER, B., PICKERING, A. & ORAM, J. D. (1964*b*). An inhibitory system—lactoperoxidase/thiocyanate/hydrogen peroxide—in raw milk. In *Microbial Inhibitors in Food*, ed. Molin, N. Stockholm: Almqvist & Wiksell.
REITER, B., PICKERING, A. & COUSINS, C.M. (1965). Report of the National Institute for Research in Dairying, pp. 104-105.
REITER, B., SHARPE, M. E. & HIGGS, T. M. (1970). Experimental infection of the non-lactating bovine udder with *Staphylococcus aureus* and *Streptococcus uberis*. *Research in Veterinary Science* **11**, 18-26.
REITER, B., BROCK, J. H. & STEEL, E. D. (1975). Inhibition of *Escherichia coli* by bovine colostrum and postcolostral milk. II. The bacteriostatic effect of lactoferrin on a serum susceptible and serum resistant strain of *E. coli*. *Immunology* **28**, 83-95.
REITER, B., MARSHALL, V. M. E., BJORCK, L. & ROSEN, C-G. (1976). The non specific bactericidal activity of the lactoperoxidase/thiocyanate/hydrogen peroxide system of milk against *E. coli* and some Gram-negative pathogens. *Infection and Immunity* **13**, 800-807.
ROGERS, H. J. (1967). Bacteriostatic effects of horse sera and serum fractions on *Clostridium welchii* Type A, and the abolition of bacteriostasis by iron salts. *Immunology* **12**, 285-301.

RUSSELL, M. W. & REITER, B. (1975). Phagocytic deficiency of bovine milk leucocytes: an effect of casein. *Journal of the Reticuloendothelial Society* **18**, 1-13.

RUSSELL, M. W., BROOKER, B. E. & REITER, B. (1976). Inhibition of the bactericidal activity of bovine polymorphonuclear leucocytes and related systems by casein. *Research in Veterinary Science* **20**, 30-35.

SCHADE, A. L. (1960). The microbiological activity of siderophilin. *Protides of the Biological Fluids, Amsterdam, Bruges* **8**, 261-263.

SCHADE, A. L. (1963). Significance of serum iron for the growth, biological characteristics and metabolism of *Staphylococcus aureus*. *Biochemische Zeitschrift* **338**, 140-148.

SCHADE, A. L. & CAROLINE, L. (1944). Raw hen eggwhite and the role of iron in growth inhibition of *Shigella dysenteriae, Staphylococcus aureus, Escherichia coli* and *Saccharomyces cerevisiae. Science, New York* **100**, 14-15.

SCHADE, A. L. & CAROLINE, L. (1946). Iron binding component in human blood plasma. *Science, New York* **104**, 340-341.

SCHADE, A. L., REINHART, R. W. & LEVY, H. (1949). Carbon dioxide and oxygen in complex formation with iron and siderophilin, the iron-binding component of human plasma. *Archives of Biochemistry* **20**, 170-172.

SCHADE, A. L., MYERS, N. H. & REINHART, R. W. (1968a). Carbohydrate metabolism and production of diffusible active substances by *Staphylococcus aureus* grown in serum at iron levels in excess of siderophilin iron saturation and below. *Journal of General Microbiology* **52**, 253-260.

SCHADE, A. L., PALLAVICINI, C. & WIESMANN, U. (1968b). Ekkrinosiderophilin of human milk. *Protides of the Biological Fluids, Amsterdam, Bruges* **16**, 619-625.

SHARPE, M. E. & REITER, B. (1972). Common antigenic determinant in a rumen organism and in Salmonellae containing the antigen 04. *Applied Microbiology* **24**, 613-617.

SHARPE, M. E., LATHAM, M. J. & REITER, B. (1969). The occurrence of natural antibodies to rumen bacteria. *Journal of General Microbiology* **56**, 353-364.

SHARPE, M. E., LATHAM, M. J. & REITER, B. (1975). The immune response of the host animal to bacteria in the rumen and caecum. *IVth International Symposium on Ruminant Physiology, 1974*, pp. 193-204.

SHEARMAN, D. J. E., PARKIN, D. M. & McCLELLAND, D. B. C. (1972). The demonstration and function of antibodies in the gastrointestinal tract. *Gut* **13**, 483-484.

SOJKA, W. J. (1971). Enteric diseases in newborn piglets. *Veterinary Bulletin* **41**, 509-579.

STADHOUDERS, J. & VERINGA, H. A. (1962). Some experiments related to the inhibitory action of milk peroxidase on lactic acid streptococci. *Netherlands Milk & Dairy Journal* **16**, 96-116.

STADHOUDERS, J. & HUP, G. (1970). Complexity and specificity of euglobulin in relation to inhibition of bacteria and to cream rising. *Netherlands Milk & Dairy Journal* **24**, 79-95.

THEODORE, T. S. & SCHADE, A. L. (1965a). Growth of *Staphylococcus aureus* in media of restricted and unrestricted inorganic iron availability. *Journal of General Microbiology* **39**, 75-83.

THEODORE, T. S. & SCHADE, A. L. (1965b). Carbohydrate metabolism of iron-rich and iron-poor *Staphylococcus aureus*. *Journal of General Microbiology* **40**, 385-395.

VIRTANEN, A. I. (1964). Über die Chemie der Brassica-Faktoren, ihre Wirkung auf die Funktion der Schilddrüse und ihr Ubergehen in die Milch. *Experientia* **17**, 241-249.

WALKER, W. A. & HAIG, R. (1973). Immunology of the gastro-intestinal tract. *Journal of Pediatrics* **83**, 517-530.

WANG, C. C. & NEWTON, A. (1969). Iron transport in *Escherichia coli*: relationship between chromium sensitivity and high iron requirement in mutants of *Escherichia coli. Journal of Bacteriology* **98**, 1135-1141.

WARING, W. S. & WERKMAN, C. H. (1942). Growth of bacteria in an iron-free medium. *Archives of Biochemistry* **1**, 303-310.

WETTSTEIN, F. O. & STENT, G. S. (1968). Physiologically induced changes in the properties of phenylalanine tRNA in *Escherichia coli*. *Journal of Molecular Biology* **38**, 25-40.

WHITTLESTONE, W. G. (1975). Discussant of Section III. Prevention of Infection. In *Proceedings of a Seminar on Mastitis Control, International Dairy Federation,* eds Dodd, F. H., Griffin, T. K. & Kingwill, R. G., p. 182.

WILSON, A. T. & ROSENBLUM, H. (1952a). The antistreptococcal property of milk. I. Some characteristics of the activity of lactenin *in vitro*. The effect of lactenin on haemolytic streptococci of the several serological groups. *Journal of Experimental Medicine* **95**, 25-38.

WILSON, A. T. & ROSENBLUM, H. (1952b). The antistreptococcal property of milk. II. The effects of anaerobiosis, reducing agents, thiamine, and other chemicals on lactenin action. *Journal of Experimental Medicine* **95**, 39-50.

WILSON, B. M. & GLYNN, A. A. (1975). Release of ^{14}C label and complement killing of *Escherichia coli*. *Immunology* **28**, 391.

WILSON, M. R. (1972). The influence of preparturient intramammary vaccination on bovine mammary secretions. Antibody activity and protective value against *Escherichia coli* enteric infections. *Immunology* **23**, 947-955.

WILSON, T. R. & HARRIS, G. M. (1961). The oxidation of thiocyanate ion by hydrogen peroxide. II. The acid-catalyzed reaction. *Journal of the American Chemical Society* **83**, 286-289.

WOLIN, A. G. & KOSIKOWSKI, F. W. (1958). Formation of bacterial inhibitory zones in whey agar by raw milk. *Journal of Dairy Science* **41**, 34-40.

WRIGHT, R. C. & TRAMER, J. (1957). The influence of cream rising upon the activity of bacteria in heat-treated milk. *Journal of Dairy Research* **24**, 174-183.

WRIGHT, R. C. & TRAMER, J. (1958). Factors influencing the activity of cheese starters. The role of milk peroxidase. *Journal of Dairy Research* **25**, 104-118.

ZEYA, H. I. & SPITZNAGEL, J. K. (1966a). Cationic proteins of polymorphonuclear leucocyte lysosomes. I. Resolution of antibacterial and enzymatic activities. *Journal of Bacteriology* **91**, 750-754.

ZEYA, H. I. & SPITZNAGEL, J. K. (1966b). Cationic proteins of polymorphonuclear leucocyte lysosomes. II. Composition properties and mechanisms of antibacterial action. *Journal of Bacteriology* **91**, 755-762.

Inactivation of Non-sporing Bacteria by Gases

A. D. RUSSELL

Welsh School of Pharmacy, University of Wales Institute of Science & Technology, Edward VII Avenue, Cardiff, Wales

CONTENTS

1. Introduction 61
2. Physical properties of gaseous disinfectants 62
3. Antibacterial activity 63
   (a) Ethylene oxide 63
   (b) Propylene oxide 67
   (c) β-Propiolactone 68
   (d) Formaldehyde 68
   (e) Other gases 70
4. Mechanism of action 72
   (a) Ethylene oxide 72
   (b) Propylene oxide 76
   (c) β-Propiolactone 76
   (d) Formaldehyde 77
   (e) Other gases 78
5. Practical uses 78
   (a) Ethylene oxide 78
   (b) Propylene oxide 80
   (c) β-Propiolactone 80
   (d) Formaldehyde 81
   (e) Other gases 82
6. Conclusions 82
7. Acknowledgement 83
8. References 83

1. Introduction

THE USE of chemicals in the gas or vapour phase to achieve disinfection has been known for many years. An early procedure was the employment of sulphur dioxide, obtained by burning sulphur, or of chlorine for fumigating sick rooms; however, as these gases are harmful to many materials, it is better to use formaldehyde instead.

A scientific basis for using gases as sterilizing or disinfecting agents has been established only comparatively recently. Nordgren (1939) described the factors influencing the activity of gaseous formaldehyde, and Phillips & Kaye (1949) reviewed the earlier work which had taken place with ethylene oxide (EO). Subsequently, a great deal of research has been published on the antimicrobial activity and uses of these, and of other, gases. Because bacterial spores are more resistant than vegetative bacteria to chemical and physical agents (Roberts &

Hitchins, 1969; Russell, 1971), it was inevitable that most of this work was directed towards the sporicidal activity of gaseous agents. The destruction of spores by such agents has been the subject of many reviews, notably those by Phillips (1952, 1961, 1968), Bruch (1961), Kelsey (1967), Kereluk & Lloyd (1969), Roberts & Hitchins (1969), Bruch & Bruch (1970), Hoffman (1971) and Russell (1971). Kereluk (1971) has dealt with lesser known gases. The activity of gaseous disinfectants against non-sporing bacteria will be considered in this paper, although it will be necessary to make occasional comments on the relative susceptibility of sporing and non-sporing bacteria to gaseous chemicals.

2. Physical Properties of Gaseous Disinfectants

Table 1 (based in part on Ernst, 1974) lists the various types of substances used as gaseous disinfectants. The chemical structures of the various substances are given in Fig. 1. Glutaraldehyde is described by Ernst (1974, see Table 1) as being one of the more important vapour phase disinfectants but in fact there is a dearth of published information on this aspect.

Table 1
Types of gaseous disinfectants

Most important	Lesser interest
Ethylene oxide	Glycidaldehyde
β-Propiolactone	Ethylene imine
Formaldehyde	Nitric oxide
Propylene oxide	Epichlorohydrin
Methyl bromide	
Peracetic acid	
Ozone	
Glutaraldehyde*	

* Few studies made with glutaraldehyde to date.
Based in part on Ernst (1974).

Physical properties of the five most widely used gases are presented in Table 2, which is based on the excellent paper of Bruch & Bruch (1970). This table shows that ethylene oxide (EO) and methyl bromide (MB) are gases at normal temperatures, whereas formaldehyde and β-propiolactone (BPL) require heating to produce the vapour form. Propylene oxide (PO) requires only mild heating. Table 2 also indicates that differences exist in diffusibility or penetrability. Ethylene oxide, for example, is freely diffusible and penetrates paper, cellophane, cardboard, fabrics and some plastics such as polyvinyl chloride, but less readily through polythene. It cannot penetrate many solid materials, particularly if they are crystalline (Abbott *et al.*, 1965; Royce &

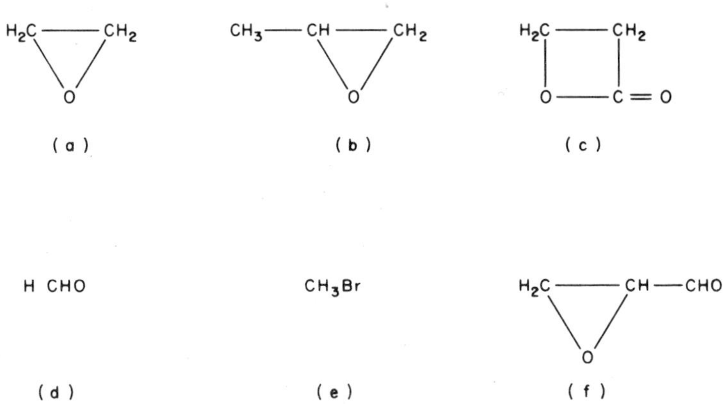

Fig. 1. Chemical structures of some gaseous disinfectants: (a) ethylene oxide, (b) propylene oxide, (c) β-propiolactone, (d) formaldehyde, (e) methyl bromide, (f) glycidaldehyde.

Bowler, 1961). β-Propiolactone has weak penetrating powers, and hydrolyses readily in water to give hydracrylic acid (β-hydroxypropionic acid), the rate of hydrolysis being temperature-dependent (Hoffman & Warshowsky, 1958; see Table 3). Propylene oxide gives propylene glycol on hydrolysis, which is non-toxic; PO hydrolyses slowly in the presence of only a small amount of moisture and thus there is no need to remove it from exposed materials (Sykes, 1965).

Ernst (1973) has stated that EO acts as a 'carrier' for moisture through non-polar and normally hydrophobic films having low moisture permeations, whereas water aids the permeation of EO through polar type films which normally allow water to permeate readily but which impede diffusion of EO.

3. Antibacterial Activity

On a molar basis (Table 2) formaldehyde and BPL are considerably more potent antibacterial agents than EO, MB and PO. However, the activity of all antimicrobial substances depends upon various factors (Russell, 1974), notably the concentration of the substance, the type and pretreatment growth conditions of organism, the temperature and the presence of organic or other modifying matter; with gases, relative humidity (RH) is an extremely important parameter. These factors will be discussed in relation to the antibacterial activity of each gas.

(a) *Ethylene oxide*

Bacterial spores are generally only some 2-10 times as resistant as non-sporing bacteria to EO gas (Phillips, 1952, 1968; Toth, 1959; Blake & Stumbo, 1970).

Table 2

Properties of the most commonly used gaseous disinfectants

Gaseous disinfectant	Molecular weight	Boiling point (°)	Solubility in water	Sterilizing concn. (mg/l)	RH requirement (%)	Penetration of materials	Microbicidal activity*	Best application as gaseous disinfectant
Ethylene oxide	44	10.4	Complete	400–1000	Non-desiccated 30–50; large loads 60	Moderate	Moderate	Sterilization of plastic medical supplies
Propylene oxide	58	34	Good	800–2000	Non-desiccated 30–60	Fair	Fair	Decontamination
Formaldehyde	30	90°/Formalin†	Good	3–10	75	None (Surface sterilant)	Excellent	Surface sterilant for rooms
β-Propiolactone	72	162	Moderate	2–5	> 70	None (Surface sterilant)	Excellent	Surface sterilant for rooms
Methyl bromide	95	4.6	Slight	3500	30–50	Excellent	Poor	Decontamination

* Based on an equimolar comparison.
† Formalin contains formaldehyde plus methanol.

Spores of the thermophilic *Bacillus stearothermophilus* and of certain other organisms may, in fact, be less resistant to EO than some vegetative bacteria such as *Staphylococcus aureus* (Thomas et al., 1959; Freeman & Barwell, 1960; Thomas, 1960), *Micrococcus radiodurans* (Kereluk & Lloyd, 1969; Kereluk et al., 1970), *Streptococcus faecalis*, including radiation-resistant strains (Kereluk et al., 1970) and *Mycobacterium smegmatis* under certain conditions (Gilbert et al., 1964).

Table 3
Rate of hydrolysis of β-propiolactone

Temperature (°)	Half-life (min)
10	1080
25	210
50	20
75	5

Hoffman & Warshowsky (1958).

Spaulding (1968) designated EO as a disinfectant rather than as a sterilizing agent but adds that EO used appropriately is a practical agent for producing sterility. Spaulding also states that, although there is a paucity of exact information, EO probably has a very high activity against vegetative bacteria, including *Myco. tuberculosis*. Kaye (1950) has shown that EO will inactivate mycobacteria, Smith (1968) states that it is mycobactericidal and Newman et al. (1955) reported that EO could be used to decontaminate articles handled by tuberculous patients.

It has been known for many years (Kaye, 1949; Kaye & Phillips, 1949; Phillips, 1949; Phillips & Kaye, 1949) that RH influences the sporicidal activity of EO. Under optimum conditions, i.e. with 'naked' spores placed on filter paper, a certain minimum amount of water is necessary, and a sporicidal effect is most rapid at c. 28-33% RH. Spores which have been dried are far more resistant to destruction by EO (Mayr, 1961; Perkins & Lloyd, 1961; Phillips, 1961, 1968; Gilbert et al., 1964). There is less information available as to the effect of RH on the activity of EO against non-sporing bacteria. Moreover, it is doubtful whether RH should be considered in isolation, and in view of what is known about the action of EO, it is proposed that the inter-relationship of four separate points should be discussed: (i) the RH of the environment, (ii) the state of hydration of the cells themselves, (iii) the adsorption of moisture by solid surfaces, and (iv) the presence of 'protecting' substances.

The optimum RH for the maximum bactericidal effect of EO is c. 33%; at high RH values EO is converted to ethylene glycol, which has a weaker antibacterial activity. However, equilibration of *Staph. aureus* and *Myco.*

smegmatis at 1% RH increases their resistance to EO at 33% RH (Gilbert *et al.*, 1964). *Salmonella senftenberg* cells preconditioned at 11, 23, 33, 53 and 73% RH and exposed to EO at the same RH values are more resistant to the gas as the RH increases (Michael & Stumbo, 1970). Once bacterial cells have been dried beyond a certain critical point they must be wetted or placed in an environment of 100% RH to become rehydrated (Royce & Bowler, 1961; Gilbert *et al.*, 1964) and it is important to realize that rehydration of cells is a slow process (Bateman *et al.*, 1962). The menstruum from which cells are dried and the surface on which they are dried will also influence their response to EO; organisms in dried broth films, for example, are not always sterilized (Royce & Sykes, 1955). *Staph. aureus* is generally more difficult to kill than bacterial spores or *Escherichia coli* when placed in dried nutrient broth films on glass surfaces, but not on filter paper discs (Royce & Bowler, 1961). Nutrient broth has been shown to be a better 'protecting' agent than serum (Royce & Bowler, 1961) despite the greater amount of protein in the latter. Gilbert *et al.* (1964) have shown that *Staph. aureus* and *Myco. smegmatis* cells grown in tryptose broth, then equilibrated at 1% RH before exposure on filter paper discs to EO gas at 33% RH were rather more resistant than *B. globigii* spores similarly grown, equilibrated and treated, and considerably more resistant than such spores which had been washed with water. However, a proportion of the desiccated cells resisted EO. *Staph. aureus* grown in tryptose broth, washed with water, placed on paper discs and exposed to EO at 33% RH were more readily killed than similarly grown, but unwashed, cells of this organism and the higher the concentration of organic matter (as represented in this case, by tryptose broth) the slower the death rate (Gilbert *et al.*, 1964). Likewise *Salm. senftenberg* cells lyophilized in egg powder are twice as resistant to EO gas as 'clean' lyophilized cells (Michael & Stumbo, 1970).

Bacteria trapped inside crystals of various substances are protected from the effects of EO (Royce & Sykes, 1955; Royce & Bowler, 1961). Cells of *Staph. aureus* and *E. coli* dried from saline are rather more resistant to EO than cells dried from broth (Royce & Bowler, 1961), and the situation is probably analogous to that pertaining with saline-dried spores which are resistant to EO (Beeby & Whitehouse, 1965). Winge-Heden (1963) found that freshly prepared test 'slips' of *Staph. aureus* and *Strep. faecalis* were considerably easier to inactivate than slips which had been stored dry, and the latter effect probably represents the desiccation effect described above. Dried *Staph. epidermidis* cells are more resistant to EO than *B. subtilis* spores (Opfell *et al.*, 1967).

Staphylococcus aureus and *E. coli* dried on hard surfaces are more resistant to EO inactivation than those dried on absorbent surfaces (Royce & Bowler, 1961; see also Blake & Stumbo, 1970 although conditions used by the latter authors did not actually represent dried cells, since exposure to EO of *Lactobacillus brevis* and *L. mesenteroides* appeared to be made immediately after inoculation).

However, not all organisms dried on absorbent surfaces can be sterilized (Royce & Bowler, 1961). Bacteria dried on non-hygroscopic surfaces are less sensitive to EO than the same bacteria dried on hygroscopic surfaces (Kereluk et al., 1970).

Miller (1972) has described the establishment of EO sterilization cycles based upon the 'natural' contamination of products to be sterilized.

(b) *Propylene oxide*

The antibacterial activity of PO has been studied less than that of EO. Whelton et al. (1946) reported that bacteria were rather more resistant to PO than were moulds and yeasts, but little attention was paid to RH. Bruch & Koesterer (1961) studied the microbicidal activity of PO against *Serratia marcescens* and especially against bacterial spores, and found that the coefficient of dilution was near unity and that the antibacterial activity of the gas decreased with an increase in RH. In contrast, Himmelfarb et al. (1962) (Tables 4 & 5) observed that their test bacteria became progressively less resistant to PO with an increase in RH up to a maximum of 65-70%. There were, however, notable differences in the techniques of the two groups of workers: Bruch & Koesterer (1961) used

Table 4
Effect of contact time on the bactericidal activity of propylene oxide at 52% RH

Contact time (h)	% survivors of	
	E. coli	Staph. aureus
0	100	100
0.5	< 0.001	25
1	< 0.0001	6.4
2	–	0.4
3	–	< 0.1

RH, relative humidity.
Propylene oxide concentration 1.0 cm^3/l, temperature 35°.
Based on the graphical data presented by Himmelfarb et al. (1962).

air-dried filter paper strips of bacteria, and did not draw a vacuum before admission of the PO vapour; Himmelfarb et al. (1962), on the other hand, preconditioned bacteria or spores at moisture levels corresponding to the RH levels during PO treatment, and furthermore, drew vacua before admission of PO vapour. It is thus likely that they were dealing with desiccated bacteria akin to those described under EO (see also Salle & Korzenovsky, 1942). Of the bacteria tested by Himmelfarb et al. (1962), B. subtilis spores were the most resistant to PO, with *Staph. aureus, Sarcina lutea, M. flavus* and *Strep. faecalis* the most resistant of the non-sporing organisms.

Table 5
Bactericidal activity of propylene oxide vapour at different RH values

Relative humidity (RH) (%)	D value (min) with	
	E. coli	Staph. aureus
1	c. 22	c. 160
52	10	60
65	5	19
80	< 1	–
98	–	5

Propylene oxide concentration 1.0 cm^3/l, temperature 35°.
Based on the graphical data presented by Himmelfarb et al. (1962).

(c) β-*Propiolactone*

The antibacterial activity of BPL is a direct function of its concentration and the temperature and RH at which it is used (Spiner & Hoffman, 1960). Of bacteria, viruses and fungi, bacterial spores are the most resistant to BPL solutions (Trafas et al., 1954; LoGrippo et al., 1955; Hazeu & Bueck, 1965) although spores are only some 4-5 times more resistant to BPL vapour than are some strains of *Staph. aureus* (Hoffman & Warshowsky, 1958; Allen & Murphy, 1960).

Most of the work with gaseous and liquid BPL has been carried out with bacterial spores, although reports of its activity with non-sporing bacteria have appeared (Allen & Murphy, 1960; Bruch & Koesterer, 1961; Toplin & Gaden, 1961). The antibacterial activity is maximal at RH levels of 75-85%, although as with EO, it is not so much the environmental moisture content that is important but the content and location of water within the bacterial cell (Hoffman & Warshowsky, 1958). Under conditions of maximum effectiveness, BPL vapour is claimed to be about 25 times more active as a vapour phase disinfectant than formaldehyde, 4000 times more active than EO and 50,000 times more active than MB (Hoffman & Warshowsky, 1958).

(d) *Formaldehyde*

Vegetative bacteria and bacterial spores are fairly readily killed by formaldehyde gas (Sykes, 1965) and the degree of resistance of spores is only some 2-3 times

(*Anon.*, 1958) or 2-15 times (Phillips, 1952) that of non-sporing bacteria. Formaldehyde in the liquid (Smith, 1968) or vapour (*Anon.*, 1958) state is mycobactericidal.

There is a linear relation between the concentration of formaldehyde and the killing rate (*Anon.*, 1958), but little effect on disinfection rate from variation in temperature over the range 0-30° has been observed (*Anon.*, 1958). However, Nordgren (1939) observed that the rate of disinfection of spores exposed to formaldehyde vapour increased as the temperature was increased from 10-70°. Organic matter, in the form of blood, sputum or soil, reduced the rate of bacterial inactivation (Nordgren, 1939) and cocci suspended in serum are more difficult to kill than those suspended in gelatin (*Anon.*, 1958).

As with other vapour phase disinfectants, the antibacterial activity of formaldehyde is dependent on RH. Various RH levels for optimum activity of formaldehyde have been proposed: an optimum RH of 80-90% but with no great increase in disinfection rate upon increasing the RH above 58% has been mentioned (*Anon.*, 1958; Baird-Parker & Holbrook, 1971). Nordgren (1939) reached a somewhat similar conclusion in that he observed an increase in the rate of bacterial kill as the RH was raised to 50% but little increase as the RH was increased from 50-90%. Spaulding (1968), however, is of the opinion that there is no bactericidal effect of formaldehyde unless the RH is at 70% or above. What can be agreed, however, is that the RH must be much higher than the level of 28-33% given as the optimum level with EO.

Formaldehyde gas may be generated in the following ways: (i) evaporation of commercial formaldehyde solution (formalin) which consists of a 40% solution of formaldehyde in water, plus 10% methanol to prevent polymerization; (ii) addition of formalin to potassium permanganate; (iii) volatilization of paraformaldehyde.

Although paraformaldehyde has been stated to be of little practical use because of its slow volatilization (Nordgren, 1939) this statement no longer holds true. Paraformaldehyde, a polymer of formaldehyde with the formula $HO(CH_2O)_n.H$, where n = 8-100, is a flake or a fine or coarse powder and is produced by evaporating aqueous solutions of formaldehyde. When heated, paraformaldehyde depolymerizes rapidly to give formaldehyde, and this is the basis of the process used by C. Kaitz (cited by Taylor *et al.*, 1969) for disseminating formaldehyde gas. This process is now widely used.

A recent and interesting paper by Tulis (1973) draws attention to the fact that certain organic resins and polymers, when exposed to elevated temperature, release potentially sterilizing amounts of gaseous formaldehyde. This evolution of formaldehyde is such that the rate of release is a function of time and temperature. Examples of such products are: (a) melamine formaldehyde (Fig. 2a) which is formed from formaldehyde and melamine under alkaline conditions; (b) urea formaldehyde products (Fig. 2b), a mixture of mono-methylol urea and di-methylol urea; (c) paraformaldehyde. Tulis (1973) considers paraformaldehyde to be an excellent source of monomeric formaldehyde gas because it can be produced in a temperature-controlled reaction, and no contaminating residues (methanol and formic acid) are produced during the evaporation of formalin solutions.

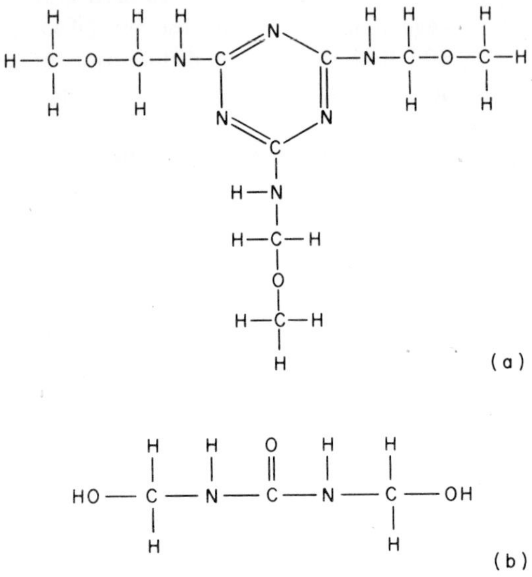

Fig. 2. (a) Melamine formaldehyde, (b) urea formaldehyde.

Paraformaldehyde-produced formaldehyde gas is lethal to spores and various non-sporing Gram positive and Gram negative bacteria (Taylor et al., 1969; Tulis, 1973) and is considerably more effective as a disinfecting and sterilizing agent than the formaldehyde-releasing resins. The inactivation process is strictly a function of the available formaldehyde gas and at various temperatures the percentage loss of formaldehyde is much greater from paraformaldehyde than from the resins (Tulis, 1973).

(e) *Other gases*

(i) *Methyl bromide*

Methyl bromide is considerably less active as an antibacterial agent than EO (Table 2: Bruch & Bruch, 1970; Kolb & Schneiter, 1950; Kelsey, 1967) or even PO (Kelsey, 1967) but has good penetrating ability. It kills sporing and non-sporing bacteria.

(ii) *Glycidaldehyde*

Glycidaldehyde vapour inactivates sporing and non-sporing bacteria (Dawson, 1962). The inactivation rate depends on temperature of treatment, gas concentration and RH, a decrease in RH resulting in an increase in the time necessary for complete inactivation of *Staph. aureus* and of spores (Dawson, 1962).

(iii) *Ozone*

The possibility of using ozone as an aerial disinfectant was described by Elford & van den Ende (1942) and Ingram & Haines (1949) showed that it inhibited bacterial growth. Gram positive cocci are considered to be more sensitive than Gram negative bacteria but less so than Gram positive bacilli, but the actual activity of ozone is highly dependent on the test method (Baird-Parker & Holbrook, 1971).

The rapid bactericidal effects of vapour phase ozone (200 p/m) (Warshaw, 1953), and of ozone solutions at 1° (Fetner & Ingols, 1959) have been described. Recently it has been observed that various Gram positive and Gram negative bacteria are rapidly inactivated by treatment with ozone in phosphate-buffered saline, but less rapidly in the presence of organic matter, and in the latter case a synergistic effect is observed with the simultaneous use of ozone and sonication (Burleson *et al.*, 1975).

(iv) *Glutaraldehyde*

Ernst (1974) lists glutaraldehyde as one of the most important gaseous disinfectants. There is, however, a surprising lack of published work on vapour-phase glutaraldehyde. This dialdehyde is very active against bacteria and their spores when it is used at alkaline pH values (Borick & Pepper, 1970). However, under conditions of high temperatures which would presumably be necessary to achieve vaporization, polymerization of the glutaraldehyde molecule (at alkaline pH) would occur (Boucher, 1974) and it is likely that the resulting vapour would have little, if any, antibacterial activity.

(v) *Nitric oxide*

Nitric oxide increases the duration of the lag phase of *E. coli* and the extent of this increase is a function of the concentration of the gas (Russell, 1965).

(vi) *Carbon dioxide*

Carbon dioxide in soft drinks inhibits the development of *Salm. typhimurium, Corynebacterium diphtheriae* and *Brucella abortus* (Dunn, 1968). The fact that it inhibits the growth of slime-producing bacteria has been known for many years. The paper by Clark & Lentz (1969) describes briefly much of the earlier work and also indicates that the growth of psychrotolerant, slime-producing bacteria is markedly inhibited by CO_2 gas in the atmosphere. This inhibition depends on (a) low temperatures, 0-5°, (b) the addition of CO_2 at an early stage before the organisms became adjusted to the environmental conditions, (c) high CO_2 concentrations of *c.* 20%.

4. Mechanism of Action

Two excellent reviews of the inactivation of sporing and non-sporing bacteria by EO and other gases have been made by Bruch & Bruch (1970) and Hoffman (1971). Even these papers, however, depend heavily for their conclusions on mechanisms of action upon: (a) studies with liquid forms of these gases, (b) investigations with non-bacterial sources. Hoffman (1971) points out that it must be assumed that the mode of action of the gaseous and liquid forms is identical or at least similar. Thus, despite extensive usage of gaseous disinfectants, it is true to state that studies on their mechanism of inactivation are comparatively few. However, some useful information has accrued, and this is considered below.

(a) *Ethylene oxide*

Phillips (1949, 1952, 1961, 1968) showed ethylene oxide to be an alkylating agent, and suggested that it inactivated bacteria or their spores by combining with the $-NH_2$, $-COOH$, $-SH$ and $-OH$ groups of proteins (Fig. 3). The general

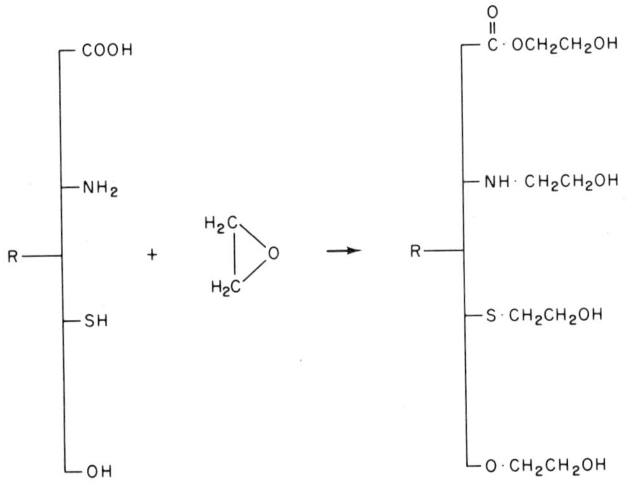

Fig. 3. Postulated interaction of ethylene oxide with proteins.

activity of similar substances paralleled their activity as alkylating agents (Fig. 4), with cyclopropane, which has a similar 3-membered ring as EO but which is not an alkylating agent, having no antibacterial activity. Alkylation is itself defined (Price, 1958) as the conversion $H-X \rightarrow R-X$, where R is an alkyl group. The biological activity of the alkylating agents is indicated by reaction with nucleophilic groups (Stacey *et al.*, 1958; Smith & Spencer, 1975).

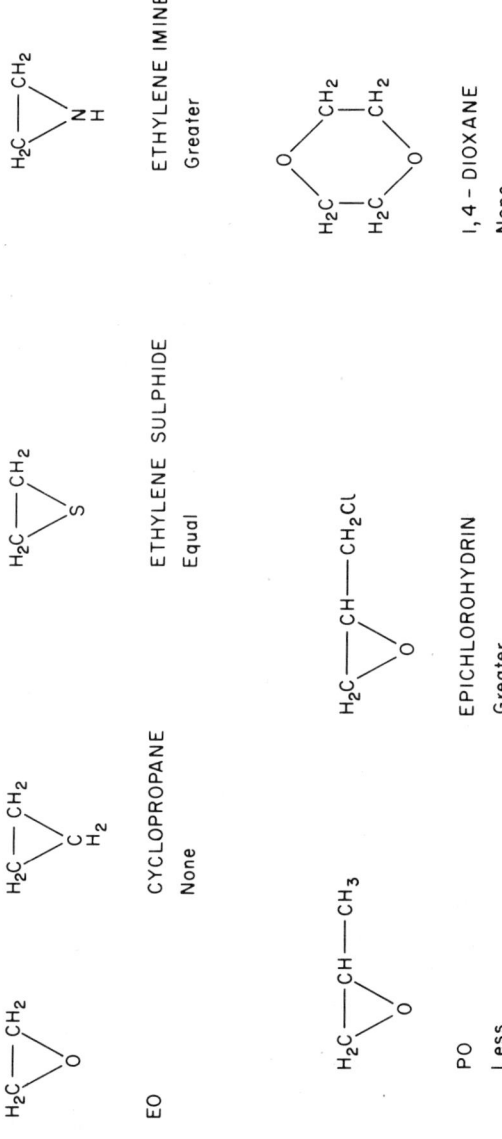

Fig. 4. General antibacterial activity of alkylating agents, compared with that of ethylene oxide (EO).

It has frequently been observed (Kaye & Phillips, 1949; Toth, 1959; Phillips, 1968; Blake & Stumbo, 1970) that bacterial spores are only a few times as resistant as vegetative bacteria to EO, in contrast to liquid disinfectants where the difference in resistance may be several ·1000-fold (Russell, 1971). This appears to be a general property of the alkylating agents.

Research in biological systems other than bacteria had earlier suggested an interaction of epoxides, of which EO is but one example, with proteins and amino acids (Ross, 1958). Certainly, EO causes hydroxyethylation of amino acids (Starbuck & Busch, 1963). EO also interacts with nucleic acids, e.g. with the phosphate group to form a tri-ester (Alexander & Stacey, 1958), although this may be only to a minor extent in comparison to reaction at other sites, and thus has been considered an unlikely lesion for the antibacterial effect of EO (see below, however).

EO also interacts with guanine to give 7-(2'-hydroxyethyl) guanine (Fig. 5a; Brooks & Lawley, 1964), and alkylating agents have a principal site of interaction at N-7 of guanine moieties in deoxyribonucleic acid (DNA), the second most reactive site being at N-3 of adenine moieties (Brooks & Lawley, 1964). Windemueller & Kaplan (1961) found that EO gas hydroxyethylates tertiary heterocyclic N atoms in adenosine and adenosine triphosphate. EO reacts almost exclusively with guanine residues of ribonucleic acid (RNA) from tobacco mosaic virus, TMV (Fraenkel-Conrat, 1961).

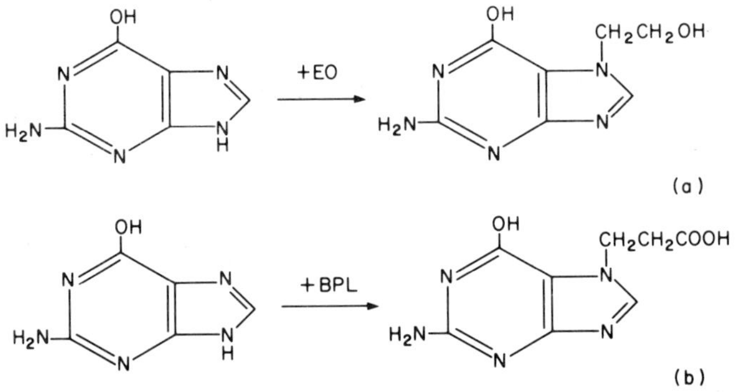

Fig. 5. Interaction with guanine of (a) ethylene oxide, (b) β-propiolactone.

The significance of these findings in relation to the mechanism of action of EO against non-sporing (or sporing) bacteria has yet to be fully realized. One important paper (Michael & Stumbo, 1970) has shed some light on this action. This work, using *Salm. senftenberg* and *E. coli* as test organisms, is summarized in Table 6. Two points must be emphasized; (i) the repair and recovery of EO-treated cells in minimal salts (MS) medium supplemented with guanine or guanosine triphosphate (GTP) but not any other supplement, (ii) the lack of

Table 6
Growth of Salmonella senftenberg *after exposure to ethylene oxide*

Experiment	Cells	Recovery medium	Result
1	Unexposed	TSY Broth	'Normal' growth
2	EO-exposed	TSY Broth	Slight lag, then 'normal' growth
3	Unexposed	MS Broth	'Normal' growth (rate much less than in Exp. 1)
4	EO-exposed	MS Broth	(a) Very long lag. (b) Rise in O.D.; cell protein, mass and RNA contents increase; glucose content decreases; DNA content and viable count on TSY agar constant. (c) Fall in O.D.
5	EO-exposed	MS Broth + guanine	Repair and reproduction
6	EO-exposed	MS Broth + other supplements*	No repair or reproduction
7	EO-exposed	MS Broth + EO-exposed guanine	Repair and reproduction
8	EO-exposed	MS Broth + EO-exposed GTP	No repair or reproduction

EO, ethylene oxide; TSY Broth and agar, trypticase soy broth or agar respectively, supplemented with 0.5% of yeast extract; MS Broth, a minimal salts + glucose liquid medium; GTP, guanosine triphosphate.
* Other supplements tested: amino acids, organic acids, base compounds of DNA and RNA, vitamins, nucleic acid sugars.
Michael & Stumbo (1970).

recovery of EO-treated cells in MS medium containing EO-treated GTP but not EO-treated guanine. These facts suggested to the authors that the alkylation of phosphated guanine in the cell was primarily responsible for the loss of its reproductive ability as a result of EO treatment. Certainly the evidence in favour of this conclusion is interesting if at present rather circumstantial.

Liquid sulphur mustard, a known alkylating agent with a mutagenic effect, produces abnormal colonies with bacteria exposed to it (Gilbert *et al.*, 1964). However, mutations do not occur with EO-treated TMV (Fraenkel-Conrat, 1961) or with EO-treated bacteria (Gilbert *et al.*, 1964; Hoffman, 1971; see also Bruch & Bruch, 1970), although a consequence of N-7 alkylation of purines could be anomalous base pairing leading to mutation and ultimately defective proteins and thereby cell death (Roxon, 1973).

Thus, the reaction(s) responsible for the death of EO-treated vegetative bacteria cannot yet be stated with any degree of certainty. As Hoffman (1971) has pointed out, death could result from a general poisoning of the cell involving alkylation of nucleic acid or proteins, or from a single reaction with a vital site in the DNA or RNA molecule. Ethylene imine has been found (Loveless et al., 1954) to induce morphological effects in growing *E. coli* cells, but no similar studies appear to have been made with EO-treated cells, despite the similarity of the two substances (Hendry et al., 1951). Overall, research on the mechanisms of action of EO against non-sporing bacteria has proved to be disappointing, with very little actual knowledge being gained. It would be instructive to examine, for example, the DNA (including possible strand breakage) of bacteria exposed under various conditions to EO.

The reasons for desiccation-induced EO resistance of a small proportion of spores and vegetative bacteria, referred to in Section 3, are not yet known with any degree of certainty. It has been shown that decreased RH can lead to mutation in *E. coli* cells (Webb, 1967) and it is thus possible that EO-resistant mutants are produced by exposing cells to low RH. However, since desiccated cells after exposure to high RH values become sensitive to EO, it is probably more likely that a reversible change occurs involving (a) removal of water, thereby reducing the likelihood of alkylation; (b) removal of water leading to conformational changes in DNA.

(b) *Propylene oxide*

The mode of action of PO against vegetative bacteria has been little studied, but it is believed to act in a similar manner, i.e. as an alkylating agent (Lawley & Jarman, 1972), to EO (Hoffman, 1971; Kereluk, 1971). Recent studies by Walles (1974) using calf thymus DNA *in vitro* have shown that PO causes an increased irreversibility of heat denaturation, the extent of the irreversibility depending on the degree of alkylation. This is indicative of single-strand DNA breaks caused predominantly by phosphate alkylation, breaks caused by guanine alkylation also occurring but at a slower rate (Rhaese & Freese, 1969).

However, there have been no significant studies made on the effect of PO on non-sporing bacteria at the molecular level and thus it is not at present possible to come to any firm conclusions as to its mode of action against such organisms.

(c) *β-Propiolactone*

BPL is a highly reactive chemical, although it owes this reactivity to the highly strained structure of the β-lactones rather than to the presence of an unsaturated bond in the lactone ring (Dickens, 1964). It reacts with various cellular groups and is considered (Phillips, 1968) to be an alkylating agent which combines with $-NH_2$, $-COOH$, $-SH$ and $-OH$ groups, although it is not certain which (if any) reaction is responsible for inactivation of bacteria. It reacts *in vitro* with guanosine, deoxyguanylic acid and RNA (Roberts & Warwick, 1963) and

Ichikawa et al. (1967) (cited by Hoffman, 1971) extracted the nucleic acids from BPL-treated *E. coli* and concluded that the N-7 guanine position (Fig. 5b) was a primary reaction site for BPL. Likewise Troll et al. (1969) have shown that BPL substitutes on the N-7 position of guanine in DNA *in vitro*. BPL causes no increase in the irreversibility of the heat denaturation of calf thymus DNA *in vitro*, although it causes a much higher degree of guanine alkylation than PO (Walles, 1974). The former finding is indicative of a lack of single-strand DNA breaks, and suggests that BPL does not alkylate the phosphate groups of DNA, since phosphate alkylation causes breaks much faster than base alkylation (Rhaese & Freese, 1969). Similar studies on bacterial DNA *in vitro* and *in vivo* would be instructive.

Iyer & Szybalski (1958) have shown that BPL induces mutations in *E. coli* and thus this appears to be a further difference between the action of BPL on the one hand and EO and PO on the other.

Dickens (1964) stated that although the reaction between BPL and guanine might prove to occur less readily than that between BPL and –SH groups, it might provide a mechanism for its interference with nucleic acid metabolism and thus with the genetic material of the cell, since BPL is known to be carcinogenic (Walpole et al., 1954).

(d) *Formaldehyde*

Formaldehyde is an extremely reactive chemical, and details of its interactions have been provided (Walker, 1964). All the published work relating to its mechanism of action has been performed with aqueous solutions acting on bacteria or with isolated protein, DNA or RNA *in vitro*, and thus it can only be assumed at present that gaseous formaldehyde acts in a similar manner.

Formaldehyde interacts with protein by combining with the primary amide, as well as with the amino groups although phenolic moieties bind little of the chemical (Fraenkel-Conrat et al., 1945). Fraenkel-Conrat (1961) subsequently showed that formaldehyde gave an intermolecular cross-linkage of protein amino groups with phenolic or indole residues. Proposals that formaldehyde acts as a mutagenic agent (Loveless, 1951) and as an alkylating agent by reaction with carboxyl, hydroxyl and sulphydryl groups (Phillips, 1952) have also been made.

Formaldehyde can also react extensively with nucleic acids. With the nucleic acid of TMV there is a biphasic reaction, an initial loose binding, followed by a firm attachment after extensive reaction (Staehelin, 1958). Formaldehyde reacts with the amino groups of bases and is much less reactive with DNA than with RNA. The second, stable product of RNA with formaldehyde results from the formation of methylene bridges with adenosine (Staehelin, 1958). With bacteria as the test system, it has been shown that sublethal concentrations of formaldehyde inhibit the synthesis of cytoplasmic and nuclear material (Neely, 1963a, b, c) but subsequently the aldehyde is metabolized to carbon dioxide, and cellular growth resumes (Neely, 1966).

Obviously, the nature of bacterial inactivation by formaldehyde in the vapour (or, indeed, liquid) state is unknown, and there is scope for a considerable amount of worthwhile research to be carried out in this area.

(e) *Other gases*

If little is known specifically about the mechanism of action of EO, PO, BPL and formaldehyde vapours, then it must be added that even less information is available about this aspect of the other antibacterial gases.

Methyl bromide for example, is considered to act by alkylation (Phillips, 1968) although there is scant, if any, evidence in support of this contention. As far back as 1950, Kolb & Schneiter postulated that three possibilities existed as to the nature of its bactericidal action; (a) an intracellular penetration of the substance, which then acts as the lethal agent, (b) an intracellular hydrolysis of methyl bromide takes place resulting in the formation of hydrobromic acid and methanol which could be the lethal agent, (c) further intracellular products might act as the toxic agents.

Glutaraldehyde in the vapour state has been little studied and thus no conclusions as to its mechanisms of action in this form can be made. In the liquid state, at alkaline pH values it has been shown to bind strongly to, and 'fix' the cell envelope of Gram positive (Russell & Vernon, 1975) and Gram negative (Munton & Russell, 1970, 1972; Russell & Haque, 1975) bacteria, although the relationship to bactericidal activity has yet to be proved.

Little is also known about the mechanisms of action of other vapour phase disinfectants.

5. Practical Uses

Apart from sterilization of various items, involving the destruction of heat-resistant and other bacterial spores as well as of non-sporing bacteria, gases have also been widely used for inactivation of vegetative bacteria only, and this section considers the latter aspect.

(a) *Ethylene oxide*

Since its introduction as a practical gaseous disinfectant or sterilizing agent, EO has been shown to have many uses, as indicated below.

(i) *Decontamination procedures*

Newman *et al.* (1955) described the use of EO as a decontaminating agent for articles made or handled by tuberculous patients; in their tests, virulent tubercle bacilli were employed. Herman (1968) advocated the use of EO as a method of controlling contaminants in microbiological laboratories.

(ii) *Sterilization of hospital blankets*

Thomas *et al.* (1959) have investigated the sterilizing effect of EO on test strips of hospital blankets and have shown this to be a useful practical procedure. Interestingly, however, *Staph. aureus* appeared to be more resistant to EO than the two spore forming organisms likewise tested. Foter (1960a, b) has likewise described the usefulness of EO in sterilizing bedding.

(iii) *Lung ventilators*

EO has been employed to sterilize lung ventilators (Baker *et al.*, 1963). The efficacy of the method was checked by deliberately infecting machines with *Staph. aureus.*

(iv) *Ureteral catheters*

An overnight exposure of catheters, deliberately contaminated with *Pseudomonas aeruginosa,* to EO gas (110 mg/l at 23°) produced sterility, whereas two commonly used antiseptic solutions for the cold sterilization of urological instruments proved to be inadequate (Kalinska *et al.*, 1966).

(v) *Ophthalmic instruments*

Various ophthalmic instruments have been sterilized, without damage, by EO (Skeehan *et al.*, 1956). Tests were made against bacterial spores and *Ps. aeruginosa.*

(vi) *Anaesthetic equipment*

EO is a practical sterilizing agent for such equipment (Snow *et al.*, 1962). Tests were carried out against bacterial spores, *Staph. aureus* and *Ps. aeruginosa.*

(vii) *Sterilization of drugs*

Kaye *et al.* (1952) showed that penicillin could be sterilized by exposure of the dry powder to EO vapour, or by adding EO to solutions of the antibiotic, and that there was no loss of potency or increase in toxicity. In contrast, there was some loss of potency, but no increase in toxicity, when streptomycin–calcium chloride was exposed to EO vapour, *E. coli* and spores of *B. subtilis* being used as indicators of sterilization. More recently, Diding *et al.* (1968) have described the treatment of crude drugs with EO; their procedure involved carrying out total bacterial and coliform counts before and after treatment with EO. Several of the crude drugs were heavily contaminated and although EO reduced the contamination of all the drugs, a few with high initial contamination levels were not sterilized by this treatment.

(viii) *Fumigation of shell eggs*

Contamination and fumigation with regard to poultry are discussed more fully in Section 5(d). However, it has been shown (Lorenz et al., 1950) that fresh, clean eggs placed deliberately in a suspension of a fluorescent *Pseudomonas* sp. and then exposed to EO vapour were afforded almost complete protection against spoilage.

(ix) *Packaged dry fruits*

EO has been used, usually mixed with *iso*-propyl formate to give a liquid at ordinary temperature, to inhibit moulds, yeasts and bacteria in packaged dry fruits (Dunn, 1968).

(x) *Spacecraft sterilization*

EO has been studied extensively for its use in spacecraft sterilization (see Bruch, 1968) who also points out that Russian workers use a mixture of EO and methyl bromide for this purpose.

In general terms, EO has many uses, particularly for delicate equipment or instruments and heat-sensitive plastic equipment and supplies (see also Macek, 1973). Bruch (1973) has described the toxicity of EO residues in plastics. However, because of its very high penetrating ability, EO is not normally used (cf. formaldehyde and BPL) for fumigating large interior spaces, as it is usually impossible to make such spaces sufficiently tight to retain EO (Phillips, 1968).

(b) *Propylene oxide*

The use of EO for the processing or treatment of foods has been discouraged. In contrast, PO, which hydrolyses to the non-toxic propylene glycol (which is apparently more acceptable than ethylene glycol—see Kereluk, 1971), has been shown (Bruch & Koesterer, 1961) to be suitable for treating powdered or flaked foods.

(c) β-*Propiolactone*

BPL is a highly active antibacterial agent, but it has low penetrative powers. Its possible carcinogenicity (Walpole et al., 1954) has obviously limited its applications, although BPL vapour has been shown to have a use in the decontamination of premises (Hoffman & Warshowsky, 1958; Spiner & Hoffman, 1960). Liquid BPL has been employed for the sterilization of culture media (Toplin & Gaden, 1961; Hazeu & Bueck, 1965; Holme, 1965), the test organisms being sporing and non-sporing bacteria. *E. coli* grows well in BPL-treated media after removal of BPL (Holme, 1965).

(d) *Formaldehyde*

Despite some critical reports as to its suitability as a gaseous disinfectant, formaldehyde is still widely used as such, and the following list describes some of its important applications.

(i) *Hospital bedding and blankets*

Vegetative bacteria are generally killed when freely exposed to formaldehyde vapour; however, when the organisms are protected by layers of blankets, they may be able to survive exposure to the aldehyde (*Anon.*, 1958; Thomas *et al.*, 1959). These findings obviously suggest that formaldehyde vapour should be unsuitable for the disinfection of blankets. However, the use of formaldehyde solutions in the penultimate rinse of laundering blankets shows evidence of a residual bactericidal activity associated with the blankets as a result of the slow evolution of formaldehyde vapour (Dickinson & Wagg, 1967; Dickinson *et al.*, 1970) which is sufficient to kill non-sporing bacteria. This persistent bactericidal action, which occurs with woollen, but not cotton, blankets (Alder *et al.*, 1971) means in practice that there is (a) a reduced rate of bacterial contamination during use, or (b) a more rapid disappearance of bacteria during storage (Alder *et al.*, 1971).

(ii) *Low temperature steam*

Alder *et al.* (1966) described a method for disinfecting heat-sensitive material by means of low-temperature steam (70-90°) with formaldehyde vapour. This procedure kills sporing and non-sporing bacteria, although the latter are undoubtedly destroyed by the steam itself.

(iii) *Disinfection of premises*

Formaldehyde vapour has a limited use in the terminal disinfection of premises (Kelsey, 1967), but it has been suggested for decontaminating microbiological laboratories (Hundemann & Holbrook, 1959).

(iv) *Poultry husbandry*

Fumigation by formaldehyde has found considerable use in poultry science. Poultry houses, for example, become progressively contaminated even when there are no signs of clinical disease in the flock, and regular treatment of such houses is necessary, preferably after emptying and before new stock is introduced. Formaldehyde is particularly well suited for this purpose because of its ease of application (*Anon.*, 1970). Nicholls *et al.* (1967), in describing hatchery hygiene manufacture, showed that samples of fluff were contaminated with various types of bacteria and indicated that fumigation generally caused a reduced level of contamination.

One of the major problems of the poultry industry is bacterial contamination of shell eggs (Graves & MacLaury, 1962). Contamination of the egg fluids occurs mainly as a result of the penetration of the shell by bacteria which are deposited on the surface of the egg after it has been laid (Harry, 1963). The shells of deep litter eggs contain several times more bacteria than the shells of battery laid eggs; furthermore, bacteria in the former case consist mainly of those types present in litter, from which they are transferred by hens to eggs or nest linings (Harry, 1963). Fumigation in the hatchery can be carried out after the eggs have been set in the incubators, although a disadvantage of this is the risk of exposing eggs containing developing embryos to toxic doses of formaldehyde; fumigation is thus now usually undertaken in a fumigation chamber, the eggs being fumigated after being set (Harry & Gordon, 1966). It is in fact, recommended that eggs should be disinfected immediately after collection, either by fumigation with formaldehyde gas or by use of a liquid disinfectant (*Anon.*, 1971). It is also important that eggs should be cleaned before disinfection: formaldehyde, for example, is unable to penetrate dirt (*Anon.*, 1971).

(e) *Other gases*

In comparison with EO, PO, BPL and formaldehyde, other gases are little used as disinfectants or sterilizing agents. In many cases this could be due to the high level of toxicity, but in others (e.g. glycidaldehyde, glutaraldehyde) there is a relative dearth of information. Glycidaldehyde would seem to be a particularly useful substance for further investigation. Methyl bromide is listed by Kereluk (1971) as being suitable for the fumigation of many products common to cereal mills, ships' holds, cargo areas and bedding whilst Bruch & Bruch (1970) state that it may be used for the decontamination of furs and hides. The use of carbon dioxide in the food industry has been adequately described by Dunn (1968).

6. Conclusions

Several factors influence the sensitivity of non-sporing bacteria to gaseous disinfectants, notably concentration of gas, time of exposure, temperature of treatment, RH, the presence of organic matter and the predrying of bacteria. The underlying reasons for this influence are not always immediately apparent, however.

Despite. oft-repeated claims that alkylation of proteins is responsible for inactivation of bacteria exposed to gases, no critical work has been published in support of this claim. There is also little information available as to the nature of the interaction of these gases with bacterial DNA and RNA, in particular on the possibility of single-strand breakage occurring and contributing to death. Finally, there is no basic information as to why non-sporing bacteria have a similar order of sensitivity to gaseous disinfectants as bacterial spores, in contrast to the enormous differences that exist in their response to liquid disinfectants.

7. Acknowledgement

I thank Dr H. J. Smith for several useful discussions.

8. References

ABBOT, C. F., COCKTON, J. & JONES, W. (1956). Resistance of crystalline substances to gas sterilization. *Journal of Pharmacy and Pharmacology* 8, 709-720.
ALDER, V. G., BROWN, A. M. & GILLESPIE, W. A. (1966). Disinfection of heat-sensitive material by low-temperature steam and formaldehyde. *Journal of Clinical Pathology* 19, 83-89.
ALDER, V. G., BOSS, E., GILLESPIE, W. A. & SWANN, A. J. (1971). Residual disinfection of wool blankets treated with formaldehyde. *Journal of Applied Bacteriology* 34, 757-763.
ALEXANDER, P. & STACEY, K. A. (1958). Comparison of the changes produced by ionizing radiations and by the alkylating agents: evidence for a similar mechanism at the molecular level. *Annals of the New York Academy of Sciences* 68, 1225-1237.
ALLEN, H. F. & MURPHY, J. T. (1960). Sterilization of instruments and materials with β-propiolactone. *Journal of the American Medical Association* 172, 1759-1763.
ANON. (1958). Disinfection of fabrics with gaseous formaldehyde. Committee on formaldehyde disinfection. *Journal of Hygiene, Cambridge* 56, 488-515.
ANON. (1970). *The Disinfection and Disinfestation of Poultry Houses.* Ministry of Agriculture, Fisheries and Food: Advisory Leaflet 514, revised 1970. London: H.M.S.O.
ANON. (1971). *Hatching Egg Hygiene for More and Better Chicks.* Ministry of Agriculture, Fisheries and Food: Agricultural Development and Advisory Services. London: H.M.S.O.
BAIRD-PARKER, A. C. & HOLBROOK, R. (1971). The inhibition and destruction of cocci. In *Inhibition and Destruction of the Microbial Cell,* ed. Hugo, W. B. London: Academic Press.
BAKER, F. J., LUCAS, B. G. & SEIBER, A. B. (1963). The sterility and sterilization of lung ventilators. *Thorax* 18, 313-315.
BATEMAN, J. B., SEVENS, C. L., MERCER, W. B. & CARSTENSON, E. L. (1962). Relative humidity and the killing of bacteria: the variation of cellular water content with external relative humidity or osmolality. *Journal of General Microbiology* 29, 207-219.
BEEBY, M. M. & WHITEHOUSE, C. E. (1965). A bacterial spore test piece for the control of ethylene oxide sterilization. *Journal of Applied Bacteriology* 28, 349-360.
BLAKE, D. F. & STUMBO, C. R. (1970). Ethylene oxide resistance of micro-organisms important in spoilage of acid and high-acid foods. *Journal of Food Science* 35, 26-29.
BOUCHER, R. M. G. (1974). Potentiated acid 1,5 pentanedial solution—a new chemical sterilizing and disinfecting agent. *American Journal of Hospital Pharmacy* 31, 546-557.
BORICK, P. M. & PEPPER, R. E. (1970). The spore problem. In *Disinfection,* ed. Benarde, M. A. New York: Dekker.
BROOKS, P. & LAWLEY, P. D. (1964). Alkylating agents. *British Medical Bulletin* 20, 91-95.
BRUCH, C. W. (1961). Gaseous sterilization. *Annual Review of Microbiology* 15, 245-262.
BRUCH, C. W. (1968). Spacecraft sterilization. In *Disinfection, Sterilization and Preservation,* eds Lawrence, C. A. & Block, S. S. Philadelphia, U.S.A.: Lea & Febiger.

BRUCH, C. W. (1973). Sterilization of plastics: toxicity of ethylene oxide residues. In *Industrial Sterilization: International Symposium, Amsterdam, 1972*, eds Phillips, G. B. & Miller, W. S. Durham, North Carolina, U.S.A.: Duke University Press.
BRUCH, C. W. & BRUCH, M. K. (1970). Gaseous disinfection. In *Disinfection*, ed. Benarde, M. A. New York: Dekker.
BRUCH, C. W. & KOESTERER, M. G. (1961). The microbicidal activity of gaseous propylene oxide and its application to powdered or flaked foods. *Journal of Food Science* **26**, 428-435.
BURLESON, G. R., MURRAY, T. M. & POLLARD, M. (1975). Inactivation of viruses and bacteria by ozone, with and without sonication. *Applied Microbiology* **29**, 340-344.
CLARK, D. S. & LENTZ, C. P. (1969). The effect of carbon dioxide on the growth of slime producing bacteria on fresh beef. *Canadian Institute of Food Technology Journal* **2**, 72-75.
DAWSON, F. W. (1962). Glycidaldehyde vapour as a disinfectant. *American Journal of Hygiene* **76**, 209-215.
DICKENS, F. (1964). Carcinogenic lactones and related substances. *British Medical Bulletin* **20**, 96-101.
DICKINSON, J. C. & WAGG, R. E. (1967). Use of formaldehyde for the disinfection of hospital woollen blankets in laundering. *Journal of Applied Bacteriology* **30**, 340-346.
DICKINSON, J. C., WAGG, R. E. & LITCHFIELD, S. (1970). Residual bactericidal action of wool blankets laundered with formaldehyde: a trial. *Journal of Applied Bacteriology* **33**, 566-573.
DIDING, N., WERGEMAN, L. & SAMUELSON, G. (1968). Ethylene oxide treatment of crude drugs. *Acta Pharmaceutica Suecica* **5**, 177-182.
DUNN, C. G. (1968). Food preservatives. In *Disinfection, Sterilization and Preservation*, eds Lawrence, C. A. & Block, S. S. Philadelphia, U.S.A.: Lea & Febiger.
ELFORD, W. S. & VAN DEN ENDE, J. (1942). Investigation of ozone as an aerial disinfectant. *Journal of Hygiene, Cambridge* **42**, 240-265.
ERNST, R. R. (1973). Ethylene oxide gaseous sterilization for industrial applications. In *Industrial Sterilization: International Symposium, Amsterdam, 1972*, eds Phillips, G. B. & Miller, W. S. Durham, North Carolina, U.S.A.: Duke University Press.
ERNST, R. R. (1974). Ethylene oxide sterilization kinetics. *Biotechnology and Bioengineering Symposium* No. 4, pp. 865-878.
FETNER, R. H. & INGOLS, R. S. (1959). Bactericidal activity of ozone and chlorine against *Escherichia coli* at $1°$ C. *Advances in Chemistry Series* **21**, 370-374.
Foter, M. J. (1960a). Disinfectants for bedding. Part I. *Soap and Chemical Specialities* **36**, 73-76.
FOTER, M. J. (1960b). Disinfectants for bedding. Part II. *Soap and Chemical Specialities* **36**, 127-133.
FRAENKEL-CONRAT, H. (1961). Chemical modification of viral RNA. I. Alkylating agents. *Biochimica et biophysica acta* **49**, 169-180.
FRAENKEL-CONRAT, H., COOPER, M. & OLCOTT, H. S. (1945). The reaction of formaldehyde with proteins. *Journal of the American Chemical Society* **67**, 950-954.
FREEMAN, M. A. R. & BARWELL, C. F. (1960). Ethylene oxide sterilization in hospital practice. *Journal of Hygiene, Cambridge* **58**, 337-345.
GILBERT, G. L., GAMBILL, V. M., SPINER, D. R., HOFFMAN, R. K. & PHILLIPS, C. R. (1964). Effect of moisture on ethylene oxide sterilization. *Applied Microbiology* **12**, 496-503.
GRAVES, R. C. & MACLAURY, D. W. (1962). The effects of temperature, vapour pressure and absolute humidity on bacterial contamination of shell eggs. *Poultry Science* **41**, 1219-1225.
HARRY, E. G. (1963). The relationship between egg spoilage and the environment of the egg when laid. *British Poultry Science* **4**, 91-100.
HARRY, E. G. & GORDON, R. F. (1966). Egg and egg hatchery hygiene. *The Veterinarian* **4**, 5-15.

HAZEU, W. & BUECK, H. J. (1965). The use of β-propiolactone for the sterilization of heat-labile materials. *Antonie van Leeuwenhoek* **31**, 295-300.
HENDRY, J. A., HOMER, R. F., ROSE, F. L. & WALPOLE, A. L. (1951). Cytotoxic agents. III. Derivatives of ethyleneimine. *British Journal of Pharmacology* **6**, 357-410.
HERMAN, L. G. (1968). Control of biological agents as contaminants in microbiological laboratories. In *Disinfection, Sterilization and Preservation*, eds Lawrence, C. A. & Block, S. S. Philadelphia, U.S.A.: Lea & Febiger.
HIMMELFARB, P., EL–BISI, H. M., READ, R. B. & LITSKY, W. (1962). Effect of relative humidity on the bactericidal activity of propylene oxide vapour. *Applied Microbiology* **10**, 431-435.
HOFFMAN, R. K. (1971). Toxic gases. In *Inhibition and Destruction of the Microbial Cell*, ed. Hugo, W. B. London: Academic Press.
HOFFMAN, R. K. & WARSHOWSKY, B. (1958). Beta-propiolactone vapour as a disinfectant. *Applied Microbiology* **6**, 358-362.
HOFFMAN, R. K., BUCHANAN, L. M. & SPINER, D. R. (1966). β-propiolactone vapour decontamination. *Applied Microbiology* **14**, 989-992.
HOLME, T. (1965). Sterilization of microbiological media with β-propiolactone. *Biotechnology and Bioengineering* **7**, 129-132.
HUNDEMANN, A. S. & HOLBROOK, A. A. (1959). A practical method for the decontamination of microbiologic laboratories by use of formaldehyde gas. *Journal of the American Veterinary Medical Association* **135**, 544-553.
INGRAM, M. & HAINES, R. B. (1949). Inhibition of bacterial growth by pure ozone in the presence of nutrients. *Journal of Hygiene, Cambridge* **47**, 146-168.
IYER, V. N. & SZYBALSKI, W. (1958). Two simple methods for the detection of chemical mutagens. *Applied Microbiology* **6**, 23-29.
KALINSKA, R. W., YELDERMAN, J. J., WEAVER, R. G. & HOEPRICH, P. D. (1966). Cold sterilization of ureteral catheters with ethylene oxide gas. *Journal of Urology* **96**, 31-35.
KAYE, S. (1949). The sterilizing action of gaseous ethylene oxide. III. The effect of ethylene oxide and related compounds upon bacterial aerosols. *American Journal of Hygiene* **50**, 289-295.
KAYE, S. (1950). Use of ethylene oxide for the sterilization of hospital equipment. *Journal of Laboratory and Clinical Medicine* **50**, 289-295.
KAYE, S. & PHILLIPS, C. R. (1949). The sterilizing action of gaseous ethylene oxide. IV. The effect of moisture. *American Journal of Hygiene* **50**, 290-306.
KAYE, S., IRMINGER, H. F. & PHILLIPS, C. R. (1952). The sterilization of penicillin and streptomycin with ethylene oxide. *Journal of Laboratory and Clinical Medicine* **40**, 67-72.
KELSEY, J. C. (1967). Use of gaseous antimicrobial agents with special reference to ethylene oxide. *Journal of Applied Bacteriology* **30**, 92-100.
KERELUK, K. (1971). Gaseous sterilization: methyl bromide, propylene oxide and ozone. In *Progress in Industrial Microbiology* Vol. 10, ed. Hockenhull, D. J. D. Edinburgh: Churchill Livingstone.
KERELUK, K. & LLOYD, R. S. (1969). Ethylene oxide sterilization. A current review of principles and practices. *Journal of Hospital Research* **7**, 7-75.
KERELUK, K. GAMMON, R. A. & LLOYD, R. S. (1970). Microbiological aspects of ethylene oxide sterilization. II. Microbial resistance to ethylene oxide. *Applied Microbiology* **19**, 152-156.
KOLB, R. W. & SCHNEITER, R. (1950). The germicidal and sporicidal efficacy of methyl bromide for *B. anthracis. Journal of Bacteriology* **59**, 401-411.
LAWLEY, P. D. & JARMAN, M. (1972). Alkylation by propylene oxide of deoxyribonucleic acid, adenine, guanosine and deoxyguanylic acid. *Biochemical Journal* **126**, 893-900.
LOGRIPPO, G. A., OVERHULSE, P. R., SZILAGYI, D. E. & HARTMAN, F. W. (1955). Procedure for sterilization of arterial homografts with beta-propiolactone. *Laboratory Investigation* **4**, 217-231.

LORENZ, F. W., STARR, P. B. & BOUTHILLET, R. (1950). Fumigation of shell eggs with ethylene oxide. *Poultry Science* **29**, 545-547.

LOVELESS, A. (1951). Qualitative aspects of the chemistry and biology of radiomimetic (mutagenic) substances. *Nature, London* **167**, 338-342.

LOVELESS, L. E., SPOERL, E. & WEISMAN, T. H. (1954). A survey of effects of chemicals on division and growth of yeast and *Escherichia coli*. *Journal of Bacteriology* **68**, 637-644.

MACEK, T. J. (1973). Biological indicators and the effectiveness of sterilization procedures. In *Industrial Sterilization: International Symposium, Amsterdam, 1972*, eds Phillips, G. B. & Miller, W. S. Durham, North Carolina, U.S.A.: Duke University Press.

MAYR, G. (1961). Equipment for ethylene oxide sterilization. In *Recent Developments in the Sterilization of Surgical Materials. Report of a Symposium.* London: Pharmaceutical Press.

MICHAEL, G. T. & STUMBO, C. R. (1970). Ethylene oxide sterilization of *Salmonella senftenberg* and *Escherichia coli*: death kinetics and mode of action. *Journal of Food Science* **35**, 631-634.

MILLER, W. S. (1972). Establishment of ethylene oxide sterilization cycles. *Bulletin of the Parenteral Drug Association* **26**, 34-40.

MUNTON, T. J. & RUSSELL, A. D. (1970). Aspects of the action of glutaraldehyde on *Escherichia coli*. *Journal of Applied Bacteriology* **33**, 410-419.

MUNTON, T. J. & RUSSELL, A. D. (1972). Effect of glutaraldehyde on the outer layers of *Escherichia coli*. *Journal of Applied Bacteriology* **35**, 193-199.

NEELY, W. B. (1963a). Action of formaldehyde on micro-organisms. I. Correlation of activity with formaldehyde metabolism. *Journal of Bacteriology* **85**, 1028-1035.

NEELY, W. B. (1963b). Action of formaldehyde on micro-organisms. II. The formation of 1,3-thiazone-4-carboxylic acid in *Aerobacter aerogenes*. *Journal of Bacteriology* **85**, 1420-1422.

NEELY, W. B. (1963c). Action of formaldehyde on micro-organisms. III. Bactericidal activity of formaldehyde in sublethal concentrations. *Journal of Bacteriology* **86**, 445-448.

NEELY, W. B. (1966). The adaptation of *Aerobacter aerogenes* to the stress of sublethal doses of formaldehyde. *Journal of General Microbiology* **45**, 187-194.

NEWMAN, L. B., COLWELL, C. A. & JAMESON, E. L. (1955). Decontamination of articles made by tuberculous patients in physical medicine and rehabilitation. *American Review of Tuberculosis and Pulmonary Diseases* **71**, 272-279.

NICHOLS, A. A., LEAVER, C. W. E. & PANES, J. J. (1967). Hatchery hygiene evaluation as measured by microbiological examination of samples of fluff. *British Poultry Science* **8**, 297.

NORDGREN, G. (1939). Investigations on the sterilizing efficacy of gaseous formaldehyde. *Acta pathologica et microbiologica scandinavica supplement XL*, pp. 1-165.

OPFELL, J. B., SHANNON, J. L. & CHAN, H. (1967). Comparison of methyl bromide and ethylene oxide resistances of *Staphylococcus epidermidis* and *Bacillus subtilis* spore population. *Bacteriological Proceedings* **67**, 13-14.

PERKINS, J. J. & LLOYD, R. S. (1961). Applications and equipment for ethylene oxide. In *Recent Developments in the Sterilization of Surgical Materials. Report of a Symposium.* London: Pharmaceutical Press.

PHILLIPS, C. R. (1949). The sterilizing action of gaseous ethylene oxide. II. Sterilization of contaminated objects with ethylene oxide and related compounds: time, concentration and temperature relationships. *American Journal of Hygiene* **50**, 280-288.

PHILLIPS, C. R. (1952). Relative resistance of bacterial spores and vegetative bacteria to disinfectants. *Bacteriological Reviews* **16**, 135-138.

PHILLIPS, C. R. (1961). The sterilizing properties of ethylene oxide. In *Recent Developments in the Sterilization of Surgical Materials. Report of a Symposium.* London: Pharmaceutical Press.

PHILLIPS, C. R. (1968). Gaseous sterilization. In *Disinfection, Sterilization and Preservation*, eds Lawrence, C. A. & Block, S. S. Philadelphia, U.S.A.: Lea & Febiger.

PHILLIPS, C. R. & KAYE, S. (1949). The sterilizing action of gaseous ethylene oxide. I. Review. *American Journal of Hygiene* **50**, 270-279.
PHILLIPS, C. R. & WARSHOWSKY, B. (1958). Chemical disinfectants. *Annual Review of Microbiology* **12**, 525-550.
PICKERILL, J. K. (1975). Practical system for steam formaldehyde disinfection. *Laboratory Practice* **24**, 401-403.
PRICE, C. C. (1958). Fundamental mechanisms of alkylation. *Annals of the New York Academy of Sciences* **68**, 663-668.
RHAESE, H. J. & FREESE, E. (1969). Chemical analysis of DNA alterations. IV. Reactions of oligodeoxyribonucleotides with nitrosoguanidine. *Biochimica et biophysica acta* **190**, 418-433.
ROBERTS, J. J. & WARWICK, G. P. (1963). The reaction of propiolactone with guanosine, deoxyguanylic acid and RNA. *Biochemical Pharmacology* **12**, 1441-1442.
ROBERTS, T. A. & HITCHINS, A. D. (1969). Resistance of spores. In *The Bacterial Spore*, eds Gould, G. W. & Hurst, A. London: Academic Press.
ROSS, W. J. C. (1958) *In vitro* reactions of biological alkylating agents. *Annals of the New York Academy of Sciences* **68**, 669-681.
ROXON, J. J. (1973). Ethylene oxide sterilization. *Australian Journal of Pharmaceutical Sciences* **NS2**, 65-76.
ROYCE, A. & BOWLER, C. (1961). Ethylene oxide sterilization—some experiences and some practical limitations. *Journal of Pharmacy and Pharmacology* **13**, 87T-94T.
ROYCE, A. & SYKES, G. (1955). A new approach to sterility testing. *Journal of Pharmacy and Pharmacology* **7**, 1046-1052.
RUSSELL, A. D. (1971). The destruction of bacterial spores. In *Inhibition and Destruction of the Microbial Cell*, ed Hugo, W. B. London: Academic Press.
RUSSELL, A. D. (1974). Factors influencing the activity of antimicrobial agents: an appraisal. *Microbios* **10**, 151-174.
RUSSELL, A. D. & HAQUE, H. (1975). Inhibition of EDTA-lysozyme lysis of *Pseudomonas aeruginosa* by glutaraldehyde. *Microbios* **13**, 151-153.
RUSSELL, A. D. & VERNON, G. N. (1975). Inhibition by glutaraldehyde of lysostaphin-induced lysis of *Staphylococcus aureus*. *Microbios* **13**, 147-149.
RUSSELL, C. (1965). The effect of nitric oxide on the growth of *Escherichia coli* M. *Experientia* **21**, 625.
SALLE, A. J. & KORZENOVSKY, M. (1942). The effect of a vacuum on the destruction of bacteria by germicides. *Proceedings of the Society for Experimental Biology and Medicine* **50**, 12-16.
SKEEHAN, R. A., KING, J. H. & KAYE, S. (1956). Ethylene oxide sterilization in ophthalmology. *American Journal of Ophthalmology* **42**, 424-430.
SMITH, C. R. (1968). Mycobactericidal agents. In *Disinfection, Sterilization and Preservation*, eds Lawrence, C. A. & Block, S. S. Philadelphia, U.S.A.: Lea & Febiger.
SMITH, H. J. & SPENCER, P. S. J. (1975). Enzyme inhibitors—active in therapy. *Manufacturing Chemist & Aerosol News* **46**(9) 57-61, (10) 43-48.
SNOW, J. C., MANGIARACINE, A. B. & ANDERSON, M. L. (1962). Sterilization of anaesthesia equipment with ethylene oxide. *New England Journal of Medicine* **266**, 443-446.
SPAULDING, E. H. (1968). Chemical disinfection of medical and surgical materials. In *Disinfection, Sterilization and Preservation*, eds Lawrence, C. A. & Block, S. S. Philadelphia, U.S.A.: Lea & Febiger.
SPINER, D. R. & HOFFMAN, R. K. (1960). Method of disinfecting large enclosures with BPL vapour. *Applied Microbiology* **8**, 152-155.
STACEY, K. A., COBB, M., COUSENS, S. F. & ALEXANDER, P. (1958). The reactions of the "radiomimetic" alkylating agents with macromolecules *in vitro*. *Annals of the New York Academy of Sciences* **68**, 682-701.
STAEHELIN, M. (1958). Reaction of tobacco mosaic virus nucleic acid with formaldehyde. *Biochimica et biophysica acta* **29**, 410-417.
STARBUCK, W. C. & BUSCH, H. (1963). Hydroxyethylation of amino acids in plasma

albumin with ethylene oxide. *Biochimica et biophysica acta* **78**, 594-605.
SYKES, G. (1965). *Disinfection and Sterilization*, 2nd Edn. London: Spon.
TAYLOR, L. A., BARBEITO, M. S. & GREMILLION, G. G. (1969). Paraformaldehyde for surface sterilization and detoxification. *Applied Microbiology* **17**, 614-618.
THOMAS, C. G. A. (1960). Sterilization by ethylene oxide. *Guy's Hospital Reports* **109**, 57-74.
THOMAS, C. G. A., WEST, B. & BESSER, H. (1959). Cleansing and sterilization of hospital blankets. *Guy's Hospital Reports* **108**, 446-463.
TOTH, L. J. (1959). The sterilizing effect of ethylene oxide vapour on different micro-organisms. *Archiv für Mikrobiologie* **32**, 409-410.
TOPLIN, T. & GADEN, E. L. (1961). The chemical sterilization of liquid media with beta-propiolactone and ethylene oxide. *Journal of Biochemical and Microbiological Technology and Engineering* **3**, 311-323.
TRAFAS, P. C., CARLSON, R. E., LOGRIPPO, G. A. & LAM, C. R. (1954). Chemical sterilization of arterial homografts. *Archives of Surgery* **69**, 415-424.
TROLL, W., RINDE, E. & DAY, P. (1969). Effect on N-7 and C-8 substitution of guanine in DNA on T_m, buoyant density and RNA polymerase priming. *Biochimica et biophysica acta* **174**, 211-219.
TULIS, J. J. (1973). Formaldehyde gas as a sterilant. In *Industrial Sterilization: International Symposium, Amsterdam, 1972*, eds Phillips, G. B. & Miller, W. S. Durham, North Carolina, U.S.A.: Duke University Press.
WALKER, J. F. (1964). *Formaldehyde*. Vol. III. New York: Van Nostrand Reinhold.
WALLES, S. A. (1974). The influence of some alkylating agents on the structure of DNA *in vitro*. *Chemico-Biological Interactions* **9**, 97-103.
WALPOLE, A. L., ROBERTS, D. C., ROSE, F. L., HENDRY, J. A. & HOMER, R. F. (1954). Cytotoxic agents. IV. The carcinogenic actions of some monofunctional ethylene amine derivatives. *British Journal of Pharmacology* **9**, 306-323.
WARSHAW, L. J. (1953). Bactericidal and fungicidal effects of ozone on deliberately contaminated 3D viewers. *American Journal of Public Health* **43**, 1558-1562.
WEBB, S. J. (1967). Mutation of bacterial cells by controlled desiccation. *Nature, London* **213**, 1137-1139.
WHELTON, R., PHAFF, H. J., MRAK, E. M. & FISHER, C. D. (1946). Control of microbiological food spoilage by fumigation with epoxides. Parts I and II. *Food Industry* **18**, 23-25, 174-176, 228.
WINDERMUELLER, H. G. & KAPLAN, N. D. (1961). The preparation and properties of *N*-hydroxyethyl derivatives of adenosine, adenosine triphosphate and nicotinamide adenosine dinucleotide. *Journal of Biological Chemistry* **236**, 2716-2726.
WINGE-HEDEN, K. (1963). Ethylene oxide sterilization without special equipment. *Acta pathologica et microbiologica scandinavica* **53**, 225-244.

The Antimicrobial Activity of SO_2—with Particular Reference to Fermented and Non-fermented Fruit Juices

S. M. HAMMOND AND J. G. CARR

University of Bristol, Research Station, Long Ashton, Bristol, England

CONTENTS

1. Introduction 89
2. Uses of SO_2 in fermented and non-fermented fruit beverages 90
3. Ionization of SO_2 in aqueous solutions 90
4. Sulphite addition compounds in fruit beverages 91
5. The active antimicrobial principle in sulphur dioxide solutions 92
6. Mechanisms of action of SO_2 94
    (a) Uptake 94
    (b) Interaction with thiol groups 95
    (c) Interaction with enzymes, co-factors and vitamins 96
    (d) Interaction with nucleic acids and mutagenesis 97
    (e) Interaction with components of the microbial 'metabolic pool' 98
    (f) Sulphur dioxide and microbial sulphur metabolism 99
    (g) Interaction with lipids 100
7. Difficulties encountered in experimentation with SO_2 100
8. Effects of SO_2 on the viability of micro-organisms present in fermented fruit juices 101
9. Conclusions 104
10. Acknowledgements 105
11. References 106

1. Introduction

THE ANCIENTS BELIEVED that the fumes of burning sulphur could purify both body and soul. According to Homer, Odysseus, on his return to Ithaca after the siege of Troy, directed his nurse,

> "Eurycleia, bring me some disinfecting sulphur and make a fire, so that I can fumigate the house."

It was perhaps because of its edificatory properties that the fumes of burning sulphur, now known as the gas sulphur dioxide (SO_2), was first used by the Egyptians and later the Romans to cleanse wine vessels. The first modern accounts of the use of SO_2 in beverages was made by a Dr Beale in John Evelyn's Pomona published in 1670.

> "As sulphur hath some use in wine, so some do lay brimstone on a rag and by a wire let it down into the cider vessel and there fire it and when the vessel is full of smoak, the liquor speedily poured in."

2. Uses of SO_2 in Fermented and Non-fermented Fruit Beverages

Sulphur dioxide is the preservative of choice for a wide range of low pH foods and beverages. This is due partly to its marked antimicrobial properties, but also to its ability to prevent enzymic and non-enzymic browning in fruit products. These latter properties will not be discussed here.

Sulphur dioxide is often added to fruit juices before fermentation to inhibit the growth of bacteria and non-fermenting yeasts, allowing subsequent controlled fermentations. Pure cultures, adapted to SO_2, can then be added to produce the desired fermentation. The origin, fruit cultivar and quality of the juice will determine the amount of SO_2 necessary to control a fermentation. For example Amerine & Joslyn (1951) suggested the addition of 75 p/m of SO_2 to musts from sound, mature acid grapes, 112 p/m to musts from fully matured grapes and 270 p/m to musts from over-ripe, diseased or damaged grapes. Generally, the addition of 50-100 p/m of SO_2 to musts and fruit juices will inhibit the growth of naturally occurring yeasts, but the sensitivity of individual species varies widely, e.g. *Pichia membranaefaciens* and *Kloeckera apiculata* are SO_2-sensitive while *Saccharomycodes ludwigii, Saccharomyces bailii* and *Brettanomyces* spp. can resist SO_2 concentrations up to 500 p/m (Reed & Peppler, 1973). The growth of many bacteria in fruit juices and musts can be inhibited by 100 p/m SO_2. It has been reported that SO_2 is usually more effective in inhibiting the growth of Gram negative rods (notably *Escherichia coli* and pseudomonads) than Gram positive rods, e.g. lactic acid bacteria (Christian, 1963; Dyet & Shelley, 1966). Freese *et al.* (1973) reported that the antibacterial effect of SO_2 against *Bacillus* spp. increased linearly with concentration.

Addition of SO_2 to grape musts or fruit juices can also inhibit the action of fruit enzymes, preventing off-flavours (Haisman, 1974).

Sulphur dioxide is also added to the fermented product to prevent refermentation or microbial spoilage. The effect of SO_2 on a red wine, the sugar content of which had been raised to 1.5% with sucrose, inoculated with 2.3×10^5 yeasts/ml, has been described by Ough & Ingraham (1960). After two months the viable yeast count was determined, when it was found that there was a linear decrease in the number of survivors with increased SO_2 concentration. Sulphur dioxide is equally effective in fruit wines and is used widely in the fermentation of ciders and berry juices to achieve biological stability in the finished product (Reed & Peppler, 1973).

3. Ionization of SO_2 in Aqueous Solutions

The antimicrobial properties of SO_2 depend upon the degree of ionization of the molecule. Sulphur dioxide is applied to foods and beverages as the liquefied gas

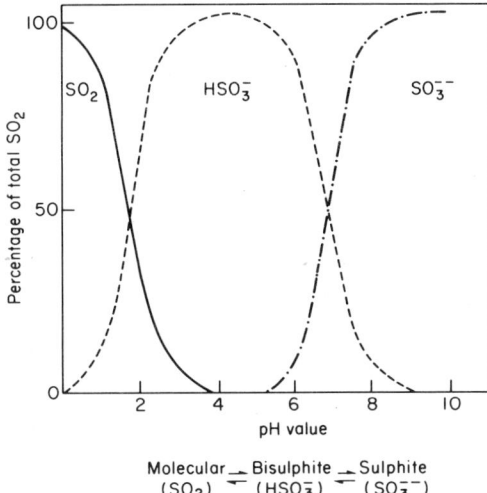

Fig. 1. Percentage distribution of sulphite, bisulphite and molecular SO_2 as a function of pH in aqueous solution.

or more commonly in the form of sulphite, bisulphite or metabisulphite salts. Sulphur dioxide or its salts set up a pH-dependent equilibrium mixture when dissolved in water. From the dissociation constants for sulphite and bisulphite (Tartar & Garretson, 1941) the proportion of each molecular species existing at any pH value can be calculated (Fig. 1). As the pH falls the proportion of sulphite ions in the mixture is decreased and the proportion of SO_2 molecules increases at the expense of bisulphite ions. The term molecular SO_2 will be used in this paper to describe the molecules existing in aqueous solutions at low pH values. The existence of sulphurous acid has been questioned by many workers. Ultra-violet, infra-red and Raman spectroscopy has failed to reveal the presence of a sulphurous acid molecule (Ley & Konig, 1938; Simon & Waldman, 1956; Jones & McLaren, 1958). Falk & Giguère (1958) suggested that since no stable sulphurous acid molecules could be detected in aqueous solution, SO_2 is dissolved in the molecular state.

4. Sulphite Addition Compounds in Fruit Beverages

When SO_2 is added as a preservative or anti-oxidant to fermented and non-fermented beverages, part of it combines with organic molecules in the product to form so-called sulphite addition compounds. Aqueous SO_2 solutions react with aldehydes (where $R'=H$) and ketones to produce hydroxysulphonates (Fig. 2).

While all aldehydes react to form hydroxysulphonates, many ketones react

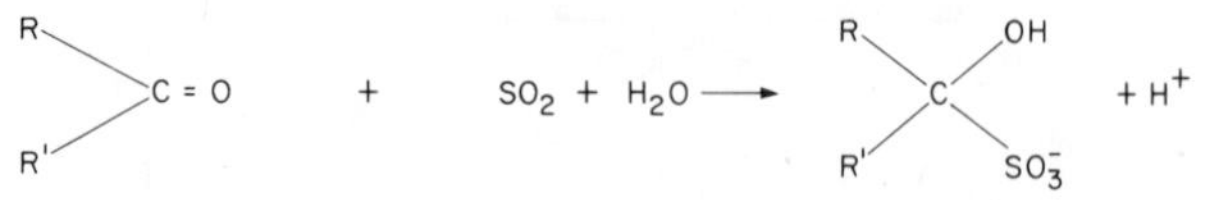

$$\underset{\text{Carbonyl compound}}{\overset{R}{\underset{R'}{>}}C=O} \;+\; SO_2 + H_2O \longrightarrow \underset{\text{Hydroxysulphonate}}{\overset{R}{\underset{R'}{>}}C\overset{OH}{\underset{SO_3^-}{<}}} \;+\; H^+$$

Fig. 2. Formation of hydroxysulphonates.

slowly if at all. With the exception of diethyl ketone, which reacts only slightly, only ketones containing either a methyl group attached to the carbonyl group (i.e. $R'=CH_3$) or having the carbonyl group as part of a ring system, as for example in keto sugars, combine appreciably with SO_2. Sulphur dioxide will also react with olefinic compounds, where addition to the double bond is sometimes as rapid as to the carbonyl group. Amines, particularly tertiary amines will react with SO_2 to form amine bisulphites (Joslyn & Braverman, 1954).

Combination of SO_2 with sugars, which may be regarded as cyclized aldehydes, is slower than with open chain aldehydes. Ingram & Vas (1950) found that galactose, mannose and arabinose readily form addition compounds with SO_2, but maltose, lactose and glucose are less active in complex formation, while fructose and sucrose are probably inactive.

Burroughs & Sparks (1964) have shown that the amount of SO_2 bound by fruit products is a function of a series of reversible chemical equilibria, each depending upon pH, temperature and active concentrations of the reactants. These workers found that in ciders it was possible to attribute bound SO_2 to eight main components. The very high SO_2-binding capacity of certain ciders and fruit juices was traced to the combined activities of moulds and bacteria, particularly *Acetomonas* spp. (*Gluconobacter*) that produce additional SO_2-binding compounds when growing in juices prepared from damaged or diseased fruit. In addition to the usual SO_2-binding agents present in ciders, i.e. sugars, acetaldehyde, xylosone, galacturonic acid, pyruvic acid and 2-oxoglutaric acid, ciders made from damaged fruit contained additional SO_2-binding compounds, including 5-oxofructose, 2-oxogluconic acid and 2,5-dioxogluconic acid. Rankine & Pocock (1969) showed that the amount of SO_2 bound by acetaldehyde, pyruvic acid and 2-oxoglutaric acid during the fermentation of three grape juices ranged from 43-89%, depending upon the yeast strain used and the juice type. Weeks (1969) reported that in experimental white wines the main SO_2-binding compounds were pyruvic acid and 2-oxoglutaric acid. Peynaud & Sapis (1972) showed that wine made from damaged and diseased grapes contained increased numbers of SO_2-binding compounds.

5. The Active Antimicrobial Principle in Sulphur Dioxide Solutions

It has been shown that the germicidal effect of SO_2 in sugar solutions was related not to the total amount added but to the unbound, or so-called free SO_2

(Ingram, 1948). Sulphite addition products show little antimicrobial activity (Neuberg, 1929; Rehm, 1964). Any germicidal activity exhibited by SO_2 bound to organic molecules is thought to be due to that small proportion of molecular SO_2 released by dissociation of the complex.

Oka (1960) showed that the antimicrobial effect of many acid antiseptics, including SO_2, varied with pH, being powerful inhibitors of microbial growth under acid conditions but relatively ineffective at neutrality. The antimicrobial activity of acid antiseptics appears to depend on the concentration of non-ionized acid molecules, which will increase with lowered pH. Equal concentrations of non-ionized molecules show an antimicrobial effect independent of pH (Rahn & Conn, 1944). Bosund (1962) explained this by suggesting that non-ionized molecules of acid antiseptics can penetrate the cell wall and membranes of micro-organisms more rapidly than ionized molecules. It has been shown that molecular SO_2 is > 1000 times as active as the bisulphite or sulphite ion against *E. coli*, 500 times more effective against yeasts and 100 times more effective against *Aspergillus niger* (Rehm & Wittmann, 1962).

When a suspension of *Saccharomyces cerevisiae* containing c. 8×10^6

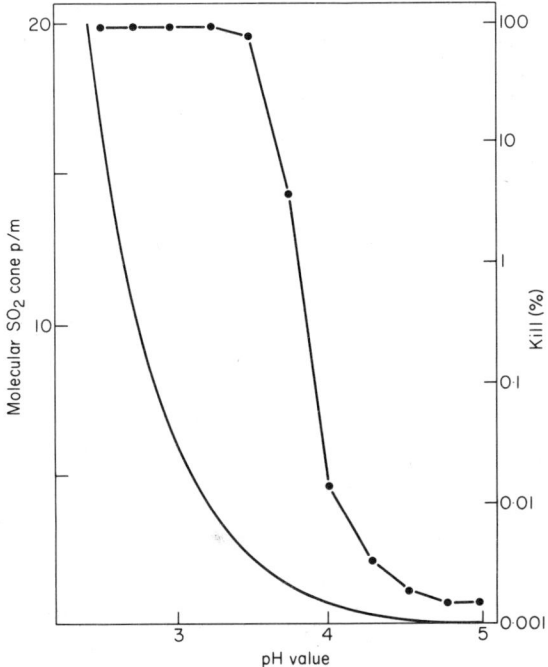

Fig. 3. Effect of 100 p/m sodium metabisulphite on the viability of a yeast suspension (8×10^6 organism/ml) at different pH values (●). Solid line shows the molecular SO_2 concentration.

yeasts/ml was treated with sodium metabisulphite at a constant concentration over a range of pH values (Fig. 3) and the viable count determined after 2 h, the greatest kill was observed at low pH. Using the dissociation constants for bisulphite in water (Tartar & Garretson, 1941), the molecular SO_2 concentration for each pH value can be calculated. The antifungal effect of SO_2 also increases with the molecular SO_2 concentration.

The antimicrobial effect of adding SO_2 to fruit beverages is, therefore, related only to the small part of the added SO_2 not bound by compounds in the product and existing in the non-ionized molecular form, i.e. the free molecular SO_2 concentration. For effective preservation with SO_2 the product must have a low pH, the more acidic the less SO_2 need be added. Sulphur dioxide is unsatisfactory for preserving products with a $pH > 4$. Palm wine, which has a natural pH of 4-5, would require > 1000 p/m of SO_2 to suppress the growth of spoilage organisms (Faparusi, 1969).

6. Mechanisms of Action of Sulphur Dioxide

Many antimicrobial agents are known to act at specific sites within the microbial cell, but this is not so with the majority of preservatives, which often exhibit multisite concentration-dependent antimicrobial activity. Uptake of sulphur dioxide may be complete within 2 min in yeasts (Macris, 1972). After penetrating the cell membrane, SO_2 could interact with components of microbial metabolism to produce stasis or death. Because SO_2 is very reactive and probably interacts with many cell components and metabolites, the actual cause of SO_2-induced inhibition or death remains debatable.

(a) *Uptake*

The first step in the action of an antimicrobial agent is its uptake by the micro-organism. Undissociated acid antiseptic molecules penetrate microbial cells more readily than ionic species (Ingram et al., 1956). Compounds in solution may enter the cell by active transport or by passive diffusion.

Macris & Markakis (1974) suggested that uptake of SO_2 by yeasts was not by passive diffusion. If SO_2 were taken up passively, uptake would correlate with applied SO_2 concentration. In fact, SO_2-uptake displayed saturation kinetics, i.e. the cell can be saturated with substrate. The temperature coefficient for uptake of material by passive diffusion is of the order of $1.4/10°$ temperature rise. Macris & Markakis (1974) examined SO_2-uptake by yeasts over a temperature range and found that the temperature coefficient did not correspond to the phenomena of passive diffusion.

Usually active uptake systems are specific, the degree of specificity depending upon stereo-specific sites within the cell membrane. It has been claimed that as microbial membranes appear to select molecular SO_2 from a mixture of sulphite and bisulphite ions, this is indicative of an active transport mechanism (Macris, 1972). However, non-ionized molecules of weak electrolytes usually diffuse into cells more readily than charged ions (Jacobs, 1940).

Transport systems which can be slowed by metabolic inhibitors are also thought to involve active processes. Pretreatment with inhibitors that block the formation or utilization of high energy phosphate bonds can prevent the accumulation of SO_2 inside yeast cells (Macris, 1972), suggesting that SO_2-uptake is carried out by an energy requiring active site within the cell membrane.

Further evidence for active uptake of SO_2 was obtained by Crompton et al. (1974) who reported that rat liver mitochondria actively took up SO_2 molecules. It appeared that the dicarboxylate carrier present in the mitochondrial membrane had an affinity for SO_2 molecules and was capable of actively transporting SO_2 into the mitochondrion.

Studies on the uptake of SO_2 by bacteria have not been reported.

(b) *Interaction with thiol groups*

In living systems thiol groups have unique properties, involving reversible oxidation to form disulphide bridges. These bridges contribute to the secondary and tertiary structure in proteins which are vital to enzymic activity. In addition many enzymes owe their activity to free thiol groups (Lynen, 1970).

Sulphur dioxide can replace the less nucleophilic sulphur group from the disulphide bond of cystine or cystine peptides (Clarke, 1932) (Fig. 4). The reaction is reversible in alkaline solutions in the absence of added oxidants. The rate of cystine sulphitolysis and the point of equilibrium is markedly affected by neighbouring groups on the protein molecule, by solution pH, temperature and ionic strength (Stricks & Kolthoff, 1951; Parker & Kharasch, 1959). The disulphide bonds of proteins exhibit differing sensitivities to SO_2 (Middlebrook & Phillips, 1942; Pechere et al., 1958; Kolthoff et al., 1959, 1960; Cecil & Leoning, 1960; Henchen, 1962), but generally interchain disulphide bonds are easily broken by SO_2 and this cleavage is reversible. Intrachain disulphide bonds are more resistant to cleavage, probably due to protection by folding of the protein molecule. Pretreatment of proteins with urea makes intrachain disulphide groups more sensitive to sulphite cleavage. Leach (1960) suggested that pretreatment of proteins with urea imposed a strain on disulphide bonds by unfolding the protein and hence reducing the effective bond strength; urea also increases the disulphide bond reactivity with SO_2. Similarly if the chains are unfolded by adding guanine, phenyl mercury hydroxide or by physical stress then intrachain disulphide bonds can be broken (Speakman, 1958; Cecil & Wake, 1962). Formation of protein sulphonates is important because of the well-recognized fact that sulphonation of disulphide bridges in enzymes, even under conditions where re-oxidation to the native state can occur, produces losses of enzyme function (Chan, 1968).

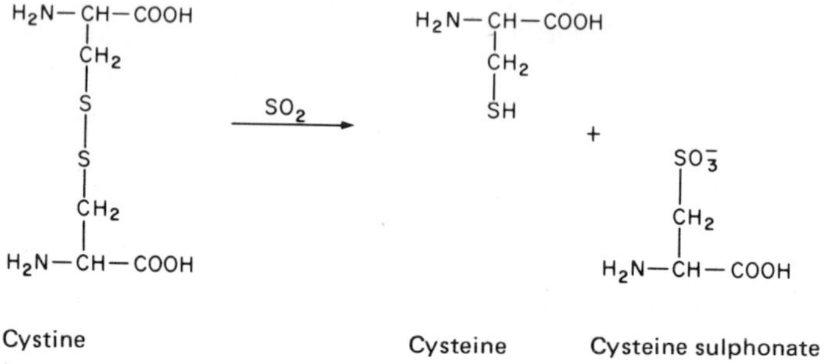

$$\begin{array}{c} H_2N-CH-COOH \\ | \\ CH_2 \\ | \\ S \\ | \\ S \\ | \\ CH_2 \\ | \\ H_2N-CH-COOH \end{array} \xrightarrow{SO_2} \begin{array}{c} H_2N-CH-COOH \\ | \\ CH_2 \\ | \\ SH \end{array} + \begin{array}{c} SO_3^- \\ | \\ CH_2 \\ | \\ H_2N-CH-COOH \end{array}$$

Cystine   Cysteine   Cysteine sulphonate

Fig. 4. Cleavage of the disulphide bond of cystine by SO_2.

In enzymes, the juxtaposition of two thiol groups often allows the formation of a dithiolane ring resulting in the distortion of the dihedral angle of the sulphur atom (from 90-26°) with an energy strain of 3.5 kcal. Relief from this strain explains the ready reducibility of the dithiolane ring and the role of dithiole in oxidative and reductive enzymes. The proximity of two thiol residues provides a ligand with a high potential affinity for a number of groups. This has led to the suggestion that their function in nature may be to bind substrates or co-factors to enzymes (Massey & Williams, 1965). Such groups would be particularly prone to SO_2 attack.

Structural proteins may be susceptible to sulphitolysis and this could cause considerable damage to the microbial cell.

(c) *Interaction with enzymes, co-factors and vitamins*

Thiamine (Vitamin B_1) on treatment with SO_2 is broken into two fragments which show no vitamin activity (Williams, 1935); see Fig. 5. Many micro-organisms have an absolute requirement for thiamine, for example yeasts of the genus *Brettanomyces* and certain lactobacilli, and addition of SO_2 may inhibit microbial growth by effectively removing thiamine from the system.

Thiamine pyrophosphate, an almost universal co-enzyme in living systems, is associated with oxidative and non-oxidative decarboxylases of oxo-acids and the formation of 2-ketols. Very probably all thiamine-dependent enzyme reactions are inhibited by SO_2 cleavage of the thiamine moiety of thiamine pyrophosphate (Haisman, 1974).

The co-enzyme folic acid, concerned in purine biosynthesis is also cleaved by SO_2 (Vanderschmitt et al., 1967). The co-enzyme glutathione contains a dithiol

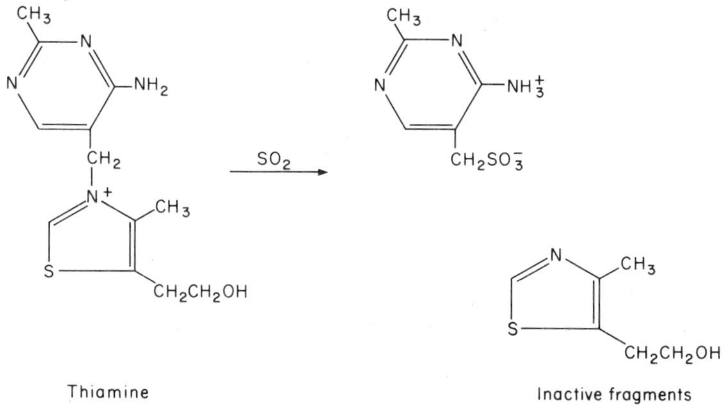

Fig. 5. Sulphitolysis of thiamine.

sensitive to SO_2 (Stricks et al., 1955; Massey & Williams, 1965). An interaction has been demonstrated between nicotinamide adenine dinucleotide (NAD) and SO_2 (Mayerhof et al., 1938). Pfleiderer et al. (1956) suggested that the SO_2-sensitivity of enzymes was in part due to the inactivation of NAD, without which the energy releasing oxidative reactions of micro-organisms could not proceed. Rehm (1964) showed that three NAD-dependent steps of glycolysis in *E. coli* and *S. cerevisiae* were strongly inhibited by SO_2.

Many enzymes and enzyme systems are inhibited by SO_2 either by inducing changes in the enzyme molecular conformation, modifying the enzyme active site or by co-enzyme destruction. There can be few microbial enzymes not affected in some way by SO_2.

(d) *Interaction with nucleic acids and mutagenesis*

Hayatsu (1969) reported changes in the nucleotide sequence of *E. coli* transfer-RNA in the presence of SO_2 and showed (Hayatsu et al., 1970) that bisulphite was capable of converting cytosine to uracil (Fig. 6). The ability of SO_2 to induce cytosine–uracil conversions has been quoted as the basis of mutagenic effects of SO_2 on micro-organisms (Inoue et al., 1972) as well as for interference in double-helix formation, transcription and inactivation of RNA in coding for protein syntheses (Shapiro & Braverman, 1972).

Hayatsu & Miura (1970) and Summers & Drake (1971) independently reported that addition of SO_2 to cultures of bacteriophages induced mutations as a result of its specific action against guanine–cytosine pairs. Dorange & Dupuy (1972) have reported SO_2-induced mutations in yeasts, while Mukai et al. (1970) reported that SO_2-induced mutants of *E. coli* showed modifications of bacterial DNA restricted to guanine–cytosine transfers.

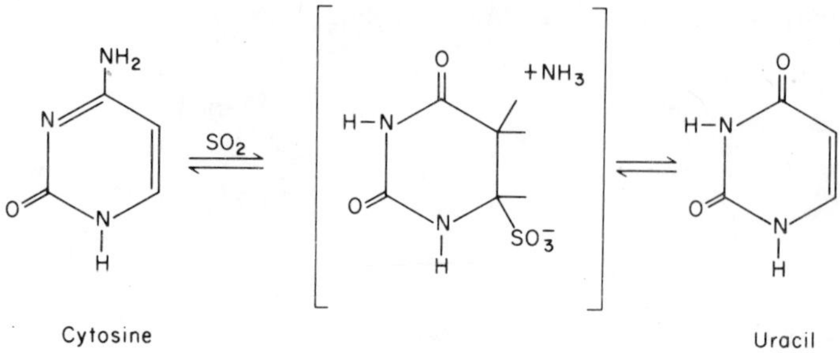

Fig. 6. Deamination of cytosine in the presence of SO_2 to form uracil.

Shapiro et al. (1973) have suggested that SO_2 can only interact with nucleic acid *in vivo*, when the nucleic acids exist as single strands, i.e. SO_2 can react with transfer or messenger RNA, but not with DNA when it exists as a double strand.

Sulphur dioxide catalyses the reaction between cytosine derivatives and amines to form 4-amino cytosine (Shapiro & Weisgras, 1970). Amines may compete with water as nucleophiles for the cytosine–bisulphite adduct under physiological conditions. The discoverers of this reaction have speculated that it may be a basis for SO_2-induced damage to nucleic acid, since cytosine residues modified by SO_2 addition could conjugate with protein molecules, or react with guanine or adenine residues of the opposite nucleic acid strand. The resulting cross-linkage could produce effects similar to those elicited by cytotoxic difunctional alkylating agents.

Although SO_2 undoubtedly reacts with nucleic acids *in vitro*, intact organisms possess defence mechanisms capable of preventing access of mutagens to the seat of the genetic code, i.e. DNA, and are capable, to some extent, of detoxifying SO_2; see Section 6(f).

Evidence suggests that SO_2 becomes more effective in transforming nucleic acid as its concentration is lowered (Inoue et al., 1972). This is an important consideration in the preservation of fruit products, where legislation restricts preservative concentration. Persistent use of low concentrations of SO_2 and partial microbial detoxification may be factors in the production of SO_2-resistant strains.

(e) *Interaction with the microbial 'metabolic pool'*

The affinity of SO_2 for a wide range of sugars, aldehydes and ketones has been mentioned previously. Rehm (1964) investigated the inhibition of yeast glycolysis by SO_2 and reported that the formation of addition products by SO_2 in the microbial cytoplasm was an important factor in the inhibitory mechanism. Complex formation was thought to begin at the glucose level and extend to

3-phosphoglycerate, 3-phosphohydroxyacetone phosphate, pyruvate, acetaldehyde oxaloacetate and 2-oxoglutarate. These compounds will then be effectively removed from participation in cellular reactions. Macris (1972) examined the distribution of radioactive-labelled SO_2 in treated yeasts by cell fractionation. The major part of bound SO_2 was found to be combined with sugars, aldehydes or ketones. Chen & Sakaguchi (1972) reported that the intermediate metabolism of *Staphylococcus aureus, E. coli* and *Proteus vulgaris* was inhibited non-specifically by SO_2.

Anraku & Sano (1966) reported that solutions of sulphite at 100° for 20 h brought about breakdown of tryptophan to give a complex mixture of products. Inoue & Hayatsu (1971) investigated the reaction between sulphite solutions and amino acids at pH 7 and room temperature. They reported breakdown of tryptophan into three products, two of which were sulphited and none of which corresponded to the products previously reported. These workers also found that methionine in the presence of sulphite and oxygen was converted to methionine sulphoxide. Other amino acids appeared not to react.

(f) *Sulphur dioxide and microbial sulphur metabolism*

The microbial biochemistry of sulphur has been extensively reviewed (Maw, 1965; Lawrence & Cole, 1968; Roy & Trudinger, 1970). Micro-organisms actively take up sulphate ions from the environment, reduce them to sulphite and then to sulphide, which can then be used in the synthesis of sulphur amino acids. It seems likely that the sulphite moiety produced by reduction of sulphate is not released into the cytoplasm but exists in a complex with a disulphide group of an enzyme. Whether sulphite exists free or bound, addition of SO_2 to the system inhibits competitively the conversion of sulphite to sulphide and may thereby affect microbial sulphur metabolism.

Sulphur dioxide can be detoxified by two routes: the enzyme sulphite oxidase catalyses the oxidation of SO_2 to sulphate, while sulphite reductase converts SO_2 to sulphide. In mammals, sulphite oxidase, present mainly in the liver but also in other tissues, rapidly converts ingested sulphite to sulphate (Fridovich & Handler, 1956; Kun, 1961) by a complex series of reactions with flavoproteins, lipoic acid and hypoxanthine as co-factors. The presence of this enzyme has been reported in micro-organisms (Heimberg et al., 1953). Sulphite reductase has been isolated from *E. coli* (Ellis, 1964) *Salmonella typhimurium* (Dreyfus & Monty, 1963), *Aspergillus nidulans* (Yoshimoto et al., 1967), *Neurospora crassa* (Seigel et al., 1965), yeasts (Nickerson, 1953; Naiki, 1965; Wainwright, 1967; Yoshimoto & Sato, 1968a, b), photosynthetic bacteria (Peck et al., 1974), algae (Schmidt et al., 1974) and a number of higher plants (Tamura, 1965; Asada, 1967). Wattiaux-de Coninck & Wattiaux (1971) found that a sulphite reductase − cytochrome *c.* complex was located in the intermembrane spaces of mitochondrial granules. Sulphite reductase preparations also possess hydroxylamine, nitrite and sometimes cytochrome reductase properties.

The ability of yeasts to produce SO_2 from sulphate, elemental sulphur or sulphur amino acids is well documented (Rankine & Pocock, 1969; Dittrich & Staudenmayer, 1970). Wurdig & Schlotter (1968) reported yeast strains capable of producing up to 130 p/m of SO_2 in fermentation broths. Sulphur dioxide formation by yeasts was found to depend upon pH, oxygen tension, temperature, medium composition and yeast strain. Premuzic *et al.* (1972) demonstrated that SO_2 was formed when grape must was fermented by *Saccharomyces fructuum* (now called *S. chevalieri*) or *S. carlsbergensis* (*S. uvarum*), while in comparison *S. oviformis* (*S. bayanus*), *S. uvarum, S. elegans* (*S. bailii*) and *S. rosei* were poor producers of SO_2. *Saccharomyces veronae* (*Kluyveromyces veronae*) did not produce SO_2.

(g) *Interaction with lipids*

Utsumi *et al.* (1973) have reported that SO_2 was able to peroxidize lipids. This could have a profound effect upon microbial metabolism and membrane function (Tappel, 1973).

7. Difficulties Encountered in Experimentation with SO_2

A little-appreciated difficulty in the study of the antimicrobial activity of SO_2 is the instability of sulphite salts and solutions. Before beginning any work with SO_2 workers would do well to read the salutary warnings of Postgate (1963). Sulphite solutions readily undergo auto-oxidation. A 0.1 M sulphite solution in physiological saline, shaken in air at 37° fell to 0.07 M after 1 h and to 0.022 M after 2½ h (Postgate, 1963). Sulphite auto-oxidation is catalysed by a wide range of metal ions including zinc, cobalt, nickel, ferrous and ferric iron, vanadium and copper. The effects of metal ions on sulphite oxidation are abolished by complexing agents. Since sulphite auto-oxidation is a free radical chain reaction it may also be inhibited by reagents which break the reaction chain, e.g. organic acids, alcohols, sugars, amines and amides. It must also be remembered that in an unbuffered system, oxidation of sulphite to sulphate will lower the pH of the medium. It is advisable that any experimental system chosen for SO_2 work be carefully examined to ensure that SO_2 concentrations do not fall during the experimental period. The actual SO_2 present should be measured directly and not inferred from the amount of sulphite added.

8. Effects of SO_2 on the Viability of Micro-organisms Present in Fermented Fruit Juices

Although SO_2 rapidly kills yeasts in fermented fruit products, viable lactobacilli are commonly isolated from ciders containing appreciable amounts of unbound SO_2 (Carr & Davies, 1971). Lactobacilli appear to grow so slowly in sulphited cider that they do not even cause a visible haze after several months' storage. This selective toxicity of SO_2 has been the subject of study at Long Ashton for some time.

Working with SO_2 imposes restrictions upon microbiological technique. The test system must be strictly anaerobic; the apparatus must be examined for SO_2-permeability and SO_2-binding compounds excluded from nutrient media. The medium of Green & Gray (1950) was adapted for SO_2 work. Fructose, which does not bind SO_2, was substituted for glucose, the media filter-sterilized, as autoclaving increases the proportion of SO_2-binding compounds, and then purged with sterile nitrogen to remove dissolved oxygen.

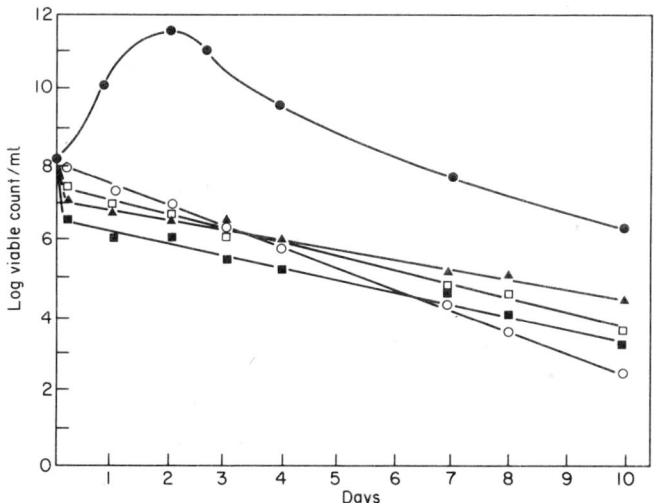

Fig. 7. Growth of *Lactobacillus plantarum* at pH 4 (●) and in the presence of 29 (○), 46 (▲), 85 (□) or 161 (■) p/m of free SO_2.

When inoculated with *Lactobacillus plantarum, L. mali, L. collinoides* or *Leuconostoc* spp. levels of free $SO_2 < 10$ p/m (equivalent to a free molecular SO_2 conc. of < 0.1 p/m) were found to prevent visible bacterial growth at 30° after ten days. Plating out on apple juice yeast extract (AJYE) agar showed viable organisms were present in these flasks.

To examine this phenomenon more closely, quantitative experiments were carried out using *L. plantarum* as a test organism. Flasks containing medium at pH 4 equilibrated with graded doses of SO_2 were inoculated with the organism and incubated at 30°. Samples were taken at regular intervals and plated out on AJYE agar after serial dilution (Fig. 7). Control flasks without SO_2 produced dense growth reaching a maximum of 10^{11}-10^{12} bacteria/ml at two days, but flasks containing SO_2 showed no visible growth. The death rate in these flasks appeared to be almost independent of SO_2 concentration in the range 30-161 p/m free SO_2. To extend the range of SO_2 concentrations examined a series of flasks was prepared, dosed with SO_2, inoculated with *L. plantarum* and incubated at 30° as before. After three days, when the controls showed dense growth, the viable count was determined. Free SO_2 concentrations of 10-390 p/m again produced an almost linear slow decline in bacterial number. It appeared that 20 p/m of free SO_2 was almost as efficient a bacteriostat as 390 p/m and even at the highest SO_2 concentration considerable numbers of survivors could be found after three days. When these experiments were repeated at pH 3.4 the cell number showed only a slow decline with increasing free SO_2 concentration, up to 90 p/m, but above this concentration the viable count began to fall more rapidly. At free SO_2 concentrations above 175 p/m no viable organisms were found after three days.

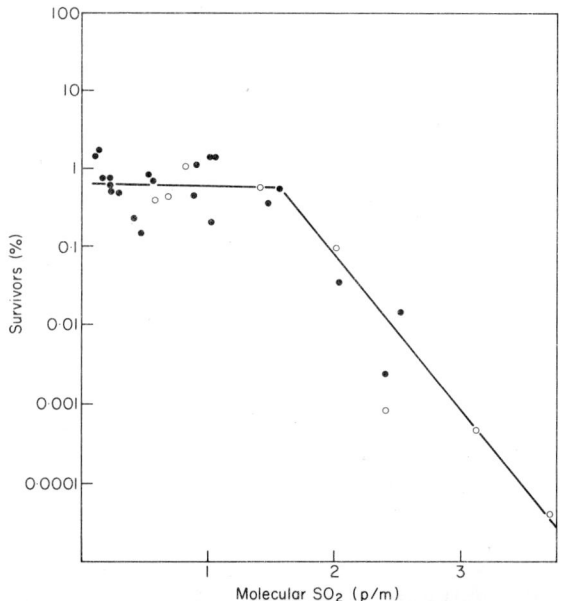

Fig. 8. Effect of molecular SO_2 concentration on the viability of *Lactobacillus plantarum* after 3 days; data collected at pH 4 (●) and pH 3.4 (○).

If these data are presented in terms of molecular SO_2, calculated from the dissociation constant to remove the effect of pH, the death curve for *L. plantarum* has two phases (Fig. 8). For molecular SO_2 concentrations up to

1.5 p/m it appears that SO_2 is bacteriostatic but above this concentration it becomes bactericidal.

In this system two possible explanations of the antibacterial activity of SO_2 are possible. Either SO_2 removes or interacts with metabolite(s) in the growth medium essential for bacterial growth, or interacts with components of bacterial metabolism within the organism. Flasks containing medium and SO_2 were inoculated with *L. plantarum* and incubated for two days. The free SO_2 in the flasks was then removed either by oxidation with a precise amount of hydrogen peroxide or complexed with a small excess of acetaldehyde. Removal of free SO_2 from the growth medium allowed bacterial growth to restart after a lag of two days and the viable count reached 10^{10} bacteria/ml after seven days (Fig. 9).

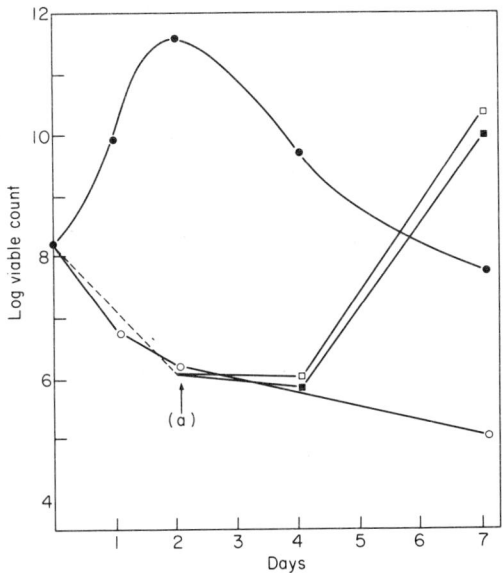

Fig. 9. Effect of 142 p/m of free SO_2 on the growth of *Lactobacillus plantarum* at pH 4 (o) and with SO_2 neutralized by acetaldehyde (□) or hydrogen peroxide (■) added at (a). Control, no SO_2 (●).

This is in contrast with the situation we have found in yeasts. relatively small concentrations of free SO_2 (< 25 p/m) do not merely prevent yeast growth; they produce a rapid reduction in the viable yeast count, i.e. SO_2 is fungicidal. Yeasts from an overnight culture of *S. cerevisiae* were inoculated (*c.* 10^5/ml into a series of SO_2 concentrations dissolved in growth medium (pH 3.4) incubated at 25°; no viable organisms could be isolated after 8 h, even at low free SO_2 concentrations (Fig. 10).

It is possible to speculate on the selective toxicity of SO_2 It seems likely that a compound as chemically reactive as SO_2, would be equally effective in inhibiting yeast or bacterial metabolism. This selectivity may be due to differential uptake of SO_2 by yeasts and bacteria.

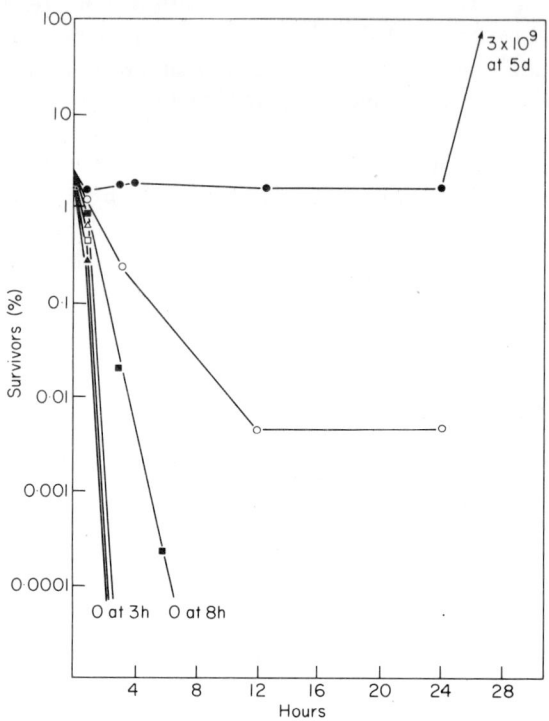

Fig. 10. Growth of *Saccharomyces cerevisiae* at pH 3.4 (●) and in the presence of 15 (○), 25 (■), 50 (△), 75 (□) and 100 (▲) p/m of free SO_2.

9. Conclusions

There is now pressure from many sources to reduce the levels of SO_2 added to foods and beverages (*Anon.*, 1967; Jaulmes, 1970; Lüthi, 1972). Since entry into Europe, the UK permitted levels for SO_2 have been reduced and are about to be reduced further (Table 1). Concern has been expressed about the consumption of SO_2 by the populations of industrial nations. An acceptable daily intake (ADI) for man was recommended by the Joint FAO/WHO Expert Committee on Food Additives (*Anon.*, 1967); 0-0.35 mg of SO_2/Kg body wt was proposed as the unconditional ADI (levels of use that are effective technologically, at least for some purposes, and can be safely employed without expert advice) and 0.35-1.5 mg of SO_2/Kg body wt as the conditional ADI (levels of use that can be employed safely but at which it is desirable that some degree of expert supervision and advice is available). The widespread use of SO_2 as a preservative and anti-oxidant in foodstuffs means that some persons receive a larger daily dose than the conditional ADI. Bigwood (1973) noted that consumption of three glasses of wine/day may exceed the ADI. However, there is still little evidence, in spite of extensive investigation, that orally administered

Table 1
Legal limits for use of SO_2 in beverages in the UK

	Permitted levels (p/m)	Interim period until August 1976 (wines only) (p/m)	
Red wines	200	250	
White wines	250	300	EEC Regulation 2592/73 *Anon.* (1973)
Special white wines	–	400	
Cider	200		
Fruit juices	350		
Beer	70		

SO_2, in moderate doses, shows any mammalian toxicity (Lockett & Natoff, 1960; Steinhoff & Marquardt, 1963; Causeret *et al.,* 1965; Cluzan *et al.,* 1965; Lantaeume *et al.,* 1965; Hugot *et al.,* 1965; Gibson & Strong, 1974). Indeed small amounts of sulphite are regularly formed in mammalian intermediary metabolism, in cystine catabolism and certain other pathways (Larson & Salisbury, 1953).

Sulphur dioxide has been added to fruit products before and after fermentation over countless generations, with little evidence of toxicity. Although we do not advocate the use of preservatives for their own sake, or to cover manufacturers' malpractices, fermented and non-fermented fruit beverages are often biologically unstable. If the legal limits for SO_2 addition are reduced much further, then the use of other preservatives in conjunction with SO_2 will increase and we may be exchanging the use of a relatively safe compound for mixtures of preservatives which may not be as satisfactory from the technological or biological point of view, and perhaps in the long term not as safe, toxicologically.

10. Acknowledgements

We thank Mrs P. A. Davies for her skilled assistance, Dr F. W. Beech and Dr L. F. Burroughs for helpful discussions and the Wine Companies Consortium for a research fellowship (S.M.H.).

11. References

AMERINE, M. A. & JOSLYN, M. A. (1951). *Table Wines: The Technology of Their Production in California.* Berkeley, California: University of California Press.
ANON. (1967). Toxicological evaluation of some antimicrobials, antioxidants, emulsifiers, stabilisers, flour treatment agents, acids and bases. F.A.O. Nutritional Meetings Report Series No. 40 A, B & C. W.H.O./Food Additives/67 29.
ANON. (1973). E.E.C. Regulation No. 2592/73. Published by order of the Council of Europe.
ANRAKU, T. & SANO, H. (1966). Decomposition of tryptophane by sulphurous acid salts. I. Isolation and identification of indigotin, indirubin and other compounds produced by decomposition of tryptophane. *Yakugaku Zasshi* **86**, 1034-1042.
ASADA, K. (1967). Purification and properties of a sulfite reductase from leaf tissue. *Journal of Biological Chemistry* **242**, 3646-3654.
BIGWOOD, E. J. (1973). Acceptable daily intake of food additives. *Critical Reviews in Toxicology* **2**, 41-93.
BOSUND, I. (1962). The action of benzoic acid and salicylic acid on the metabolism of micro-organisms. *Advances in Food Research* **11**, 331-353.
BURROUGHS, L. F. & SPARKS, A. H. (1964). The identification of SO_2-binding compounds in apple juices and ciders. *Journal of the Science of Food and Agriculture* **15**, 176-185.
CARR, J. G. & DAVIES, P. A. (1971). Lactic acid bacteria in juices and fermenting ciders. *Annual Report Long Ashton Research Station* for 1970, 133.
CAUSERET, J., HUGOT, D., LHUISSIER, M., BIETTE, E., LECLERC, J. & CLUZAN, R. (1965). L'utilization des sulfites en technologie alimentaire: quelques aspects toxicologiques et nutritionnels, *Fruits, Paris* **20**, 109-115.
CECIL, R. & LEONING, U. E. (1960). The reaction of the disulphide groups of insulin with sodium sulphite. *Biochemical Journal* **76**, 146-155.
CECIL, R. & WAKE, R. G. (1962). The reactions of inter- and intra-chain disulphide bonds in protein with sulphite. *Biochemical Journal* **82**, 401-406.
CHAN, W. (1968). A method for complete S–sulphonation of cystine residues. *Biochemistry* **7**, 4247-4254.
CHEN, R. & SAKAGUCHI, O. (1972). Food hygiene studies of sulfite. II. Effect of sulfite on the intermediate metabolism of bacteria. *Nippon Eisengaku Zasshui* **26**, 475-480.
CHRISTIAN, J. H. R. (1963). Preservation of minced meat with SO_2. *CISRO Food Preservation Quarterly* **23**, 30-36.
CLARKE, H. T. (1932). Action of sulfite on cystine. *Journal of Biological Chemistry* **97**, 235-247.
CLUZAN, R., CAUSERET, J. & HUGOT, D. (1965). Long term study of the toxicity of potassium metabisulphite in rats. *Annales de biologie animales biochemie et biophysique* **5**, 267-281.
CROMPTON, M. PALMIERI, F., CAPANO, M. & QUAGLIARIELLO, E. (1974). The transport of sulphate and sulphite in rat liver mitochondria. *Biochemical Journal* **142**, 127-137.
DITTRICH, H. H. & STAUDENMAYER, T. (1970). Über die Zusammenhänge zwischen der Sulfit-Bildung und der Schwefelwasserstoff–Bildung bei *Saccaromyces cerevisiae*. *Zentralblatt für Bakteriologie, Parasitenkunde, Infektionskrankheiten und Hygiene*, Abt II **124**, 113-118.
DORANGE, J. L. & DUPUY, P. (1972). Mise en évidence d'une action mutagène du sulfite de sodium sur la levure. *Compte rendu hebdomadaire des séances de l'Academie des Sciences, Paris* **274**, 2798-2800.
DREYFUS, J. & MONTY, K. J. (1963). The biochemical characterisation of cysteine requiring mutants of *Salmonella typhimurium. Journal of Biological Chemistry* **238**, 1019-1024.
DYET, E. J. & SHELLEY, D. (1966). The effect of sulphite preservative on British fresh sausage. *Journal of Applied Bacteriology* **29**, 439-446.

ELLIS, R. (1964). The site of end product inhibition of sulphate reduction in *E. coli*. *Biochemical Journal* **93**, 19P.
EVELYN, J. (1670). Pomona, or an appendix concerning fruit trees. In relation to ciders the making and several ways of ordering it. Supplement, An advertisement concerning Cider by Beale. Printed by J. Martyn & J. Allestry.
FALK, M. & GUIGUÈRE, P. A. (1958). The nature of sulphurous acid. *Canadian Journal of Chemistry* **36**, 1121-1125.
FAPARUSI, S. I. (1969). Effect of pH on the preservation of palm wine by SO_2. *Applied Microbiology* **18**, 122-123.
FREESE, E., SHEU, C. W. & GALLIERS, E. (1973). Function of lipophilic acids as antimicrobial food additives. *Nature, London* **241**, 321-325.
FRIDOVICH, I. & HANDLER, P. (1956). Hypoxanthine as a co-factor for enzymic oxidation of sulfite. *Journal of Biological Chemistry* **221**, 323-331.
GIBSON, W. B. & STRONG, F. M. (1974). Accumulation of ingested sulphite and sulphate sulphur and utilization of sulphited protein by rats. *Food and Cosmetic Toxicology* **12**, 625-640.
GREEN, S. & GRAY, P. P. (1950). A differential procedure applicable to bacteriological investigation in brewing. *Wallerstein Laboratories Communications* **13**, 357-368.
HAISMAN, D. R. (1974). The effect of SO_2 on oxidising enzyme systems in plant tissues. *Journal of the Science of Food and Agriculture* **25**, 803-810.
HAYATSU, H. (1969). The oxygen catalysed reaction between 4-thiouridene and sodium sulphite. *Journal of the American Chemical Society* **91**, 5693-5694.
HAYATSU, H. & MIURA, A. (1970). The mutagenic action of sodium bisulfite. *Biochemical and Biophysical Research Communications* **39**, 156-160.
HAYATSU, H., WATAYA, Y. & KAI, K. (1970). The addition of sodium bisulfite to uracil and to cystosine. *Journal of the American Chemical Society* **92**, 724-726.
HEIMBERG, M., FRIDOVICH, I. & HANDLER, P. (1953). The enzymic oxidation of sulfite. *Journal of Biological Chemistry* **204**, 913-926.
HENCHEN, A. (1962). On the preparation and properties of sulfitolysed fibrinogen and fibrin. *Acta chemica scandanavica* **16**, 1037-1038.
HOMER, The Odyssey XXII. Translated by E. U. Rieu. London: Penguin Books.
HUGOT, D., CAUSERET, J. & LECLERC, J. (1965). Effect of sulphite on the excretion of calcium by rats. *Annales de biologie animales biochemie et biophysique* **5**, 53-59.
INGRAM, M. (1948). The germicidal effects of free and combined SO_2. *Journal of the Society for Chemistry and Industry* **67**, 18-21.
INGRAM, M. & VAS, K. (1950). Combination of SO_2 with concentrated orange juice. I. Equilibrium states. *Journal of the Science of Food & Agriculture* **1**, 21-27.
INGRAM, M., OTTOWAY, F. J. H. & COPPOCK, J. B. M. (1956). The preservative action of acid substances in food. *Chemistry and Industry*, 1154-1163.
INOUE, M. & HAYATSU, H. (1971). The interaction between bisulfite and amino acids. The formation of methionine sulfoxide from methionine in the presence of oxygen. *Chemical and Pharmaceutical Bulletin* **19**, 1286-1289.
INOUE, M., HAYATSU, H. & TANOOKA, H. (1972). Concentration effect of bisulfite on the inactivation of transforming activity of DNA. *Chemico-Biological Interactions* **5**, 89-95.
JACOBS, M. H. (1940). Some aspects of cell permeability to weak electrolytes. *Symposium on Quantitative Biology, Cold Spring Harbour* **8**, 30-39.
JAULMES, P. (1970). État actuel des techniques pour le remplacement de l'anhydride sulfureux. *Bulletin de l'Office international de la Vigne et du Vin* **43**, (478), 1320-1333.
JONES, L. H. & McLAREN, E. (1958). Infra red absorption of SO_2 and CO_2 in aqueous solution. *Journal of Chemical Physics* **28**, 995.
JOSLYN, M. A. & BRAVERMAN, J. B. S. (1954). The chemistry and technology of the pretreatment and preservation of fruit and vegetable products with SO_2 and sulfites. *Advances in Food Research* **5**, 97-160.
KOLTHOFF, I. M., ANASTASI, A. & TAN, B. H. (1959). Reactivity of sulfhydryl and disulfide in proteins. IV. Reaction between disulfide and sulfite in bovine serum

albumin denatured in guanidine hydrochloride and urea solutions. *Journal of the American Chemical Society* **81**, 2047-2052.

KOLTHOFF, I. M., ANASTASI, A. & TAN, B. H. (1960). Reactivity of sulfhydryl and disulfide in proteins. V. Reversal of denaturation of bovine serum albumin in 4M guanidine hydrochloride or 8M urea and of splitting of disulfide groups in 4M GHCl. *Journal of the American Chemical Society* **82**, 4147-4150.

KUN, E. (1961). The metabolism of sulfur containing compounds. In *Metabolic Pathways* **2**, ed. Greenberg, D. New York: Academic Press.

LARSON, B. L. & SALISBURY, G. W. (1953). The reactive reducing components of semen. *Journal of Biological Chemistry* **201**, 601-608.

LANTAEUME, M. T., RAMEL, P., GIRARD, O., JAULMES, P., GASQ, M. & RANAN, J. (1965). Effets physiologiques à long terme de l'anhydride sulfureux ou des sulfites utilisés pour le traitement des vins rouge. *Annales des falsifications et de l'expertise chimique* **58**, 16-31.

LAWRENCE, W. C. & COLE, E. R. (1968). Yeast sulfur metabolism and the formation of H_2S in brewery fermentations. *Wallerstein Laboratories Communications* **31**, 95-105.

LEACH, S. (1960). The reaction of thiol and disulphide groups with mercuric chloride and methyl mercuric iodide. I. Simple thiols and soluble proteins. *Australian Journal of Chemistry* **13**, 520-546.

LEY, H. & KONIG, E. (1938). Die Lösungsspektrum von wichtigeren Saüren der Elemente de Schwefelgruppe. *Zeitschrift für physikalische Chemie* **41B**, 365-387.

LOCKETT, M. F. & NATOFF, I. L. (1960). A study of the toxicity of sulphite. *Journal of Pharmacy and Pharmacology* **12**, 48-496.

LÜTHI, H. R. (1972). To what extent can SO_2 be replaced in wine treatment? *Third International Oenological Symposium*. Stellenbosch, S. Africa, pp. 1-18.

LYNEN, F. (1970). Functional sulphydryl groups in enzyme catalysis. In *Chemical Reactivity and Biological Role of Functional Groups in Enzymes*, ed. Smellie, R. M. S. London: Academic Press.

MACRIS, B. (1972). Transport and toxicity of SO_2 in the yeast *S. cerevisiae*. Ph.D. Thesis, University of Michigan.

MACRIS, B. & MARKAKIS, P. (1974). Transport and toxicity of SO_2 in *S. cerevisiae* var. *ellipsoideus*. *Journal of the Science of Food and Agriculture* **25**, 21-29.

MASSEY, V. & WILLIAMS, C. H. (1965). On the reaction mechanism of yeast glutathione reductase. *Journal of Biological Chemistry* **240**, 4470-80.

MAYERHOF, O., OHLMEYER, P. & MOHLE, W. (1938). Über die Koppelung zwischen Oxyreduktion und Phosphatveresterung bei der anaeroben Kohlenhydratspaltung. 2. Die Koppelung als Gleichgewichtsreaction. *Biochemische Zeitschrift* **297**, 113-133.

MAW, G. A. (1965). The role of sulfur in yeast growth and in brewing. *Wallerstein Laboratories Communications* **28**, 49-68.

MIDDLEBROOK, W. R. & PHILLIPS, H. (1942). Action of sulphites on the cystine disulphide linkages of wool. *Biochemical Journal* **36**, 428-437.

MUKAI, F. HAWRYLUK, I. & SHAPIRO, R. (1970). The mutagenic specificity of sodium bisulfite. *Biochemical and Biophysical Research Communications* **39**, 983-988.

NAIKI, N. (1965). Some properties of sulfite reductase from yeast. *Plant and Cell Physiology, Tokyo* **6**, 179-194.

NEUBERG, N. (1929). Über das Verhalten des Glucose–SO_2 Natriums zu Hefe und damit zusammenhangende Fragen. *Biochemische Zeitschrift* **212**, 477-517.

NICKERSON, W. J. (1953). Sulphite reduction by *Candida* species. *Journal of Infectious Disease* **93**, 43-56.

OUGH, C. S. & INGRAHAM, J. L. (1960). Use of sorbic acid and SO_2 in sweet table wines. *American Journal of Enology and Viticulture* **11**, 117-122.

OKA, S. (1960). Studies on transfer of antiseptics to microbes and their toxic effect. I. Accumulation of acid antiseptics in yeast cells. *Bulletin of the Agricultural Chemical Society of Japan* **24**, 59-65.

PARKER, A. J. & KHARASCH, N. (1959). The scission of the S–S bond. *Chemical Reviews* **59**, 583-628.

PECHERE, J. F., DIXON, G. H., MAYBURY, R. H. & NEURATH, H. (1958). Cleavage of disulphide bonds in trypsinogen and α-trypsinogen. *Journal of Biological Chemistry* **233**, 1364-1372.
PECK, H. D., TEDRO, S. & KAMEN, M. D. (1974). Sulfite reductase activity in extracts of photosynthetic bacteria. *Proceedings of the National Academy of Sciences of the U.S.A.* **71**, 2404-2406.
PEYNAUD, E. & SAPIS, J. C. (1972). New discoveries about SO_2 combinations and measures for its economisation. 4th International Enological Symposium, Valencia, pp. 88-93.
PFLEIDERER, G., JEKEL, D. & WEILAND, T. (1956). Uber die Einwirkung von Sulfit auf einige DPN hydrierende Enzyme. *Biochemische Zeitschrift* **328**, 187-194.
PREMUZIC, D., LOVRIC, T., SOFAR, O. & JOURIC, V. (1972). Production of SO_2 during fermentation of must as a result of the metabolism of some yeast strains and their effect on wine colour. *Kemija u Industrija* **21**, 9-21.
POSTGATE, J. R. (1963). The examination of sulphur auxotrophs: a warning. *Journal of General Microbiology* **30**, 481-484.
RAHN, O. & CONN, J. E. (1944). Effect of increase in acidity on antiseptic efficiency. *Industrial and Engineering Chemistry* **36**, 185-187.
RANKINE, B. C. & POCOCK, K. F. (1969). Influence of yeast strain on binding of SO_2 in wines and its formation during fermentation. *Journal of the Science of Food and Agriculture* **20**, 104-109.
REED, G. & PEPPLER, H. (1973). *Yeast Technology.* Westport, Connecticut: Avi Publishing Co.
REHM, H. J. (1964). The antimicrobial action of sulphurous acid. In *Microbial Inhibitors in Food,* ed. Molin, N. Uppsala: Almquist & Wiksells.
REHM, H. J. & WITTMANN, H. (1962). Beitrag zur Kenntnis der antimikrobeillen Wirkung der Schwefligen Saüre. I. Übersicht über einflussnehmende Faktoren auf die antimikrobeillen Wirkung der Schwefligen Saüre. *Zeitschrift für Lebensmitteluntersuchung und -forschung,* **118**, 413-425.
ROY, A. B. & TRUDINGER, P. A. (1970). The biochemistry of inorganic compounds of sulphur. Cambridge University Press.
SCHMIDT, A., ABRAMS, W. R. & SCHIFF, J. A. (1974). Reduction of adenosine-5-phosphosulphate to cysteine in extracts from *Chlorella* and mutants blocked for sulphate reduction. *European Journal of Biochemistry* **47**, 423-434.
SEIGEL, L. M., LEINWEBER, F. J. & MONTY, K. J. (1965). Characterisation of the sulfite and hydroxylamine reductases of *Neurospora crassa. Journal of Biological Chemistry* **240**, 2705-2711.
SHAPIRO, R. & BRAVERMAN, B. (1972). Modification of polyuridylic acid by bisulfite: effect on double helix formation and coding properties. *Biochemical and Biophysical Research Communications* **47**, 544-550.
SHAPIRO, R. & WEISGRAS, J. M. (1970). Bilsulfite catalysed transformation of cytosine and cytidene. *Biochemical and Biophysical Research Communications* **40**, 839-843.
SHAPIRO, R., BRAVERMAN, B., LOUIS, J. B. & SERVIS, R. E. (1973). Nucleic acid reactivity and conformation. II. Reaction of cytosine and uracil with sodium bisulfite. *Journal of Biological Chemistry* **248**, 4060-4064.
SIMON, A. & WALDMAN, K. (1956). Über die Lonen def Schwefligen Saüre in wassringer Losung. *Zeitschrift für anorganische und allgemeine Chemie* **283**, 359-364.
SPEAKMAN, P. T. (1958). The mechanism of the reaction between cystine in keratin and sulphite–bisulphite solutions at 50°C. *Biochimica et biophysica acta* **28**, 284-293.
STEINHOFF, D. & MARQUARDT, P. (1963). Kombination von Kaliumpyrosulfit und Äthylalkohol in Trankungsversuch an Ratten. *Arzneimittel–Forschung 13,* 237-283.
STRICKS, W. & KOLTHOFF, I. M. (1951). Equilibrium constants of the reactions of sulfite with cystine and with dithioglycolic acid. *Journal of the American Chemical Society* **73**, 4569-4579.
STRICKS, W., KOLTHOFF, I. M. & KAPOOR, R. C. (1955). Equilibrium constants of the reaction between sulfite and oxidised glutathione. *Journal of the American Chemical*

Society 73, 2057-2061.
SUMMERS, G. A. & DRAKE, J. W. (1971). Bisulphite mutagenesis of the bacteriophage T4. *Genetics* 68, 603-607.
TAMURA, G. (1965). Studies on the sulfite reducing system of higher plants. II. Purification and properties of sulfite reductase from *Allium odorum*. *Journal of Biochemistry,* Tokyo 57, 207-214.
TAPPEL, A. L. (1973). Lipid peroxidation damage to cell components. *Federation Proceedings* 32, 1870-1874.
TARTAR, H. V. & GARRETSON, H. H. (1941). Thermodynamic ionisation constants of sulphurous acid at 25° C. *Journal of the American Chemical Society* 63, 808-816.
UTSUMI, K., HASEGAWA, T. & OGATA, M. (1973). Bisulfite induced lipid peroxidation and the inhibition of the reaction by radical scavengers. *Kano Shikiso* 83, 31-36.
VANDERSCHMITT, D. J., VITOS, K. S., HUENEKENS, F. M. & SCRIMGEOUR, K. G. (1967). Addition of bisulfite to folate and dihydrofolate. *Archives of Biochemistry and Biophysics* 122, 448-493.
WAINWRIGHT, T. (1967). Yeast sulphite reductase. *Biochemical Journal* 103, 56P.
WATTIAUX-DE CONNINCK, S. & WATTIAUX, R. (1971). Subcellular distribution of sulfite cytochrome-*c* reductase in rat liver tissue. *European Journal of Biochemistry* 19, 552-556.
WEEKS, C. (1969). Production of SO_2-binding compounds and of SO_2 by two *Saccharomyces* yeasts. *American Journal of Enology and Viticulture* 20, 32-39.
WILLIAMS, R. R. (1935). Structure of vitamin B. *Journal of the American Chemical Society* 57, 229-230.
WURDIG, G. & SCHLOTTER, H. A. (1968). SO_2 bildung durch Sulfatreduktion während der Garung. I. Versuche und Beobachtungen in der Praxis. *Weinwissenschaft* 23, 356-371.
YOSHIMOTO, A. & SATO, R. (1968*a*). Studies on yeast sulfite reductase. I. Purification and characterisation. *Biochimica et biophysica acta* 153, 555-575.
YOSHIMOTO, A. & SATO, R. (1968*b*). Studies on yeast sulfite reductase. II. Partial purification of genetically incomplete sulfite reductases. *Biochimica et biophysica acta* 153, 576-588.
YOSHIMOTO, A., NAKAMURA, T. & SATO, R. (1967). Isolation from *Aspergillus nidulans* of a protein catalysing the reduction of sulfite by reduced viologen dyes. *Journal of Biochemistry,* Tokyo 62, 756-766.

Inactivation by Cold

M. INGRAM AND B. M. MACKEY

*ARC Meat Research Institute, Langford, Bristol
England*

CONTENTS

1. Introduction 111
2. Minimum temperature for growth 112
   - (a) Nutrient status 112
   - (b) Water activity 112
   - (c) Acidity 114
   - (d) Cell damage 118
   - (e) Discussion 119
3. Subminimal temperatures above freezing 121
   - (a) No cold shock 121
   - (b) Reversible inactivation 123
   - (c) Cold shock 123
   - (d) Discussion 125
4. The subzero zone 128
   - (a) Changes during freezing 128
   - (b) Incidence of growth below zero 129
   - (c) Rates of growth 129
   - (d) The nature of the flora 132
   - (e) Discussion 134
5. Freezing 135
   - (a) Rate of freezing 135
   - (b) Age of cells 136
   - (c) The effect of salts 137
   - (d) Protective agents 137
   - (e) Response of different species 138
   - (f) Reversible inactivation – cell damage 138
   - (g) Time/temperature relations during frozen storage . 139
   - (h) Effects of pH and added solutes during frozen storage . 139
   - (i) Resistance of different species to frozen storage . 140
   - (j) Reversible inactivation during frozen storage . . . 140
   - (k) Thawing 141
   - (l) Discussion 141
6. Some practical implications 143
7. Conclusion 145
8. References 146

1. Introduction

THE WELL-KNOWN EFFECT of increasing cold on micro-organisms is gradually to slow down, and eventually to stop, growth. With many organisms growth is stopped at temperatures above the freezing point of their growth medium. There is thus a temperature range in which the organism, though not frozen, is unable to grow because of the low temperatures. On the other hand certain micro-organisms (herein called 'psychrotrophs' following Eddy, 1960) are capable of growth down to and, to varying degrees, below freezing point. Only at temperatures considerably below freezing point is a fourth zone reached where growth is not possible. It is common to regard events at cool temperatures above freezing point, in the first two zones, as different from those below freezing in the latter two; hence it is convenient, in this paper, to treat the above four zones separately. Nevertheless, the paper will provide indications that these distinctions are artificial, and unfortunate in tending to obscure connections between phenomena above and below freezing point.

2. Minimum Temperature for Growth

As cold gradually slows down growth and finally prevents it, there is generally a common effect of increasing lag period and generation time. Observations showing this, as regards the responses of lag phase and of generation rate to combinations of temperature and a_w, are in Fig. 1. The pattern of response is virtually the same for both. The minimum growth temperature can be regarded as the point when either lag period or generation time becomes infinite. Hence it seems permissable here to limit discussion to terms of generation time (the lag phase usually being equivalent to relatively few generations).

(a) *Nutrient status*

The minimum temperature is believed to be lower if suitable nutrients are present in the substrate. This has been demonstrated most dramatically with thermophiles, by Campbell & Williams (1953) and Long & Williams (1959). On the other hand, vitamin requirements were the same at 20° as at 0° for the psychrophilic *Bacillus* spp. of Adams & Stokes (1968); these species, however, can grow at temperatures considerably below 0° (see Section 4), so 0° is not strictly to be regarded as their minimum.

For present purposes, it suffices to note the possibility that minimum temperatures for growth may depend on the nutritive status of the suspending medium. For instance, the observations of Schmidt *et al.* (1961) recorded growth of *Clostridium botulinum* type E down to 3.3° (38°F) only in beef stew; the limit was higher in other media, e.g. 7.8° (46°F) in the case of peptone water.

(b) *Water activity*

Figure 1 plainly indicates an important influence of water activity (a_w) on minimum growth temperature. If the a_w was $\geqslant 0.965$, the minimum temperature was $< 15°$ presumably near 5°; with a_w 0.955 the minimum was

> 20°; with a_w 0.950, > 30°. The presence of a considerable concentration (> 1.0 M) of dissolved substance (in this case KCl) sufficed to raise the minimum temperature more than 10°.

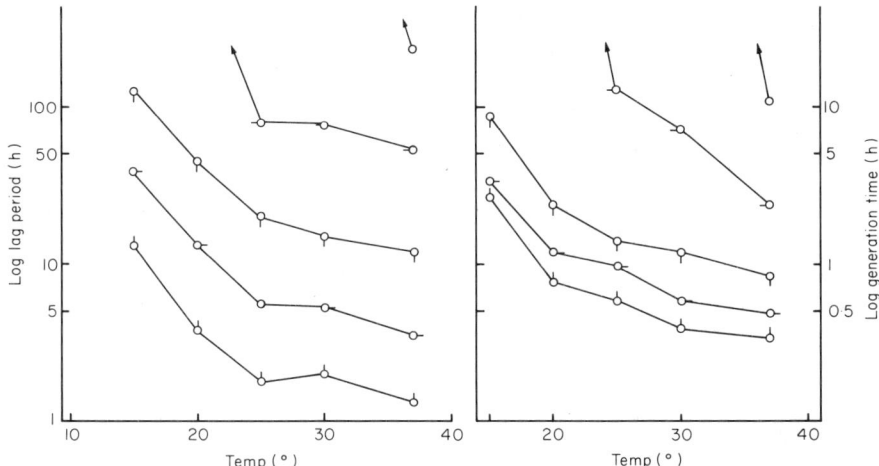

Fig. 1. Effect of temperature and a_w on lag phase and generation time of *Aerobacter aerogenes* at pH 7.0. a_w values are: ●, 0.999; ○, 0.975; ○, 0.965; ○, 0.955; ○, 0.950. (Data of Wodzinski & Frzier, 1961a.)

This phenomenon was appreciated by mycologists before bacteriologists. In a paper well ahead of its time, Tomkins (1929) showed that the effect of sub- or supraoptimal temperatures is to restrict the range of humidities at which growth is possible; or, regarded conversely, that the effect of reducing water activity is to restrict growth temperatures. The minimum temperature of *Alternaria citri* was raised from c 0° at an equilibrium relative humidity (RH) (a measure of a_w) of 100%, to c. 15° at 95%, and to nearly 25° at 92%. Tomkins' general observation was confirmed later. With the osmophilic mould *Aspergillus glaucus*, Heintzeler (1959) found that the limiting a_w was 0.71 at 30° but 0.73 at 20°; Stille (1948) observed a value of 0.78 at 10°; that is to say, at a_w 0.78 the minimum temperature is between 10 and 20°, at 0.71 between 20 and 30°. Similar observations with osmophilic yeasts have been made by Kroemer & Krumbholz (1931, 1932) and Ingram (1958).

There are several more recent publications which demonstrate an effect of sodium chloride in raising the minimum temperature for growth of *Cl. botulinum* type E, normally regarded as 3.3°. Segner *et al.* (1966) found that 4.0% NaCl, which had little effect on growth at 30°, led to extremely long lag periods at 8°; and with 4.5% of NaCl growth did not take place at 8° but did so at 16°. Ohye *et al.* (1966) likewise recorded growth up to 5.8% NaCl at 30°, 5.1% at 20°, and 4.3% at 15°; i.e. the minimum temperature was in the range

20-30° with c. 5.5% NaCl, and in the range 15-20° with 4.5%. More detailed confirmatory observations with type E were made by Emodi & Lechowich (1969b) using various solutes besides NaCl to control a_w; though there were differences between solutes in detail, the same general relation held throughout. It was also observed with three salmonellae by Matches & Liston (1972a): with *Salmonella heidelberg*, for example, growth was possible up to only 2% NaCl at 8°, to 6% at 12°, and to > 8% at 22-41°; similar observations were made at slightly lower salt concentrations with *Salm. derby* and *Salm. typhimurium*.

It is interesting to note a converse effect with the obligate halophile *Halobacterium halobium*, which has a minimum temperature of 15-20° with 15% salt, but 10-15° with 20 or 25% salt (Ingram, 1957). Apparently, where the organism prefers a high salt concentration for growth, this lowers the minimum temperature instead of raising it. This may be a reason for past confusion about the relations between temperature and the preservative effect of salt (Ingram & Kitchell, 1967).

Wodzinski & Frazier (1961c, d) examined the effects on these relations of higher carbon dioxide and lower oxygen tensions respectively. These effects were virtually negligible. That of carbon dioxide was if anything to raise the minimum temperature: for example, with *Pseudomonas fluorescens* and a medium of a_w 0.975, growth was possible below 15° in 5% CO_2, but only above 15° in 10% CO_2. Similarly, diminished oxygen tension accentuated slightly the effect of diminished a_w in raising minimum temperature, with *Ps. fluorescens* and *Aerobacter aerogenes*; diminution from 10-0.1% oxygen did not, however, affect the temperature/a_w relations of *Lactobacillus viridescens*, perhaps because this species is micro-aerophilic.

(c) *Acidity*

There have been various isolated reports that a suboptimal pH leads to a higher minimum growth temperature. For example, Barnes *et al.* (1963) concluded that the minimum temperature for growth of *Cl. perfringens* in meat is higher in meat of relatively low pH. Similarly, Baird-Parker & Freame (1967) observed that spores of *Cl. botulinum* type B grew at 30° but not 20° at pH 5.0, i.e. the minimum temperature was raised from a value normally near 10° to one of 20-30° at pH 5; and they recorded similar behaviour with type A. A particularly good example of this relation between acidity and growth temperature is given by Emodi & Lechowich (1969a, their Table 1) for *Cl. botulinum* type E. A narrowing of the pH range, near the minimum temperature for growth, was observed with three *Salmonella* serotypes by Matches & Liston (1972b), with the implication of a higher minimum temperature at low pH, but here the effect appeared relatively small.

That, corresondingly, acidity greatly influences the effect of water activity

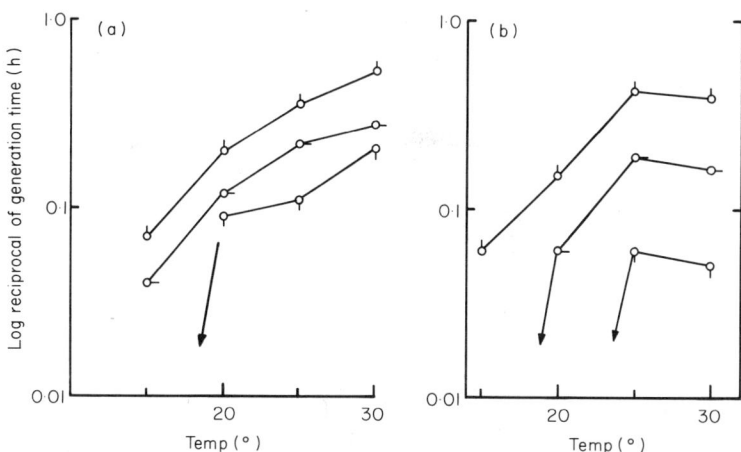

Fig. 2. The effect of a_w and pH on generation time of Ps. fluorescens, a_w values are: ⬤, 0.999; ○, 0.975; ⊘, 0.970, (a) at pH 7.0 (b) at pH 5.4 (Calculated from data of Wodzinski & Frazier, 1960).

was clearly shown by Wodzinski & Frazier (1961a, b) with Ps. fluorescens (Fig. 2). At an a_w of 0.975, the minimum temperature was < 15° at pH 7, but between 15 and 20° at pH 5.4; and similarly at a_w 0.970, a minimum temperature of 15-20° at pH 7 was raised to 20-25° at pH 5.4. A similar influence of acidity with A. aerogenes is displayed more simply in Table 1. At a_w > 0.992, where the minimum growth temperature is normally < 10° at pH 7, it was raised above 15° at pH 3.9; and at a_w 0.975, where the minimum was < 15° at pH 7, it was raised above 30° at pH 3.9. Similar observations were

Table 1

Influence of pH on limiting conditions of temperature and for growth of Aerobacter aerogenes

a_w	pH 3.9 Growth at (°)					pH 7.0 Growth at (°)					pH 8.4 Growth at (°)				
	15	20	25	30	37	15	20	25	30	37	15	20	25	30	37
> 0.99	−	+	+	+	+	+	+	+	+	+					
0.975	−	−	−	−	+	+	+	+	+	+					
0.965	−	−	−	−	−	+	+	+	+	+	+	+	+	+	+
0.955						−	−	+	+	+	−	+	+	+	+
0.950						−	−	−	−	+	−	+	+	+	+
0.945						−	−	−	−	−	−	−	−	−	−

+, growth; −, no growth.
Wodzinski & Frazier (1961a).

also made by Wodzinski & Frazier (1961b) with *L. viridescens* in the range 0.965-0.945 and between pH 7 and 4.5, this organism being less sensitive to salts and acidity.

Whereas with *Ps. fluorescens* and *A. aerogenes* the effect of increased a_w at pH 7 was to raise minimum temperature, with *L. viridescens* the effect of marginal a_w was to lower maximum and optimum temperature. Any effect on minimum temperature must at pH 7 have been below $15°$—outside the range of observation. Only at pH 4.5 was the effect on minimum temperature observable with *L. viridescens*. This difference in behaviour is probably not due to an acid-tolerant character, in which respect *L. viridescens* is not dissimilar from *A. aerogenes*. It would be interesting to establish whether it relates to the difference in Gram reaction.

Similar, unpublished, observations with *Ps. fluorescens* by Marshall & Scott, were mentioned by Scott (1961), who commented that the rates of growth in the experiments of Fig. 2 seem unusually low for that species. Experiments by Christian & Waltho (1962) revealed an exactly similar effect of high NaCl concentration (> 1.0 M) in raising the minimum growth temperature of *Staphylococcus aureus*; and a similar synergistic effect of acidity (Table 2). Subsequent work from Scott's laboratory, by Ohye & Christian (1966), revealed

Table 2
Combined effects of temperature, salt concentration and pH on growth of Staphylococcus aureus

Salt concentration (M)	pH 5.0 Growth at (°)					pH 7.0 Growth at (°)					pH 9.0 Growth at (°)				
	10	20	30	37	45	10	20	30	37	45	10	20	30	37	45
0.0	+	+	+	+	+										
1.0		+	+	+	+	+	+	+	+	+	+	+	+	+	+
1.5		+	+	+	+	+	+	+	+	+		+	+	+	+
2.0		+	+	+	+		+	+	+	+			+	+	
2.5			+	+			+	+	+	+					
3.0			+	+				+	+						

+, growth; −, no growth.
Christian & Waltho (1962).

corresponding interactions relating to *Cl. botulinum* types A, B and E, especially the last. It was also shown by Alford & Palumbo (1969) with *Salm. derby, Salm. enteritidis* and *Salm. thompson* that less salt is needed to inhibit at lower temperatures—i.e. salt raises minimum temperature—and that acidity acentuates this.

Similar phenomena are revealed at lower temperatures in unpublished work by J. Mabb & T. A. Roberts with *Streptococcus faecalis* var. *liquefaciens*

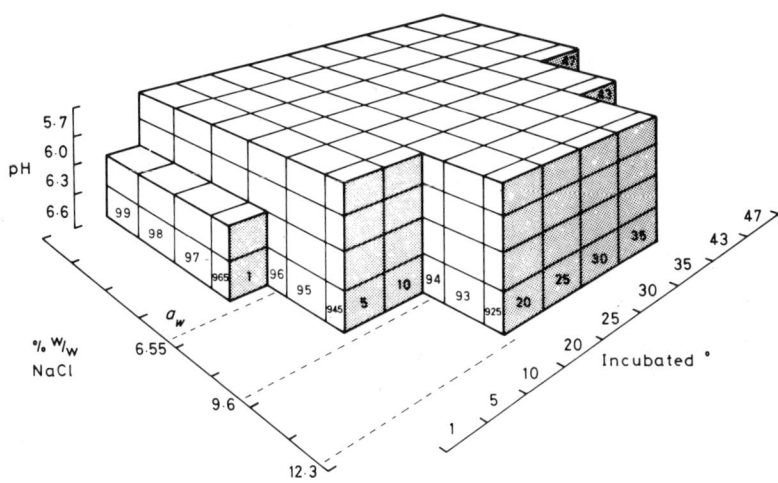

Fig. 3. Growth of *Strep. faecalis* var. *liquefaciens* in tryptone-yeast extract-dextrose medium adjusted to the indicated a_w and pH with NaCl and HCl, respectively. a_w values are indicated on the block faces pointing towards the left; e.g. 99 equals a_w of 0.99. After inoculation the bottles of media were examined at intervals up to three months. (From the unpublished data of J. Mabb & T. A. Roberts.)

(Fig. 3). As is well known, this species can grow at 1°; but this was true only if the pH was > 6.3 and the salt concentration < 6.5 g/100 g. Growth was still possible at 5°, at pH 5.7 (the lowest investigated) and 9.6% salt. With higher salt concentrations, growth was possible only above 20°.

While the effect of acidity seems fairly consistent in these experiments, that of alkalinity is not. Often alkalinity acts like acidity, in accentuating the effect of low a_w; sometimes (e.g. in Table 1) it has a protective effect. It would be interesting to follow these relations with species whose pH optima are further removed from neutrality.

Nitrite, a preservative dependent on acidity, can apparently have similar effects. It was shown by Pivnick & Barnett (1965), and has been amply confirmed since (Roberts, 1973), that the interactions between acidity, sodium chloride and nitrite, in controlling growth of *Cl. botulinum* and *Staph. aureus*, are temperature dependent so that the inhibitory concentrations of salt, nitrite, or H ion, or combinations of these, are less if the temperature is lower. That is to say that marginal values permit growth at the higher temperature but not at the lower, i.e. the minimum temperature has been raised and nitrite is similar in this respect to salt and acidity. It has recently been confirmed directly (Fig. 4) that the minimum growth temperature is higher with high nitrite concentrations. It would be useful to establish whether other preservatives, acidity-dependent or not, have the same effect as nitrite.

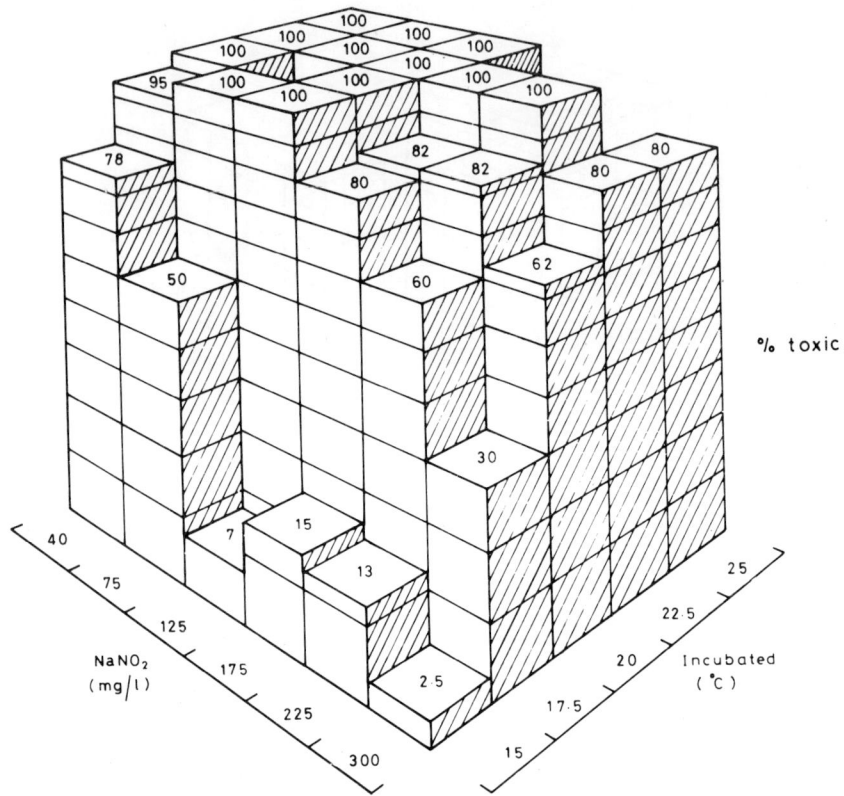

Fig. 4. The effect of sodium nitrite and storage temperature on growth and toxin production by *Cl. botulinum* in a meat slurry at pH 6.0, containing 3.5% w/v salt (in water phase) over a six month storage period. The presence of a block indicates the production of botulinal toxin at the concentrations of sodium chloride and sodium nitrite indicated after incubation at the stated temperature to spoilage, or up to six months in its absence. The height of the column of blocks shows the % of bottles containing toxic slurry which number is shown at the top of each column. (From the data of T. A. Roberts, B. Jarvis and Annette C. Rhodes, in the press.)

(d) *Cell damage*

Cell damage is another factor which profoundly affects minimum growth temperature. This can be illustrated by Fig. 2a of Beuchat & Lechowich (1968). *Streptococcus faecalis* normally grows down to 0°, hence the number of normal cells recovered is independent of incubation temperature over a wide range. But this was no longer the case below 20°, after strain S21 was heated for 20 min at 60°. About 100,000 cells (of the original 2×10^7) were still capable of growing at temperatures between 20 and 40°, but only c. 10,000 at 15° and 100 at 10°. This means that among the damaged cells, 999/1000 had minimum temperatures for growth $> 10°$, and 900/1000 had minimum temperatures $> 15°$. Heating twice as long produced a still greater effect; then, major proportions of the survivors had minimum temperatures even above 30°. Despite differences in detail, Figs 2b and c (*loc. cit.*) reveal with other strains an exactly similar large increase in minimum temperatures with increasing cell damage; they also indicate a similar effect in reducing maximum temperature.

These figures of Beuchat & Lechowich (1968) indicate that the heat resistance is lower if survivors are incubated below (or above) an optimum temperature, which may be considerably lower than the optimum for growth; in their Fig. 2b, for example, the decimal reduction time is c. 20 min at 20-30°, but c. 10 min at 5°. This aspect of the matter was noticed long ago: e.g. with *Escherichia coli* by Elliker & Frazier (1938), and for spores by Williams & Reed (1942). It has recently been re-emphasized for *Bacillus* spores by Prentice & Clegg (1974) who published diagrams exactly similar to Figs 2b and c of Beuchat & Lechowich (1968). These publications do not mention the implications for minimum growth temperature, however, the work not having extended to low enough temperatures to display the phenomena fully. (There are of course various reports that optimum temperature is lowered by sublethal heating or irradiation—see e.g. Harris, 1963—but this does not necessarily imply a similarly lowered minimum temperature, about which there is no particular information in most references.)

Radiation damage apparently produces similiar effects. It is reported (Ando *et al.*, 1968) that when spores of *Cl. botulinum* type E are irradiated, the apparent D value is less if the irradiated spores are incubated at lower temperatures.

(e) *Discussion*

Erroneous estimates of minimum temperature are obviously likely to arise through observations for too short a time or at too wide temperature intervals. Setting such mistakes aside, it is clear that various environmental or processing factors can raise the minimum growth temperature, sometimes very greatly in marginal conditions. Combinations of up to at least four factors, often act synergistically. Similar relations appear to hold for spores and for vegetative cells, perhaps because vegetative multiplication is the limiting step. But there is a need for more exploration of these relations at temperatures near the minimum for growth.

There are occasions when no similar effect has been observed. In such cases, it may be that the range of conditions explored was not wide enough to reach limiting conditions; for instance, that the temperatures involved were too far above the minimum. In particular, if no effect were observed at temperatures near 0°, the true minimum might be well into the subzero range of temperature.

Some general rationalization of these complex interactions is possible, on the basis repeated with variations by various authors (e.g. Michener & Elliot, 1964), that tolerance to one unfavourable factor (low temperature in the present context) is greater the more nearly optimal are all the related factors. However, this does not help much in understanding the phenomena.

Long ago, Foter & Rahn (1936) suggested that inactivation by cold must have a different basis in mesophiles from that in psychrophiles (i.e. psychrotrophs); because in the latter, inactivation apparently occurs when the system freezes, whereas this is evidently not the case in the former (or, still more, for thermophiles), where minimum temperatures for growth may be far above freezing point. If so, the modifying effects of ancillary factors on minimum temperature might well be different for the different types of organism. But the observations described above show that these effects are apparently similar in psychrotrophs or mesophiles. Indeed, by suitable choice of conditions, the minimum temperature of the psychrotroph can readily be raised to a level characteristic of mesophiles. This makes it seem likely that the inactivating effects of cold are essentially similar on mesophiles or psychrotrophs, except that with the latter the accident of freezing normally restricts the lower end of their range.

The effect of reducing a_w is to raise the minimum temperature of growth, continuously and without irregularity. If, for example, limitation of growth at low temperature were due to specific effects such as changes in the permeability or fluidity of the cell membrane (Farrell & Rose, 1967, 1968) then a reduction of a_w might apparently cause these changes to take place at higher temperatures than usual. But the same applies to various other factors above indicated e.g. suboptimal pH or sublethal damage by heat or radiation. It is hard to believe that all these could produce essentially the same biochemical changes in the cell; hence their effect in raising the minimum temperature is not likely to be directly related to the processes normally responsible for preventing growth at low temperatures.

If, however, cessation of growth were due to a more general phenomenon e.g. failure of the cellular homeostatic mechanisms (metabolic imbalance) at low temperatures, then it is possible that, in the presence of inhibitors, this same temperature-dependent imbalance might occur at a higher temperature. In this way biochemical comparisons of growth limitation in a psychrophile and mesophile might be made at the same temperature, by suitable adjustment of factors like pH and solute content of the medium; a useful addition to investigations involving comparison of mutant strains having altered temperature characteristics for growth.

3. Subminimal Temperatures above Freezing

Micro-organisms gradually die when held under conditions where they cannot grow; here concern is with the loss of viability which occurs if the temperature is too low to permit growth. Evidently, the matter is mainly about mesophiles. Starvation of psychrophiles in this temperature zone, roughly 10-0°, is possible but seems not to have been investigated. Section 2 has indicated conditions which might permit investigation of psychrophiles in the same range as mesophiles.

For vegetative cells, the phenomena have two apparently different aspects. With stationary phase cells, or exponential phase cells taken slowly to low temperature, there is a steady and gradual decline in viability at the lower temperature. With exponential cells cooled rapidly, there is a loss of viability related to 'cold shock', discovered by Sherman & Albus (1923).

(a) *No cold shock*

Riley & Solowey (1958) diluted a centrifuged paste of cells from a mature *Serratia* culture, with equal volumes of various diluents, all nutrient except perhaps gelatin/phosphate. On storage at 8° rates of decline in the diluted suspensions were about half that in the parent concentrate. Mead *et al.* (1960), using similarly dense suspensions of *Pasteurella* stored at +5° (10^{10} cells/ml), found that sugars were at first protective, but that later a lowering of pH through anaerobic fermentation accelerated inactivation; admission of oxygen delayed inactivation, supposedly by preventing fermentation. In both these papers, having regard to the temperatures, the rates of inactivation seem remarkably low, of the order of 10%/month; it seems likely that there was compensatory growth, as noted by Mead *et al.* (1960).

Much work has been done on the survival of bacteria in media not supporting growth, at temperatures within the normal range for growth (e.g. Postgate & Hunter, 1962). It is doubtful how far this may be relevant to inactivation below the minimum growth temperature. For example nutrients have usually been lacking: whereas Meynell (1958) showed that cold shock is virtually the same in nutrient broth as in distilled water; while the work of Jackson (1974) illustrates a steady decline at subminimal temperature in an otherwise suitable growth medium—it was not established whether it had any protective effect.

Jackson (1974) used cells from a culture of *Staph. aureus* incubated 18 h at 37°, presumably well beyond the exponential phase, and stored at +5°, well below the minimum for growth, so they may illustrate the phenomena in absence of cold shock. He observed that the loss of viability was much more rapid under acid conditions; which is the opposite of what happened with *Aerobacter* deprived of nutrients at growth temperatures (Postgate & Hunter, 1962).

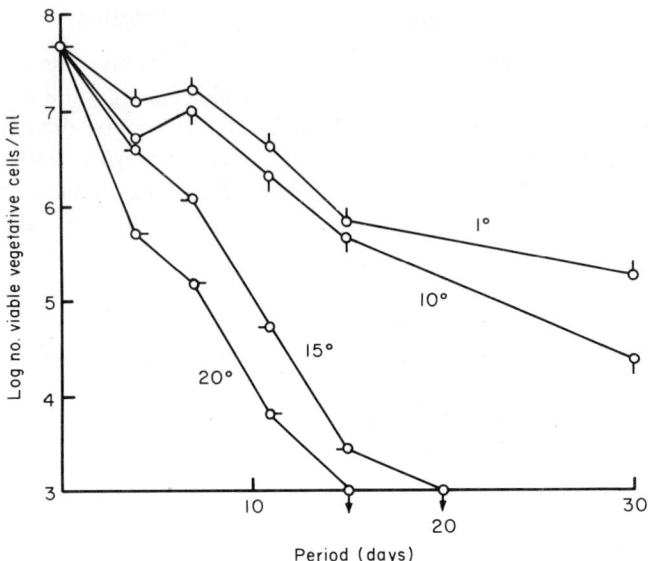

Fig. 5. Gradual death of vegetative cells of *Cl. perfringens,* held in a maintenance medium at temperatures below the minimum for growth, ♂, 1°; ♀, 10°; -o, 15°; o-, 20° (Unpublished data of S. M. El Sanousi & T. A. Roberts).

The effect simply of holding at temperatures below the minimum is also illustrated in unpublished observations of S.M. El Sanousi & T. A. Roberts, with stationary phase cells of *Cl. perfringens,* cooled slowly from 37° to the temperatures indicated (Fig. 5). The decline in viability was roughly exponential with time, down to about five decimal reductions. Decimal reduction times were c. 2 days at 20°, 4 days at 15°, 8 days at 10° (or 5°) and 12 days at 1°. It is noteworthy that survival was much better at lower temperatures, the most harmful being those just not permitting growth. (These data refer to suspensions in a rather poor maintenance medium in which the minimum temperature for growth exceeded 20°, an illustration of the effect of nutritional deficiency in raising minimum temperature cf. Section 2.)

The data suggest different rates of decline for different species. It is often supposed that vegetative cells of clostridia are especially sensitive to cold; and indeed, a decimal reduction time of about 8 days at 5° seems distinctly less than that of more than 20 days indicated for *Staph. aureus* from the data of Jackson (1974). But such figures probably depend so much on ancillary factors that comparisons have little value when (as in the above instance) those factors were not similarly controlled.

In this temperature zone, the effect of salts appears much less than at higher temperatures. to judge from the little information available. Shipp (1957) showed that in meat extract/peptone with 25% NaCl, where *Salm. enteritidis* survived less than a day at 37° and less than a week at 20°, it survived 8-10 weeks at 5°. Survival seems relatively little affected by salt concentration at lower temperatures. In Shipp's experiments, there was virtually no difference between 20% and 30% salt; and the time course of mortality for *Salm. enteritidis* in 25% NaCl at 5°, as given by Shipp (1957), is virtually the same as that given by Matches & Liston (1972a) for *Salm. derby* or *Salm. typhimurium* in 4% NaCl at 8°.

(b) *Reversible inactivation*

As the decline in viability progresses, the cells become increasingly 'damaged' in a sense analogous to that in heat, radiation or freezing damage: for example, they become increasingly sensitive to the presence of salt in the plating medium. Jackson's data suggest that, for this reason, counts of staphylococci in frozen fruit or vegetables might be too low by several powers of 10, if mannitol-salt were used as a selective medium. The damage was however restored within 2 h of incubation at 37° in a good medium (Jackson, 1974, his Fig. 5).

(c) *Cold shock*

Where exponential phase cells are transferred suddenly to appreciably lower temperature, the so-called cold shock, survival is lower at temperatures near 0° than near 20°, the opposite of what occurs with stationary phase cells, as is illustrated by Fig. 6. Sudden transfer from 37° to water or buffer at 0°, and holding at 0°, was accompanied by a progressive decline in proportion of viable cells. Meynell (1958) presented exactly similar data for *E. coli* transferred from 37° to ¼-strength Ringer solution at 4°. Whereas, in unshocked cells, the loss of viability may continue exponentially for days and become large, here it was relatively small and the impression is that it ceased in an hour; but this may be the result of the short periods of observation. Postgate & Hunter (1962) emphasised that such processes do not necessarily proceed exponentially.

Cold shock may depend on the growth medium (Strange & Ness, 1963; Houghtby & Liston, 1965), on the suspending menstruum (Meynell, 1958; Sato & Takahashi, 1968a), concentration of cells (Strange & Dark, 1962), as well as on the temperature regime. Farrell & Rose (1968) showed that it occurs in psychrophiles, and at temperatures above the minimum for growth. It has now been demonstrated with Gram positive spore-formers and streptomycetes as well as Gram negative bacteria. *Staphylococcus aureus*, however, resists cold shock (Gorrill & McNeil, 1960).

Meynell (1958) found that cold shock was prevented if the cells of *E. coli* were chilled in solutions of appreciable osmotic strength, e.g. 0.3 M sucrose. This

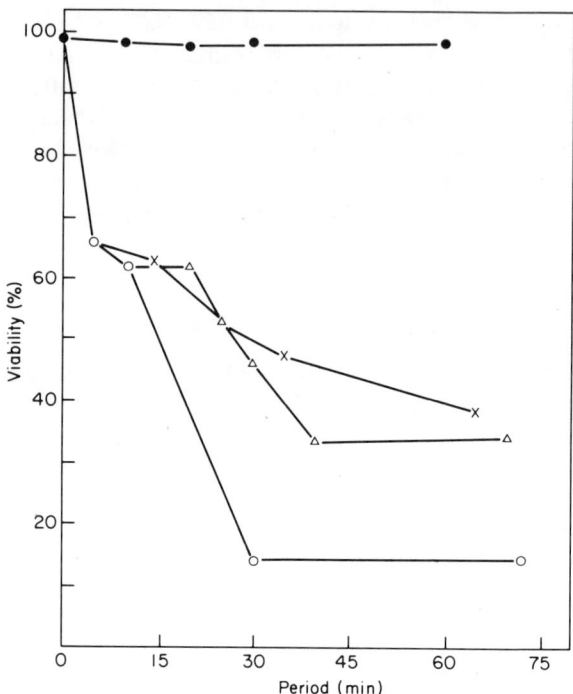

Fig. 6. Survival of exponential phase *Aerobacter aerogenes* chilled at 0° for short periods and then held at 18°. Viability at 18° immediately after chilling (●), after chilling for 5 min (x) and after chilling for 10 min (△); viability of control suspension throughout at 0° (o). (From the data of Strange & Dark, 1962: reproduced by permission of the authors and Cambridge University Press.)

has recently been confirmed by Leder (1972), with several other *E. coli* strains.

Accounts of the effect of cold shock differ, however, in detail. Traci & Duncan (1974) with *Cl. perfringens* reported an immediate mortality of 95% on transfer from 37° to 4°, followed by a nearly exponential decline with a decimal reduction time about 75 min (observed for 90 min). Strange & Dark (1962), with *A. aerogenes* reported no immediate mortality on transfer to 0°, loss of viability only appearing with storage at 0°. Similarly, while there is agreement that cells can recover from cold shock, accounts of this phenomenon seem inconsistent. According to Strange & Dark (1962) prompt return of *Aerobacter* in buffered saline to 18° after shock at 0° entirely nullified the shock, progressively longer intervals of delay at 0° progressively weakening this restorative effect at 18° (Fig. 6). But Traci & Duncan (1974) found no recovery of *Cl. perfringens* returned from 10° to 20° in water.

Exposure to low temperature causes cell damage in conditions where cold shock occurs, as in its absence. Meynell (1958) showed that cold-shocked cells of

E. coli, 4 h after transfer to 4° were sensitive to detergents in the plating medium, which might reduce counts 100- to 1000-fold. Similarly, Traci & Duncan (1974) showed that cells of *Cl. perfringens*, cold shocked in 0.1% peptone for 10 min at 10°, were sensitive to sulphite-neomycin agar, which gave counts *c*. 10 times less than a non-selective medium. Traci & Duncan observed that this sensitivity disappeared in less than an hour of incubation in TY broth at temperatures near 30°; but the apparent number of survivors was never raised above that on non-selective medium immediately after cold shock.

(d) *Discussion*

The earlier literature discusses cold shock in relation to the temperature at which the cells were grown. It now seems that a more relevant factor is the incubation/cold shock temperature difference, which the incubation temperature of course measures directly if a single shock temperature is used. In an attempt to discover which is the more important, Traci & Duncan (1974) exposed *Cl. perfringens* cells grown at 37° to 3 min at 43°, before shocking to

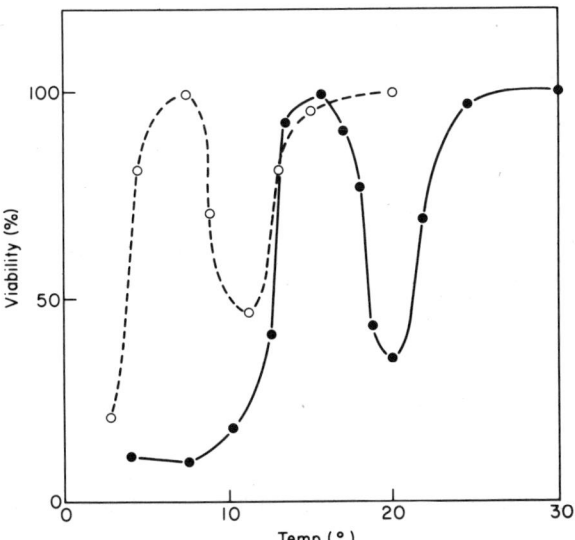

Fig. 7. Two critical temperatute zones in cold shock. A washed cell suspension of *Escherichia coli* was divided into two portions. One was kept at 30° and the other at 20°. A portion of each suspension was cold shocked in tris buffer maintained at the temperatures indicated. Viabilities were determined and plotted versus temperatures at which the cold shocks were performed. ● — ●, shock from 30°; ○ - - - ○, shock from 20°. (From the data of Sato & Takahashi, 1968: reproduced by permission of the authors and the Journal of General and Applied Microbiology.)

15 or 20°; observing that those transferred briefly to 43° acquired sensitivity similar to that of cells grown at 43°, they concluded that the difference rather than the growth temperature is critical. On the other hand Houghtby & Liston (1965) reported that, although lowering the growth temperature of *E. coli* from 35° to 22° made the cells more resistant to cold shock, rapid warming of a 22° culture to 35° just before cold shock did not make that culture sensitive. The importance of temperature difference appears more clearly in the work of Sato & Takahashi (1968*a, b*), who used cultures grown at two different temperatures, suddenly transferred to various lower temperatures. Over a range of shock temperatures, they observed a maximum and two minima of viability; and found that, comparing cultures grown at two different temperatures, these maxima/minima were repeated at an interval in shock temperature equal to the difference in growth temperatures (Fig. 7). This was demonstrated for *Ps. fluorescens* and *Bacillus subtilis* (Sato & Takahashi, 1968*b*) as well as for *E. coli*.

Traci & Duncan (1974) discuss the size of the critical temperature zone, and whether it is single or double. Remembering the well known importance of rate of freezing (cf. below), it is remarkable that no recent attempts have been made to relate cold shock phenomena to rate, for Meynell (1958) showed long ago that rate can be critical: where immediate transfer from 37° to 4° led to large mortality of bacteria, there was no effect if the transfer was gradual over 30 min. That is to say that, where a chilling rate *c*. 1°/min produced no effect, a rate perhaps 1000 times higher produced a large one. There is a suggestive parallel with the effects of similar freezing rates (cf. Fig. 11).

Except as regards cold shock, there is a surprising lack of information about the loss in viability of cells held near 0°, despite its obvious importance to those who keep cultures in refrigerators. The apparently very different responses (e.g. to holding temperature), between exponential phase cells subject to cold shock and stationary phase cells, makes one wonder about the relations between the two processes. How does one response turn into the other as cells age? What is the behaviour of mixed populations of cells, such as one usually has to deal with in practice?

It appears that the effect of cold shock is to kill a proportion of cells, more or less promptly; and to leave damaged survivors, the damage being rapidly repairable in a good medium near optimum temperature. This resembles the effects of freezing, as does the protection by osmotically active solutes (cf. Section 5).

Meynell (1958) concluded that the lethal effect of cold shock is caused by a loss of the semi-permeability of the cell membrane which affects water movement, a view supported by Strange & Dark (1962). Consistent with this view are the reports by Traci & Duncan (1974) that recovery from cold shock is not affected by nalidixic acid (inhibits DNA synthesis), rifampin (inhibits RNA polymerase), nor chloramphenicol (inhibits protein synthesis). Farrell & Rose (1968) suggested that susceptibility to cold shock depends on the unsaturated fatty acid composition of the membrane lipids, and relation to crystallization of membrane lipids was suggested by Leder (1972); but such views are not readily compatible with the conclusion of Sato & Takahashi (1968b) that what matters is the magnitude of the temperature drop, rather than the temperature at which the cells were grown and therefore the range over which the drop takes place.

The death of cold-shocked cells is preceeded and accompanied by loss of breakdown products of cell constituents–protein, nucleic acid, polysaccharides. These products are supposed to have some relation to the loss in viability, because (with Gram negative species) one cell suspension can be protected by the exudate from another and sufficiently dense suspensions are self-protecting (e.g. Strange & Dark, 1962; Farrell & Rose, 1968). Nevertheless, there is no direct relation with viability: for example, cold shock leads to release of UV-absorbing materials, but the release is much more rapid than the ensuing loss of viability, and it is not delayed by addition of Mg^{2+} ions which restores viability (Sato & Takahashi, 1968; Farrell & Rose, 1968).

Mg-mediated recovery from cold shock has been observed with *E. coli, Ps. fluorescens,* and vegetative cells or germinating spores of *B. subtilis*. This process was inhibited by 2,4-DNP, the inhibition being reversible by NAD + ATP; further, the radiation-resistant mutant B/r of *E. coli* recovers much better than the radiation sensitive mutant Bs-1. These observations led to investigation of DNA repair, showing that the DNA of cold-shocked cells contained more single-strand breakages and was more susceptible to breakdown by endogenous nucleases than that of unshocked or recovered cells (Sato & Takahashi, 1970). The authors concluded that the DNA breakages in cold-shocked cells were a consequence of loss of activity of DNA ligase which, under normal circumstances, joins together the small DNA fragments synthesized during DNA replication. The DNA of stationary phase cells (which are not susceptible to cold shock) would contain no such unjoined fragments and hence would be less susceptible to breakdown by nucleases. This proposed mechanism awaits further investigation. Postgate & Hunter (1961) remarked incidentally that *Aerobacter* cells grown in a chemostat were not susceptible to cold shock. Though such cells are regarded as growing exponentially, they did so at a rate depressed by nutrient limitation, where fewer replication points and hence fewer DNA breaks per cell might be expected (Cooper & Helmstetter, 1968) and the conditions resemble those for cells entering the stationary phase.

It is still uncertain how far the various phenomena associated with cold shock in Gram negative bacteria apply to the Gram positive bacteria. Differences appear in the degree of initial mortality and its reversibility, and in the protective effect of exudates.

4. The Subzero Zone

As a basis for this, and Section 5 it seems necessary to state briefly the changes which take place on freezing. Strictly speaking, the upper limit of this zone is not 0°, but the temperature at which the substrate freezes.

(a) *Changes during freezing*

Freezing commonly begins at −1 to −3°, depending on the concentration and nature of the soluble substances present. Although the material may appear to be frozen solid, there is actually a considerable amount of liquid water present which does not freeze until lower temperatures are reached. The lower the temperature, the more becomes frozen. For meat and fish the point at which all the water is frozen may be as low as −50 to −70°, though for vegetables and fruit the corresponding temperatures are −16 to −21° (Michener & Elliott, 1964). At their usual storage temperature, therefore, and especially at

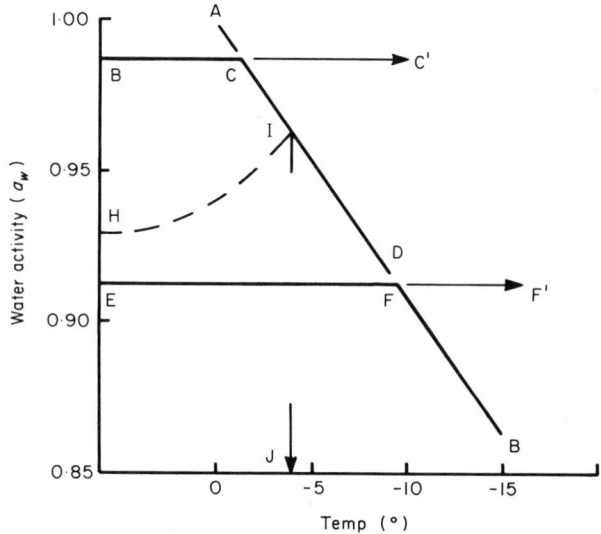

Fig. 8. The relationsip of physical state of the substrate to temperature and water activity. The relationship of a_w to temperature in a partially frozen system is shown by line AB. The effect on a_w of lowering the temperature of an initially liquid medium is shown by lines BCD (low solute medium) and EFB (high solute medium) when freezing occurs at C and F, respectively; or by the lines BCC′, EEF′, if supercooling takes place instead. The dashed line, HI, shows the minimum a_w tolerated by a hypothetical bacterium capable of growth below 0°; in the absence of supercooling, growth of this organism is impossible below the temperature J at which the dashed line intercepts line AB. (A similar diagram is given by Scott, 1961.)

temperatures not far below 0°, frozen substrates contain appreciable amounts of unfrozen water which is potentially available for the growth of micro-organisms. It has generally been accepted that growth of micro-organisms below freezing point depends on the existence of this unfrozen phase (Ingram, 1951; Scott, 1961; Michener & Elliott, 1964).

As the temperature is reduced, and progressively more of the water is converted into ice, the concentration of dissolved solids rises in that portion of the water remaining unfrozen, hence the activity of that water falls correspondingly. These changes are more fully discussed by Leistner (1975 this symposium). Although still debated (e.g. Hill & Sunderland, 1967; Storey & Stainsby, 1970), it is usually supposed that the aqueous activity is uniquely determined by the temperature, regardless (within limits) of the nature of the food or of substances added to it. Figure 8 places the relationship quoted by Scott (1961) into a diagram of the form used by Michener & Elliott (1964, their Fig. 3). That figure is unfortunately questionable as regards its indications for bacterial growth: mesophiles are indicated as growing well into the subzero range, in sufficiently dilute solutions; psychrophiles are indicated as generally more tolerant of concentrated solutions than mesophiles).

The pH of frozen foods may change appreciably during storage in the frozen state as the different salts which contribute to the indigenous buffer systems are successively precipitated out. This has been studied by van den Berg (1968) who developed a special electrode to measure pH in the frozen state. He found that, during freezing, foods with a high initial pH and relatively low protein content decreased in pH, whereas foods with a high protein content, especially those of pH above 6, increased in pH.

The microbiological significance of these relationships will now be considered.

(b) *Incidence of growth below zero*

Michener & Elliott (1964) catalogued many instances where growth of bacteria, yeasts and moulds was reported in foodstuffs and microbiological media at subzero temperatures. It appears that growth is common at temperatures down to −5 or −7°, at lower temperatures progressively less so, and below −10 to −12° it occurs very rarely. Since reports of growth below −10° are so rare (and in some cases questionable) one may regard the effective lower limit of the subzero growth zone as c. −12°. The inference is that many micro-organisms—bacteria, yeasts and moulds—have minimum growth temperatures as low as −7°, and a few even lower.

(c) *Rates of growth*

Figure 9 compares Arrhenius plots for different organisms with minimum growth temperatures near zero. Several points of interest emerge. (i) The temperature characteristic (activation energy, μ) may be slightly different for different species (or conditions). (ii) There is a continuous spectrum of minimum temperatures down to c. $-6°$, beyond which point observations are hindered by the occurrence of freezing. (iii) At temperatures near the minimum for growth, μ tends to become larger, i.e. small decreases in temperature produce relatively greater increases in generation time (Ingraham, 1958). (iv) With a true

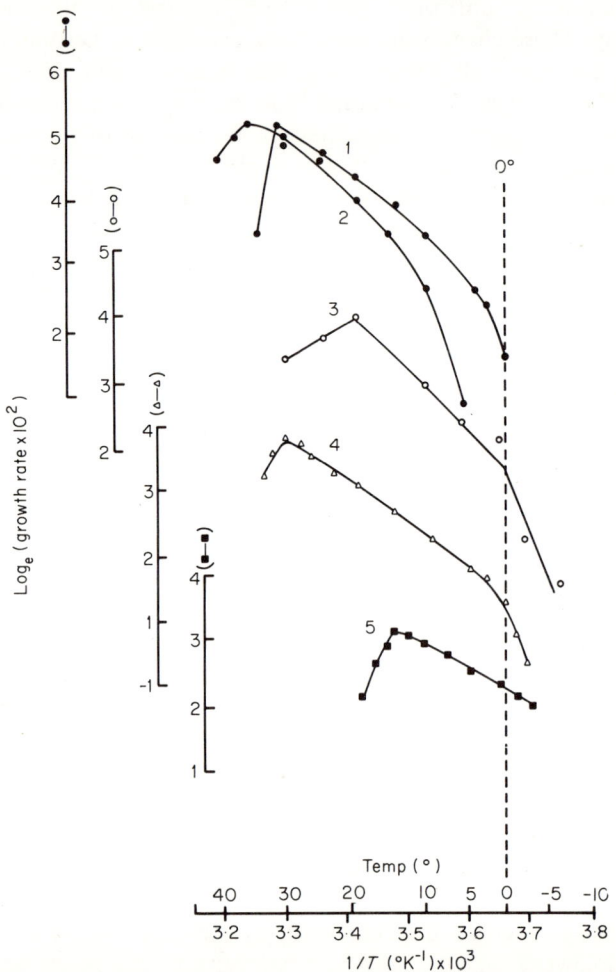

Fig. 9. Arrhenius plot of the relationship between growth rate and temperature for various bacteria capable of growing near $0°$. Curve 1, *Cl. botulinum* type E (Ohye & Scott, 1957); Curve 2, *Ps.* 21-3c (Ingraham, 1958); Curve 3, *B. psychrophilus* W16A (Larkin & Stokes, 1968; Kim & Larkin, 1973); Curve 4, *Pseudomonas* L9; Curve 5, *Pseudomonas* L12 (Harder & Veldkamp, 1971).

Table 3
Growth rates of bacteria below 0°

Organism	Temperature (°)	Generation time (h)	Reference
Achromobacter	−4 ± 1.5	30.7 (average)	Kiser (1944)
Pseudomonas fluorescens	−3	56.8	Hess (1934)
Bacillus psychrophilus T3A	−2	55	Larkin &
	−4.5	168	Stokes (1966)
Bacillus psychrophilus W16A	−2	103	Larkin &
	−5 to −7	204	Stokes (1968)
Bacillus globisporus	−2	84	Larkin &
	−4.5	156	Stokes (1968)
Bacillus insolitus	−5 to −7	252	Larkin & Stokes (1968)

psychrophile, this did not happen before freezing occurred. According to Harder & Veldkamp (1971), this behaviour has also been observed by others. Several occasions were recorded by Ingram (1951, his Table 2) where temperature coefficients from 0° to −2° were no lower than those just above 0°. The implication, in such cases, is that the minimum growth temperature would have been considerably below −2°, had observations been continued into that zone.

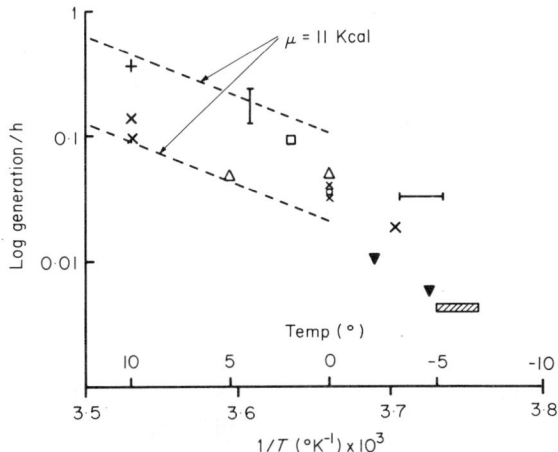

Fig. 10. Arrhenius plot of the relationship between growth rate and temperature, drawn from values given in Table 3. ⊢⊣, *Achromobacter* (Kiser, 1944); X, *P. fluorescens* (Hess, 1934); ⊺, *P. fluorescens* (Olsen & Jezeski, 1963); +, *Psuedomonas* sp (Ingraham, 1958); □, *Pseudomonas* 92 (Frank *et al.*, 1972); △, Strain 82 (Upadhyay and Stokes, 1962); ▽, ▨ *Bacillus* sp (Larkin & Stokes, 1968); 0, *M. cryophilus* (Tai & Jackson, 1969).

Observations of growth rate in the freezing zone are few and doubtfully comparable with observations under more normal conditions. Values collected in Table 3 are compared in Fig. 10 with corresponding observations above −1° as tabulated by Morita (1975). The latter are consistent with an average μ value of c. 11 kcal, believed by Harder & Veldkamp (1971) to be representative of organisms growing near 0° (though there are evident individual variations). It appears from Fig. 10 that growth in the subzero zone may be slower than might be expected on that basis.

For such relatively slower growth, two reasons can be suggested: (i) the lower the temperature, the more likely it becomes that an organism will be growing near its (subzero) minimum limit, which, as we have just seen above (e.g. Fig. 9), leads to relatively slow rates of growth; (ii) a second reason could be related to the activity of the water, in a system where freezing takes place. Reducing temperature below freezing, besides being likely to take organisms near their minimum growth temperature, steadily increases the concentration of solutes to which they are subjected. From behaviour at normal temperatures, described in Section 2, the inhibitory effects of high solute concentration and low a_w might be expected to be more profound if the organism is growing near its minimum temperature. Hence rates of growth below freezing point are often likely to be slower than if low temperature alone were the limiting factor.

(d) *The nature of the flora*

The organisms which grow on frozen foods are frequently not typical of the flora at higher temperatures. Thus at above-zero temperatures meat normally undergoes spoilage by bacteria, whereas in the frozen state moulds are usually responsible. Similarly, the usual bacterial spoilage flora of peas is often replaced by yeasts on frozen peas. The data compiled by Michener & Elliott (1964) indicate that this trend, i.e. the substitution of bacteria by yeasts or moulds as the temperature of storage is decreased, is common to a variety of foods, implying that reducing the temperature produces some change which selectively affects the growth of micro-organisms.

It has been suggested that this change might be the progressive reduction of water activity with lower temperature. Bacteria, yeasts and moulds are on the whole, respectively, able to tolerate successively lower water activities (e.g. Nickerson & Sinskey, 1972); though there are exceptions to this generalization e.g. in the halophilic bacteria and osmophilic yeasts. The common spoilage bacteria, not tolerating $a_w < c.\ 0.95$, should be excluded if only for this reason from frozen substrates at temperatures below $c.\ -5°$ (cf. Fig. 8).

This effect of temperature, and hence a_w, should however be considered in relation to other selective factors which influence the balance between bacteria, yeasts and moulds, such as the composition of the food, and environmental factors. For instance, bacteria almost never grow on frozen fruits even at temperatures which would permit growth on other foodstuffs; on the other hand, spoilage of frozen fish is more frequently caused by bacteria (even at relatively low temperatures), perhaps partly because of the naturally high incidence of salt-tolerant psychrotrophic bacteria on fish.

The pH value of the food might apparently be of major importance, for neutral foods tend to carry bacteria, highly acid fruits, moulds, and foods of intermediate acidity, yeasts, at normal temperatures. If these general relations were to persist into the subzero zone, they might confound the suggested relation with temperature. Table 4 illustrates this possibility, by showing a comparison of records for foods of low, intermediate and high acidity. Although more data would be desirable, there is an evident importance of the substrate: bacteria occurred more frequently on the low-acid group and moulds on the highly acid group of frozen foods, at both the higher and lower temperatures. At the lower temperatures, moreover, there was an apparently greater predominance of yeasts and moulds over bacteria on the neutral foods, and of moulds over yeasts on the acid foods. The foregoing is based, however, on the assumption that pH differences at normal temperature continue when the systems are frozen, which appears doubtful; the observations of van den Berg (1968, e.g. his Figs 9 & 11) suggest that the pH difference between vegetables and meats might have largely disappeared after about a month below $-7°$.

Observations made under practical conditions, like many of those in Table 4,

Table 4
The influence of the acidity of the food on the incidence of micro-organisms found on frozen foods at different storage temperatures

Foods	Acidity	Incidence of micro-organisms on food stored at					
		$0°$ to $-5°$			below $-5°$		
		B*	Y	M	B	Y	M
Meat, fish, milk	Low	10	5	4	8	1	5
Vegetables, raw or brined	Intermediate	4	0	2	1	0	7
Fruits	High	1	10	9	0	3	6

* B, Y, M indicates records of bacteria, yeast or mould, respectively.
Compiled from citations by Michener & Elliott (1964).

might be further complicated by casual environmental factors. Exclusion of air might hinder the appearance of moulds; for example, Berry (1933) reported that development of moulds on frozen berries was prevented in sealed containers. On the other hand, moulds might be encouraged by surface desiccation caused by imperfect refrigeration technology, often developed during long periods of frozen storage. There are as yet insufficient observations under defined conditions to warrant analysis of data on this basis.

(e) *Discussion*

At temperatures not too far below zero, two different kinds of behaviour occur. Some organisms grow, though slowly, but, under exactly the same conditions, others die—the familiar lethal effect of freezing. Though it is customary to regard these two patterns of behaviour separately, the relation between them is clearly a matter of special interest. Two factors now seem particularly important.

First, many organisms will be below their minimum temperature for growth and might be expected to die in a manner similar to that described in Section 3. Others have sufficiently low minimum growth temperatures. True psychrophiles may have unusually low minimum temperatures and hence might be more likely to grow at subzero temperatures. If this were so, the temperature precautions attending manipulation of true psychrophiles would be desirable in making isolations from material held at subzero temperatures; there has, however, been no indication of this hitherto.

Second, growth should be favoured by ability to tolerate reduced a_w. The fact that many common spoilage organisms, e.g. psychrotrophic pseudomonads, do not grow on frozen foods may be due to their well-known lack of this ability. Growth of organisms at or near $-10°$ would only be expected in species which, besides having sufficiently low minimum growth temperatures, were adapted for growth at high solute concentrations (e.g. halphiles or osmophilies—Ingram, 1951). Given sufficient information about the limiting water activities for growth at low temperatures, it would seem possible to predict the lowest temperature at which growth would be possible in the partially frozen state (Fig. 8). Predictions should, however, be based only on values of limiting a_w determined at appropriately low temperatures. There might conceivably be synergistic effects of pH, similar to those at normal temperatures discussed in Section 2.

At present there is very little information of this kind, doubtless due partly to the practical problems of controlling a_w at temperatures below zero. The true effects of temperature alone could only be studied in supercooled systems where no ice separates, and the composition of liquid system and hence its water activity does not change, as indicated in Fig. 8. Supercooling is occasionally possible, e.g. of culture media (Larkin & Stokes, 1968) or eggs (Hale, 1950), but only to a limited degree. Further, such systems are highly unstable, and the presence of microbial cells is likely to precipitate ice crystallization. To prevent freezing by adding solutes is by no means the same thing. Freezing is normally prevented because the freezing point is lowered, to a degree corresponding to the reduced water activity caused by the addition of solute. Larkin & Stokes (1968) added various compounds to bacteriological culture media in attempts to prevent freezing at $-7°$ but the concentrations of glycerol, sodium chloride, ethylene glycol, lactose and dimethyl sulphoxide needed to prevent freezing were inhibitory to growth even at 20°. Growth at −7° was only possible when glycerol promoted supercooling at concentrations which were not inhibitory for growth. With their data in Fig. 9, accordingly, the growth rate is higher than that which might be expected on the basis of the equilibrium water activity.

The lower limit of this zone may be set by a change in the mechanism of freezing. Intracellular water usually remains unfrozen and becomes supercooled until the temperature reaches −5 to −10°. Above c. −10°, freezing occurs only in the external solution, and a cell which can make the necessary osmotic adjustment quickly enough will escape injury other than mechanical through diminution of volume (Meryman, 1974). Below c. −10°, the cell membrane is supposed to fail to act as a barrier to proliferation of the ice already formed outside the cell, so that crystals suddenly develop inside the cell until only c. 10% of the cellular water remains unfrozen, and the cell dies (Mazur, 1966). While this corresponds with the general inability of micro-organisms to grow below −10° approximately, it appears not to correspond with the common observation that survival in the frozen state is improved at lower temperatures (see Section 5).

5. Freezing

Some inactivation occurs immediately on freezing, and it is followed by further inactivation during storage in the frozen state. Study of these phenomena is of course only possible after cells have been thawed. The subject is large, and has been extensively reviewed (Mazur, 1966; Ray & Speck, 1973; Meryman, 1974); here only points of special interest are considered.

The immediate effects of freezing are thought to be due to ice formation and solute concentration, and are affected by rate of freezing, pH, added solutes, and the permeability of the cell membrane to water. Storage death is a function of time and temperature, and may also be related to pH, solute concentration and ice crystallization (Ray & Speck, 1973). When investigating the phenomena, by plating thawed cells, more cells are usually recovered on rich than on poor, or selective, media. The difference represents cells which were damaged or injured, where the inactivation was reversible or repairable. Reports in the literature are often vague as to how long cells have been stored after being frozen, and to the difference between reversible and irreversible injury.

(a) *Rate of freezing*

The importance of cooling rate on the death of micro-organisms during freezing is now well established. Figure 1 shows how survival of various frozen cells depends on the cooling rate. Though different types of cell have different optimum rates, the existence of such an optimum implies that survival depends on at least two factors which are oppositely affected by rate of cooling. This is explained as follows (Mazur, 1966): in slowly cooled cells, freezing is extracellular and the main cause of cell damage is exposure to concentrated solutes of the substrate (cf. Section 4). At more rapid cooling rates the period of exposure to these concentrated solutes becomes progressively shorter and survival increases to a maximum. Beyond this, when cooling is so rapid that intracellular ice begins to form, survival falls; the necessary cooling rates are however rather high ($> 10°$/min) and would seldom occur in commercial practice.

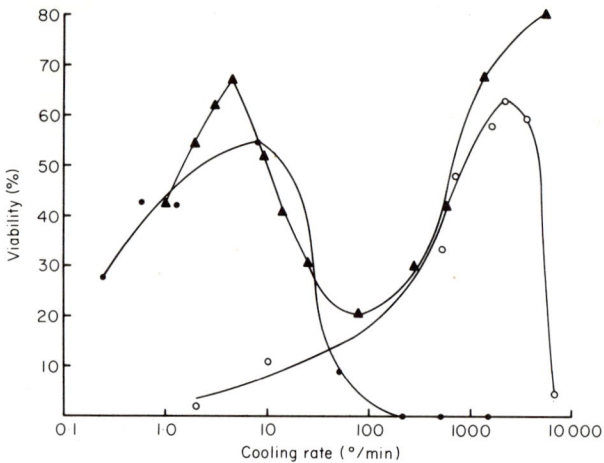

Fig. 11. Effect of cooling rate on the survival of various cells. (●) *Sacch. cerevisiae*; (○) Human red cells (From the data of Mazur, 1970); (▲) *E. coli* (From the data of Calcott & MacLeod, 1974).

The microscopic appearance of cells frozen at different rates accords with the above hypothesis. Slowly frozen cells (which have undergone osmotic loss of water) appear shrunken and distorted, whereas rapidly frozen cells retain their original outline but contain cavities representing the sites of intracellular ice crystals (Nei, 1960; Mazur, 1966; Nei, Araki & Matsusaka, 1969).

(b) *Age of cells*

Most workers have used stationary phase cells. Limited comparisons with exponential phase cells suggest that the latter may be much more susceptible to the effects of freezing. The difference disappears at high rates of freezing. The proportion of reversibly damaged cells does not vary greatly with the age of the cells (Ray & Speck, 1973).

(c) *The effect of salts*

The presence of NaCl markedly increases the lethal effect of freezing though micro-organisms differ in their susceptibility. Most appear to be rather sensitive (e.g. *E. coli, Salmonella, Aerobacter* and *Lactobacillus* spp.), though others such as *Staph. aureus* and the yeast *Saccharomyces cerevisiae* (Mazur, 1966) are almost immune to its damaging effects. Other salts, e.g. LiCl, KCl, Na_2SO_4 and K_2SO_4 (Calcott & MacLeod, 1974) are equally harmful to bacteria though some, such as phosphates, may be marginally protective (Moss & Speck, 1966; Postgate & Hunter, 1961). The effect of the added salts is to lower the freezing point, with the corollary that until the system actually freezes, the cells are exposed to more concentrated solutions and greater osmotic stress than with no addition (cf. Fig. 8).

(d) *Protective agents*

A large number of compounds, many of practical importance in foods (e.g. milk solids, sodium glutamate), protect against freezing. Some, such as sucrose, also protect against cold shock (see Section 3). The protective effect is manifest even at high freezing rates where intracellular ice is the presumed cause of damage (Fig. 12). The basis for this protection is not known though in some cases

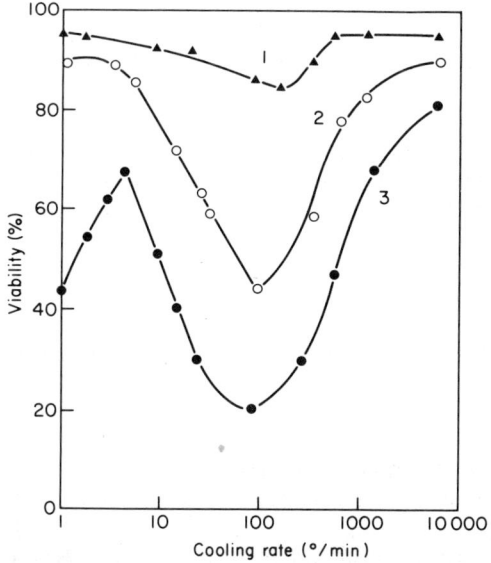

Fig. 12. Protection of *E. coli* by sucrose from freeze-thaw damage in distilled water. Organisms were frozen in distilled water +5% sucrose (curve 1), +25% sucrose (curve 2) or distilled water alone (curve 3). (From the data of Calcott & MacLeod, 1974, reproduced by permission of the National Research Council of Canada.)

penetration of the cell is apparently unnecessary (Postgate & Hunter, 1961; Meryman, 1974).

Although cryoprotectants reduce death due to freezing, the percentage of damaged cells is usually high, whence it appears that cryoprotectants do not protect correspondingly against reversible injury (Ray & Speck, 1973).

(e) *Response of different species*

Although survival rates after freezing and thawing are known for many bacterial species, in general, the experimental conditions have usually been too diverse to permit meaningful comparison. The few comparative studies which have used standardized conditions of freezing and thawing have seldom included representatives of the four major bacterial types: i.e. rods and cocci, Gram positive and Gram negative. Despite these limitations, general impressions emerge. Gram positive cocci seem particularly resistant to freezing and usually survive much better than rods. *Saccharomyces cerevisiae* appears to survive better than bacteria. There is little information for fungi, except that conidia are relatively resistant especially when air-dry (Mazur, 1966).

The survival of Gram negative bacteria (*Escherichia, Salmonella, Serratia*) increases with initial concentration of cells; that of Gram positive genera (*Lactobacillus, Bacillus, Micrococcus*) seems much less affected by cell concentration (Major *et al.*, 1955). The beneficial effect is supposed to be due to protection by leakage products from the injured cells. A similar protective effect has also been noted (see Section 3) in cold shock of Gram negative species, though whether this is true for Gram positive species is not yet known. Packer *et al.* (1965) found during freezing that even dilute suspensions of *E. coli* could be protected by filtrates of stationary phase cultures though the identity of the active material was not established.

(f) *Reversible inactivation – cell damage*

After freezing and immediate thawing of *E. coli, A. aerogenes, Shigella sonnei* and *Strep. lactis,* more bacteria were recovered in rich than in poor media. *Saccharomyces cerevisiae* does not exhibit this type of injury (Mazur, 1966).

The manifestations of this injury were recently reviewed by Ray & Speck (1973). (i) Leakage of cellular materials. In Gram negative bacteria, leakage of UV-absorbing material has been related to loss of viability; the material appears predominantly to be RNA of low mol. wt. Short chain peptides have also been implicated. (ii) Increased sensitivity to surfactants etc. It is presumed that this arises because of damage to the cell membrane. (iii) Increased nutritional need.

Trypticase, added to minimal media, allows damaged cells to recover; this effect was associated with certain active peptides.

Repair is possible in absence of cell division. Using frozen and immediately thawed *Salm. anatum*, Ray et al. (1972) showed that repair required phosphate or ATP; was assisted by energy sources, and $MgSO_4$; was inhibited by KCN or dinitrophenol; but was not prevented by actinomycin D, chloramphenicol or D-cycloserine (inhibitors of synthesis of RNA, proteins, and cell-wall mucopeptides, respectively). Similar observations were made with *E. coli* by Ray & Speck (1972); kanamycin and streptomycin inhibited repair, apparently because they removed Mg^{2+} ions. It was suggested that energy metabolism is necessary for repair of the cell membrane.

(g) *Time/temperature relations during frozen storage*

After freezing, death occasionally follows a roughly exponential course with the passage of time at constant temperature, e.g. in the experiments of Kiser (1943). More usually, death is initially more rapid, but gradually becomes relatively slow until in the later stages viable numbers remain almost constant. The progressive decline in viability is thought to be due mainly to continued exposure to concentrated solutes, and so might represent an extension of the processes responsible for immediate death on freezing (at low freezing rates); the nature of the solutes may however change with time, owing to differential precipitation (van den Berg, 1968). Alternatively, the change in size and distribution of the ice crystals, known to occur through 'grain growth' (Luyet, 1966; Kent, 1975), might be important.

Generally, the lower the storage temperature (and the less it fluctuates), the lower the rates of inactivation during storage. In particular, death at -2 to $-5°$ is often much more rapid than at -10 to $-20°$; this behaviour resembles that of unfrozen cells stored below their minimum growth temperature (Section 3). Mazur (1966) notes various exceptions to this generalization. All are agreed that death is very slow below $-60°$.

(h) *Effects of pH and added solutes during frozen storage*

A low pH greatly accelerates death during frozen storage (Georgala & Hurst, 1963); as in unfrozen storage below the minimum temperatures for growth (Jackson, 1974). However, the real pH values in the frozen state may not be the same as the (initial) measured values, because of changes during frozen storage, already noted.

Cryoprotectants usually also provide good protection against inactivation during frozen storage. Sugars, however, which usually afford protection against immediate death on freezing, are said to be deleterious during frozen storage (Ray & Speck, 1973, but see Section 5l).

(i) *Resistance of different species to frozen storage*

As in their immediate response to freezing, Gram positive species seem more resistant than Gram negative to inactivation during frozen storage. Larkin *et al.* (1955) found that, after 147 days storage in frozen orange juice at $-23°$ the number of faecal streptococci did not significantly decline whereas survival of *E. coli* was reduced to $10^{-2}\%$. Similarly, a study by Kereluk & Gunderson (1959) showed that, after 481 days storage in frozen chicken gravy, 30% of the original number of enterococci survived but only $10^{-3}\%$ of coliforms. The more comprehensive investigation of Major *et al.* (1955) showed that Gram positive species of *Bacillus, Lactobacillus, Chromobacterium* — an organism called *C. orangium,* regarded by Gilman (1953) as a variety of a strain called *C. chocolatum* which is Gram positive and not properly called *Chromobacterium* — *Microbacterium* and *Micrococcus* survived frozen storage at $-22°$ for four weeks better than the Gram negative *E. coli, Salm. gallinarum* and *Serratia marcescens.*

Under laboratory conditions, Arpai (1962) demonstrated a statistically highly significant difference between the freezing survival of a psychrotrophic pseudomonad and *E. coli*, although the differences were very slight in comparison with the effect of the suspending menstruum. In a practical situation, Zawadski (1973) found that, after nine months' storage at -18 to $-21°$, the superficial flora of frozen beef became dominated by psychrotrophs; although, much earlier, Lochhead & Jones (1936) found that after eight months' storage at $-17.8°$ in vegetables, organisms able to grow at $4°$ experienced a greater percentage reduction in viability than those growing at 20 or $37°$. The ability of psychrotrophs to survive frozen storage merits further investigation.

Although yeasts and moulds are in general better able to grow at subzero temperatures, it is doubtful that they are more resistant to frozen storage. The early experiments of Lochhead & Jones (1936) and van Eseltine *et al.* (1947) indicated no marked differences in survival between the three groups of micro-organisms after storage at $-23°$ for five months.

In examining relationships between groups of bacteria and resistance to freezing, it should be borne in mind that quite large differences have been noted between strains of the same species, e.g. *Strep. lactis* (Moss & Speck, 1963). In practice, the conditions of freezing, storage and thawing are likely to have as great an influence on survival as the type of organisms originally present.

(j) *Reversible inactivation during frozen storage*

Though there is general agreement that frozen storage contributes to non-lethal injury in potentially surviving cells, this does not necessarily keep pace with lethal injury, for the proportion of injured cells among the survivors may vary greatly. Straka & Stokes (1959) and Morichi (1969) observed that this proportion decreased with storage time; Postgate & Hunter (1963) found a large increase; Moss & Speck (1963) and Ray & Speck (1973, Table 23) found no substantial change.

Though superficially similar, it is not clear whether the reversible injury developed during frozen storage is the same as that caused only by freezing. Investigators of repair processes have worked with cells frozen and thawed immediately, presumably because more convenient.

(k) *Thawing*

Concerning thawing, only two things are noted here. The rate of thawing is not important except after very rapid freezing when rapid thawing results in greater survival than slow thawing. One explanation of this is that, during slow thawing, the very small intracellular ice crystals formed as a result of rapid freezing grow in size and in doing so cause greater damage to the cell. Rapid thawing prevents this crystal growth and also minimizes the time spent in contact with concentrated solutes. If this explanation is correct, one would expect the beneficial effect of rapid freezing to diminish after increasing periods of frozen storage, especially at not too low temperatures; but this seems not to have been investigated.

After thawing, the interval before plating may be important. Postgate & Hunter (1963) found with *A. aerogenes* that fewer survivors were recovered when the period of plating after thawing was increased. Ray & Speck (1973) reported that plate counts on non-selective medium were reduced by 40% within 2 h after thawing cells of *Salm. anatum* NF3 in water, but not with *E. coli*. These observations refer, however, to thawed cells transferred to temperatures near 40°, and are to be compared with 'starvation' phenomena rather than the cold inactivation described in Section 3.

(l) *Discussion*

Much evidence indicates damage to the semi-permeable properties of the cell membrane as the main cause of death. More recent work however (Swartz, 1971; Alur & Grecz, 1975) suggests that single-strand breakages of DNA may also be involved.

One might expect that cold shock, however caused, could contribute to the mortality which occurs on freezing. As it happens, most work on the effects of freezing has been done with stationary phase cells, which resist cold shock (see Section 3); hence, the possible contribution of cold shock has not yet become clear.

The hypothesis that freezing injury results mainly from osmotic damage to the cell membrane invites comparison with the similar hypothesis regarding cold shock. The following similarities have been noted above and are listed here. (i) Exponential phase cells are more sensitive than stationary phase cells. (ii) Both with freezing and with cold shock, rates of cooling of $c.$ $1°$/min are much less damaging than rates of the order of $100°$/min. (iii) Sugars are strongly protective in both situations. (iv) Gram positive cocci, in particular *Staph. aureus,* resist both freezing and cold shock. (v) With Gram negative species, there is in both situations an auto-protective effect of dense suspensions. (vi) The distinction between lethal and non-lethal injury in freezing resembles the similar differences in cold shock of *Cl. perfringens* described in Section 3. (vii) The associated leakage phenomena are superficially similar, and appear to depend on damage to the cell membrane in both cases. (viii) The repair phenomena are independent of the major synthetic processes; and the stimulating effects of Mg^{2+} ions are similar, though at present differently interpreted in the two cases.

These similarities, being supposed to depend on effects of solute concentration, might presumably not exist when freezing injury is due mainly to direct effects of ice cystallization e.g. with high rates of freezing. Such similarities invite further cross comparisons: for example, are Gram positive species on the whole more resistant than Gram negative to cold shock, as they are to freezing; what is the effect of non-osmotic cyroprotectants towards cold shock?

Alur & Grecz (1975) demonstrated a direct relationship between mortality on freezing and number of DNA breaks/cell, which compares with the interpretation of cold shock damage by Sato & Takahashi (1970). However the Mg^{2+}-mediated recovery of viability of cold-shocked cells (estimated on non-selective, nutrient medium) has not been reported to occur in frozen and thawed cells.

It may be significant that staphylococci, which are resistant to freezing, possess an unusual degree of osmotic plasticity. It would be interesting to know if Gram positive cocci were also resistant to rapid freezing in view of the observations of Luyet (1961) that no intracellular ice formation took place in *Strep. lactis* even at very rapid rates of freezing. This was attributed to the very large surface:volume ratio which enabled intracellular water to be lost sufficiently rapidly to prevent intracellular freezing. So far no systematic study has been made of freezing resistance in osmotolerant or halotolerant species, which normally have high intracellular solute concentrations.

The changes developing during frozen storage, in solute concentration and pH, or in ice pattern, may account for the often relatively irregular decline of numbers with time. Where the decline is most rapid at temperatures just below freezing point, the pattern resembles that with cells held above freezing but below their minimum temperature for growth (cf. Section 3); moreover, a similar relatively high temperature coefficient, Q_{10} $c.$ 4, occurs in both situations.

It appears that the effects of frozen storage could be mainly an extension of the immediate effects of freezing, especially at low rates of freezing when exposure to concentrated solutes is a probable cause of inactivation in both phases. The statement of Ray & Speck (1973) that sugars may be much more effective against immediate death than against death during storage implies that these processes are not the same. But the statement appears questionable. Only the work of Postgate & Hunter (1961) suggests a harmful effect of sugars, but, unfortunately it did not include a control free of protectants; and the observation is in marked contrast to the strongly protective effects of sucrose on organisms suspended in fruit juice (McFarlane, 1942).

Inactivation during delay after thawing has not been particularly investigated for holding temperatures near 0° such as are likely to be significant in practice. It would be interesting to see whether the inactivation under these conditions, at temperature below the minimum for growth, is similar to that with cells not frozen and thawed. More rapid inactivation might perhaps be expected with the former, because nearly all the cells will be injured at the outset.

6. Some Practical Implications

Besides their theoretical interest, these effects of various factors on inactivation by low temperature have practical implications at all temperature levels.

The minimum growth temperature, for particular food-borne pathogens, may determine regulations designed to keep food safe: for example, recommendations that smoked (Olson, 1968) or irradiated (*Anon.,* 1966) fish should be stored below 3° are based on the observation of Schmidt *et al.* (1961) that growth of *Cl. botulinum* type E occurred down to 3.3°. Different authors however sometimes report different minimum temperatures. This may be illustrated from Table 1 of Michener & Elliott (1964), which indicates unusually high minimum temperatures for toxin formation and presumably growth by *Cl. botulinum* type A in liver sausage and chicken à la king, and by type B in fruit. Reasons for such anomalies can now be suggested. The liver sausage contained 2.5% NaCl and only *c.* 45% water — i.e. the effective salt concentration was near 5% (besides small amounts of nitrate). The chicken à la king, besides having *c.* 1% salt, had a final pH mostly near 5.0. The fruit was probably still more acid. As already indicated, these factors are likely to raise the minimum temperature for growth. The essential point was well taken by Michener & Elliott in their statement

> "It also follows from these considerations that the true minimum temperature for the growth of an organism could be determined only if all other factors affecting growth were optimal."

This statement of Michener & Elliott has a noteworthy corollary, in the indication that, as the basis for legislation, it may not always be necessary to take the lowest growth temperature ever recorded. To the extent that accessory factors are not optimal, the effective minimum growth temperature will be higher, perhaps very much higher if other conditions are far from optimal; it is these interactions which permit safe storage of semi-preserves at apparently dangerous temperatures.

Further, the same general qualification evidently applies to the limiting values of other factors when temperature is not optimal; differences in composition of medium etc. which would normally have little effect may suffice to prevent growth at temperatures not far above the minimum. For the same reason, there may be unusually erratic responses in the marginal zone.

There is an obvious practical importance in the ability of cells to survive for long periods, without growing, at chill temperatures, especially as regards food-poisoning species accidentally introduced into refrigerated foods. The resistance of cells under those conditions to salts means that species which cannot grow, and indeed die rapidly in strongly salted foods or brines at normal temperatures, can survive for much longer periods if such foods are refrigerated.

The damage which occurs to cells held below the minimum temperature for growth might obviously be important in microbiological examination of chilled foods. Counts may easily be $>$ 10 times too low where selective media are used. But a normal resuscitation treatment, say 2 h near optimum temperature in a good medium, should apparently suffice to prevent this.

Cold shock also might conceivably lead to erroneous data in routine viable counting. Traci & Duncan (1974) note the likelihood, in connection with *Cl. perfringens*, of transferring exponential phase cells at 37-43° to diluents near 15° for appreciable periods, apparently sufficient to reduce the proportion of cells viable nearly 100-fold. While it is not yet clear whether diluents near 15° may be similarly significant for other mesophiles, diluents chilled to near 5° apparently could be. It is disconcerting to realize that cold shocks of as little as 10° may suffice to reduce viability by 50%, and it seems urgently desirable to establish how widespread the phenomenon may be.

Various writers on food preservation have remarked on the relatively large extensions of storage life achieved by reductions in temperature near freezing point, and have compared this with a correspondingly accentuated retardation of bacterial growth at such temperatures (Ingram, 1951). But it now appears that this may depend largely on the fortuitous fact that most of the bacteria involved have been psychrotrophs with minimum temperature limits near 0°.

The lower limit of the subzero zone of growth, -10 to $-12°$, corresponds broadly with the widespread use of 0°F, *c.* $-18°$C, as a common temperature for storing foods, representing nearly the highest level at which freedom from microbial spoilage can be assured over long periods of time.

Although freezing may greatly diminish bacterial numbers, under certain circumstances, this seldom happens in practice. Bacterial mortality in frozen foods is frequently reduced by indigenous cryoprotectants.

Although Gram negative bacteria are considered sensitive to freezing, this cannot guarantee their absence from frozen products. *Salmonella*, for example, has been recovered from frozen cheese after seven years' storage. On occasions, freezing may even protect against the bactericidal effects of certain foods; as in the case of *Eberthella typhosa* (*Salm. typhi*), present in strawberries, which was killed in 6 h at room temperature but survived up to six months if stored at $-18°$ (McClesky & Christopher, 1941). A similar protective effect of above-zero holding temperatures was shown in Section 3.

The inability of cells damaged by freezing to grow on selective media may (as with cold-shocked cells) lead to erroneously low viable counts from frozen food samples, particularly in the case of Gram negative organisms such as *Salmonella* spp. and *E. coli*. Fortunately, resuscitation techniques are being developed which should overcome this problem (Speck *et al.*, 1975).

Thawing, which in this paper has had little consideration, may in practice be the most important aspect. Thawing conditions often permit a multiplication of surviving micro-organisms which far more than offsets mortality during freezing and storage. This, however, is outside the present subject.

7. Conclusion

The preceding sections have indicated various similarities between phenomena occurring under different temperature conditions, which are usually regarded separately. The arbitrary nature of the separation between temperatures above and below freezing should now be evident. There are apparent relations between phenomena associated with the minimum temperature for growth, and the ability of organisms to grow on frozen substrates; and there are apparent relations between phenomena associated with cold shock and those related to inactivation by freezing. The separation has the unfortunate effect of obscuring such relations.

The importance of the relationships between electrolytes, pH and membrane permeability at very different temperatures has been established and it is suggested that reversible damage to the membrane may be a preliminary to more permanent damage to the DNA-replicating system.

The apparent similarities between the various manifestations of reversible injury raise the hope that a single resuscitation procedure may suffice for repair of damage inflicted over a wide range of possible temperatures. We are not, of course, justified in suggesting that the phenomena are in all the cases the same. The value of such comparisons is in transferring ideas from one field to another, to suggest how techniques illuminating one aspect of the subject might hopefully be tried elsewhere.

8. References

ADAMS, J. C. & STOKES, J. L. (1968). Vitamin requirements of psychrophilic species of *Bacillus*. *Journal of Bacteriology* **95**, 239-240.

ALFORD, J. A. & PALUMBO, S. A. (1969). Interaction of salt, pH and temperature on the growth and survival of salmonellae in ground pork. *Applied Microbiology* **17**, 528-532.

ALUR, M. D. & GRECZ, N. (1975). Mechanism of injury of *Escherichia coli* by freezing and thawing. *Biochemical and Biophysical Research Communications* **62**, 308-312.

ANDO, Y., KARASHIMADA, T., ONO, T. & IIDA, M. (1968). Toxin production by *Clostridium botulinum* type E in radiation-pasteurised fish. In *Proceedings of the First U.S.-Japan Conference on Toxic Microorganisms*, Ed. Herzberg, M. Washington: U.S. Dept. of the Interior.

ANON. (1966). Panel Proceedings. *Microbiological Problems in Food Preservation by Irradiation*. Vienna: International Atomic Energy Agency.

ARPAI, J. (1962). Nonlethal freezing injury to metabolism and motility of *Pseudomonas fluorescens* and *Escherichia coli*. *Applied Microbiology* **10**, 297-301.

BAIRD-PARKER, A. C. & FREAME, B. (1967). Combined effect of water activity, pH and temperature on the growth of *Clostridium botulinum* from spore and vegetative cell inocula. *Journal of Applied Bacteriology* **30**, 420-429.

BARNES, E. M., DESPAUL, J. E. & INGRAM, M. (1963). The behaviour of a food poisoning strain of *Clostridium welchii* in beef. *Journal of Applied Bacteriology* **26**, 415-427.

BERRY, J. A. (1933). Destruction and survival of microorganisms in frozen pack foods. *Journal of Bacteriology* **26**, 459-470.

BEUCHAT, L. R. & LECHOWICH, R. V. (1968). Survival of heated *Streptococcus faecalis* as affected by phase of growth and incubation temperature after thermal exposure. *Journal of Applied Bacteriology*, **31**, 414-419.

CALCOTT, P. H. & MACLEOD, R. A. (1974). Survival of *Escherichia coli* from freeze-thaw damage: a theoretical and practical study. *Canadian Journal of Microbiology* **20**, 671-686.

CAMPBELL, L. L., Jr. & WILLIAMS, O. B. (1953). The effect of temperature on the nutritional requirements of facultative and obligate thermophilic bacteria. *Journal of Bacteriology* **65**, 141-145.

CHRISTIAN, J. H. B. & WALTHO, J. A. (1962). The water relations of staphylococci and micrococci. *Journal of Applied Bacteriology* **25**, 369-377.

COOPER, S. & HELMSTETTER, C. E. (1968). Chromosome replication and the division cycle of *Escherichia coli* B/r. *Journal of Molecular Biology* **31**, 519-540.

EDDY, B. P. (1960). The use and meaning of the term "psychrophilic". *Journal of Applied Bacteriology* **23**, 189-190.

ELLIKER, P. R. & FRAZIER, W. C. (1938). Influence of time and temperature of incubation on heat resistance of *Escherichia coli*. *Journal of Bacteriology* **36**, 83-97.

EMODI, A. S. & LECHOWICH, R. V. (1969a). Low temperature growth of type E *Clostridium botulinum* spores. 1. Effects of sodium chloride, sodium nitrite and pH. *Journal of Food Science* **34**, 78-81.

EMODI, A. S. & LECHOWICH, R. V. (1969b). Low temperature growth of type E *Clostridium botulinum* spores. 2. Effects of solutes and incubation temperature. *Journal of Food Science* **34**, 82-87.

FARRELL, J. & ROSE, A. H. (1965). Low temperature microbiology. *Advances in Applied Microbiology* **7**, 335-378.

FARRELL, J. & ROSE, A. H. (1967). Temperature effects on microorganisms. *Annual Review of Microbiology* **21**, 101-120.

FARRELL, J. & ROSE, A. H. (1968). Cold shock in a mesophilic and a psychrophilic Pseudomonad. *Journal of General Microbiology* **50**, 429-439.

FOTER, M. J. & RAHN, O. (1936). Growth and fermentation of bacteria near their minimum temperature. *Journal of Bacteriology* **32**, 485-497.

FRANK, H. A., REID, A., SANTO, L. M., LUM, N. A. & SANDLER, S. T. (1972). Similarity in several properties of psychrophilic bacteria grown at low and moderate temperatures. *Applied Microbiology* **24** 571-574.
GEORGALA, D. L. & HURST, A. (1963). The survival of food poisoning bacteria in frozen foods. *Journal of Applied Bacteriology* **26**, 346-358.
GILMAN, J. P. (1953). Studies on certain species of bacteria assigned to the genus *Chromobacterium*. *Journal of Bacteriology* **65**, 48-52.
GORRILL, R. H. & MCNEIL, E. (1960). The effect of cold diluent on the viable count of *Pseudomonas pyocyanea*. *Journal of General Microbiology* **22**, 437-442.
HALE, H. P. (1950). The production of 'amber whites' in shell eggs by freezing and thawing. *Journal of the Science of Food and Agriculture* **1**, 46-48.
HARDER, W. & VELDKAMP, H. (1971). Competition of marine psychrophilic bacteria at low temperatures. *Antonie van Leeuwenhoek* **37**, 51-63.
HARRIS, N. D. (1963). The influence of the recovery medium and incubation temperature on the survival of damaged bacteria. *Journal of Applied Bacteriology* **26**, 387-397.
HEINTZELER, I. (1939) Das Wachstum der Schimmelpilze in Abhängigkeit von den Hydraturverhältnissen unter verschiedenen Aussenbedingungen. *Archiv für Mikrobiologie* **10**, 92-132.
HESS, E. (1934). Effects of low temperatures on the growth of marine bacteria. *Contributions to Canadian Biology and Fisheries, Series C* **8**, 491-505.
HILL, J. E. & SUNDERLAND, J. E. (1967). Equilibrium vapour pressure and latent heat of sublimitation for frozen meats. *Food Technology Champaign* **21**, 1276-1278.
HOUGHTBY, G. & LISTON, J. (1965). Lethal cold shock of *Escherichia coli* K-12. *Bacteriological Proceedings* G. 36, 19.
INGRAHAM, J. L. (1958). Growth of psychrophilic bacteria. *Journal of Bacteriology* **76**, 75-80.
INGRAM, M. (1951). The action of cold on micro-organisms, in relation to food. *Proceedings of the Society for Applied Bacteriology* **14**, 243-260.
INGRAM, M. (1957). Micro-organisms resisting high concentrations of sugars or salts. In *Microbial Ecology*, eds Williams, R. E. O. & Spicer, C. C. The Seventh Symposium of the Society for General Microbiology. Cambridge: University Press.
INGRAM, M. (1958). Les propriétés physiologiques des levures osmophiles. *Revue des fermentations et des industries alimentaires* **14**, 23-33.
INGRAM, M. & KITCHELL, A. G. (1967). Salt as a preservative for foods. *Journal of Food Technology* **2**, 15.
JACKSON, H. (1974). Loss of viability and metabolic injury of *Staphylococcus aureus* resulting from storage at 5° C. *Journal of Applied Bacteriology* **37**, 59-64.
KENT, M. (1975). Fish muscle in the frozen state: time dependence on its microwave dielectric properties. *Journal of Food Technology* **10**, 91-102.
KERELUK, K. & GUNDERSON, M. F. (1959). Studies on the bacteriological quality of frozen meat pies. IV. Longevity studies on the coliform bacteria and enterococci at low temperature. *Applied Microbiology* **7**, 327-328.
KIM, K. J. & LARKIN, J. M. (1973). Production of mesophilic mutants from a psychrophilic *Bacillus*. *Canadian Journal of Microbiology* **19**, 1452-1454.
KISER, J. S. (1943). A quantitative study of the rate of destruction of an *Achromobacter* sp. by freezing. *Food Research* **8**, 323-326.
KISER, J. S. (1944). Effects of temperatures approximating 0° C upon growth and biochemical activities of bacteria isolated from mackerel. *Food Research* **9**, 257-267.
KROEMER, K. & KRUMBHOLZ, G. (1931). Untersuchungen über osmophile Sprosspilze I. Beiträge zur Kenntnis der Gärungsvorgänge und der Gärungserreger der Trockenberrenauslesen. *Archiv für Mikrobiologie* **2**, 352-410.
KROEMER, K. & KRUMBHOLZ, G. (1932). Untersuchungen über osmophile Sprosspilze. V. Das Verhalten von Sprosspilzen in Nährlösungen mit hohen Neutralsalzkonzentrationen. *Archiv für Mikrobiologie* **3**, 384-396.
LARKIN, J. M. & STOKES, J. L. (1966). Isolation of psychrophilic species of *Bacillus*. *Journal of Bacteriology* **91**, 1667-1671.

LARKIN, J. M. & STOKES, J. L. (1968). Growth of psychrophilic microorganisms at subzero temperatures. *Canadian Journal of Microbiology* **14**, 97-101.
LARKIN, E. P., LITSKY, W. & FULLER, J. S. (1955). Fecal streptococci in frozen foods. III. Effect of freezing storage on *Escherichia coli, Streptococcus faecalis* and *Streptococcus liquefaciens* inoculated into orange concentrate. *Applied Microbiology* **3**, 104-106.
LEDER, I. G. (1972). Interrelated effects of cold shock and osmotic pressure on the permeability of the *Escherichia coli* membrane to permease accumulated substrates. *Journal of Bacteriology* **111**, 211-219.
LOCHHEAD, A. G. & JONES, A. H. (1936). Studies of numbers and types of microorganisms in frozen vegetables and fruits. *Food Research* **1**, 29-39.
LONG, S. K. & WILLIAMS, O. B. (1959). Growth of obligate thermophiles at $37°C$ as a function of the cultural conditions employed. *Journal of Bacteriology* **77**, 545-547.
LUYET, B. J. (1962). Recent development in cryobiology and their significance in the study of freezing and freeze-drying of bacteria. *Proceedings of the Low Temperature Microbiology Symposium*. Camden, N. J.: Campbell Soup Co, 1962, 63-87.
LUYET, B. J. (1966). Anatomy of the freezing process in physical systems. In *Cryobiology*, ed. Meryman, H. T. London: Academic Press.
MAJOR, C. P., McDOUGAL, J. D. & HARRISON, A. P., Jr. (1955). The effect of the initial cell concentration upon survival of bacteria at $-22°C$. *Journal of Bacteriology* **69**, 244-249.
MATCHES, J. R. & LISTON, J. (1972a). Effects of incubation temperature on the salt tolerance of *Salmonella*. *Journal of Milk and Food Technology* **35**, 39-44.
MATCHES, J. R. & LISTON, J. (1972b). Effect of pH on low temperature growth of *Salmonella*. *Journal of Milk and Food Technology* **35**, 49-52.
MAZUR, P. (1966). Physical and chemical basis of injury in single-celled micro-organisms subjected to freezing and thawing. In *Cryobiology*, ed. Meryman, H. T. London: Academic Press.
MAZUR, P. (1970). Cryobiology: The freezing of biological systems. *Science*, New York **168**, 939-949.
McCLESKY, C. S. & CHRISTOPHER, W. N. (1941). Some factors influencing the survival of pathogenic bacteria in cold-pack strawberries. *Food Research* **6**, 327-333.
McFARLANE, V. H. (1942). Behaviour of microorganisms in fruit juices and in fruit juice-sucrose solutions stored at $-17.8°C$. *Food Research* **7**, 509-518.
MEAD, D. D., WESSMAN, G. E., HIGUCHI, K. & SURGALLA, M. J. (1960). Stability of cell suspensions of *Pasteurella pestis* at 5C and $-23C$. *Applied Microbiology* **8**, 55-60.
MERYMAN, H. T. (1974). Freezing injury and its prevention in living cells. *Annual Review of Biophysics and Bioengineering* **3**, 341-363.
MEYNELL, G. G. (1958). The effect of sudden chilling on *Escherichia coli*. *Journal of General Microbiology* **19**, 380-389.
MICHENER, H. D. & ELLIOTT, R. P. (1964). Minimum growth temperatures for food-poisoning, fecal-indicator, and psychrophilic microorganisms. *Advances in Food Research* **13**, 349-396.
MORICHI, T. (1969). Metabolic injury to frozen *Escherichia coli*. In *Freezing and drying of microorganisms*, ed. Nei, T. Baltimore: University Park Press.
MORITA, R. Y. (1975). Psychrophilic bacteria. *Bacteriological Reviews* **39**, 144-167.
MOSS, C. W. & SPECK, M. L. (1963). Injury and death of *Streptococcus lactis* due to freezing and storage. *Applied Microbiology* **11**, 326-329.
MOSS, C. W. & SPECK, M. L. (1966). Release of biologically active peptides from *Escherichia coli* at subzero temperatures. *Journal of Bacteriology* **91**, 1105-1111.
NEI, T. (1960). Effects of freezing and freeze-drying on micro-organisms. In *Recent research in freezing and drying*, eds Parkes, A. S. & Smith, A. U. Oxford: Blackwell.
NEI, T., ARAKI, T. & MATUSAKA, T. (1969). The mechanism of cellular injury by freezing in microorganisms. In *Freezing and drying of microorganisms*, ed. Nei, T. Baltimore: University Park Press.
NICKERSON, J. T. & SINSKEY, A. J. (1972). *Microbiology of Foods and Food Processing*.

New York: American Elsevier Publ. Co.
OHYE, D. F. & CHRISTIAN, J. H. B. (1966). Combined effects of temperature, pH and water activity on growth and toxin production by *Clostridium botulinum* types A, B and E. In *Botulism 1966. Proceedings of the Fifth International Symposium of Food Microbiology, Moscow,* eds Ingram, M. & Roberts, T. A. London: Chapman & Hall.
OHYE, D. F. & SCOTT, W. J. (1957). Studies in the physiology of *Clostridium botulinum* type E. *Australian Journal of Biological Sciences* 10, 85-94.
OHYE, D. F., CHRISTIAN, J. H. B. & SCOTT, W. J. (1966). Influence of temperature on the water relations of growth of *Clostridium botulinum* type E. In *Botulism 1966*. In *Proceedings of the Fifth International Symposium of Food Microbiology, Moscow,* eds Ingram, M. & Roberts, T. A. London: Chapman & Hall.
OLSON, J. C., Jr. (1970). U.S. Regulatory administration for control of microbiological health hazards in foods. In *Toxic Micro-organisms,* ed. Herzberg, M. Washington: United States Department of the Interior.
OLSEN, R. H. & JEZESKI, J. J. (1963). Some effects of carbon source, aeration and temperature on growth of a psychrophilic strain of *Pseudomonas fluorescens. Journal of Bacteriology* 86, 429-433.
PACKER, E. L., INGRAHAM, J. L. & SCHER, S. (1965). Factors affecting the rate of killing of *Escherichia coli* by repeated freezing and thawing. *Journal of Bacteriology* 89, 718-724.
PIVNICK, H. & BARNETT, H. (1965). Effect of salt and temperature on toxigenesis by *Clostridium botulinum* in perishable cooked meats vacuum packed in air-impermeable plastic pouches. *Food Technology, Champaign* 21, 1164-1167.
POSTGATE, J. R. & HUNTER, J. R. (1961). On the survival of frozen bacteria. *Journal of General Microbiology* 26, 367-378.
POSTGATE, J. R. & HUNTER, J. R. (1962). The survival of starved bacteria. *Journal of General Microbiology* 29, 233-363.
POSTGATE, J. R. & HUNTER, J. R. (1963). Metabolic injury in frozen bacteria. *Journal of Applied Bacteriology* 26, 405-414.
PRENTICE, G. A. & CLEGG, L. F. L. (1974). The effect of incubation temperature on the recovery of spores of *Bacillus subtilis* 8057. *Journal of Applied Bacteriology* 37, 501-513.
RAY, B. & SPECK, M. L. (1972). Metabolic process during the repair of freeze-injury in *Escherichia coli. Applied Microbiology* 24, 585-590.
RAY, B. & SPECK, M. L. (1973). Freeze-injury in bacteria. *CRC Critical Reviews in Clinical Laboratory Sciences* 4, 161-213.
RAY, B., JANSSEN, D. W. & BUSTA, F. F. (1972). Characterisation of the repair of injury induced by freezing *Salmonella anatum. Applied Microbiology* 23, 803-809.
RILEY, J. M. & SOLOWEY, M. (1958). Survival of *Serratia marcescens* in concentrates and suspensions stored at 5° C. *Applied Microbiology* 6, 233-235.
ROBERTS, T. A. (1974). Inhibition of bacterial growth in model systems in relation to the stability and safety of cured meats. In *Proceedings of an International Symposium on Nitrite in Meat Products, Zeist, 1973.* Wageningen, Holland: Pudoc.
ROBERTS, T. A. & HOBBS, G. (1968). Low temperature growth characteristics of clostridia. *Journal of Applied Bacteriology* 31, 75-88.
SATO, M. & TAKAHASHI, H. (1968a). Cold shock of bacteria. I. General features of cold shock in *Escherichia coli. Journal of General and Applied Microbiology* 14, 417-428.
SATO, M. & TAKAHASHI, H. (1968b). Two critical temperature zones in the cold shock of *Bacillus subtilis* and *Pseudomonas fluorescens. Agricultural and Biological Chemistry* 32, 259-260.
SATO, M. & TAKAHASHI, H. (1970). Cold shock of bacteria. IV. Involvement of DNA ligase reaction in recovery of *Escherichia coli* from cold shock. *Journal of General Applied Microbiology* 16, 279-290.
SCHMIDT, C. F., LECHOWICH, R. V. & FOLINAZZO, J. F. (1961). Growth and toxin production by type E *Clostridium botulinum* below 40° F. *Journal of Food Science* 26, 626-630.

SCOTT, W. J. (1962). Available water and microbial growth. *Proceedings of the Low Temperature Microbiology Symposium.* Camden, N. J.: Campbell Soup Co., 1962, 89-105.
SEGNER, W. P., SCHMIDT, C. F. & BOLTZ, J. K. (1966). Effect of sodium chloride and pH on the outgrowth of spores of type E *Clostridium botulinum* at optimal and suboptimal temperatures. *Applied Microbiology* **14**, 49-54.
SHERMAN, J. M. & ALBUS, N. R. (1923). Physiological youth in bacteria. *Journal of Bacteriology* **8**, 127-139.
SHIPP, H. L. (1957). Survival of *Salmonella enteritidis* in brine. *Proceedings of a Symposium on the The Microbiology of Fish and Meat Curing Brines.* London: H.M.S.O.
SPECK, M. L., RAY, B. & READ, R. B. (1975). Repair and enumeration of injured coliforms by a plating procedure. *Applied Microbiology* **29**, 549-550.
STILLE, B. (1948). Grenwertze der Relativenfeuchtigkeit und des Wassergehaltes getrockneter Lebensmittel für den mikrobiellen Befall. *Zeitschrift für Lebensmittel untersuchung und -forschung* **88**, 9-12.
STOREY, R. M. & STAINSBY, G. (1970). The equilibrium vapour pressure of frozen cod. *Journal of Food Technology* **5**, 157-163.
STRAKA, R. P. & STOKES, J. L. (1959). Metabolic injury to bacteria at low temperature. *Journal of Bacteriology* **78**, 181-185.
STRANGE, R. E. & DARK, F. A. (1962). Effect of chilling on *Aerobacter aerogenes* in aqueous suspension. *Journal of General Microbiology* **29**, 719-730.
STRANGE, R. E. & NESS, A. G. (1963). Effect of chilling on bacteria in aqueous suspension. *Nature, London* **197**, 819.
SWARTZ, H. M. (1971). Effect of oxygen on freezing damage. II. Physical chemical effects. *Cryobiology* **8**, 255-264.
TAI, P.-C. & JACKSON, H. (1969). Mesophilic mutants of an obligate psyhrophile, *Micrococcus cryophilus. Canadian Journal of Microbiology* **15**, 1145-1150.
TOMKINS, R. G. (1929). Studies on the growth of moulds. *Proceedings of the Royal Society, London, Series B* **105**, 375-401.
TRACI, P. A. & DUNCAN, C. L. (1974). Cold shock lethality and injury in *Clostridium perfringens. Applied Microbiology* **28**, 815-821.
UPADHYAY, J. & STOKES, J. L. (1962). Anaerobic growth of psychrophilic bacteria. *Journal of Bacteriology* **83**, 270-283.
VAN DEN BERG, L. (1968). Physiochemical changes in foods during freezing and subsequent storage. In *Low Temperature Biology of Foodstuffs.* eds Hawthorn, J. & Rolfe, E. J. London: Pergamon Press.
VAN ESELTINE, W. P., NELLIS, L. F., LEE, F. A. & HUCKER, G. J. (1947). Effect of rate of freezing on bacterial content of frozen vegetables. *Food Research* **13**, 271-280.
WILLIAMS, O. B. & REED, J. M. (1942). The significance of the incubation temperature of recovery cultures in determining spore resistance to heat. *Journal of Infectious Diseases* **71**, 225-227.
WODZINSKI, R. J. & FRAZIER, W. C. (1960). Moisture requirements of bacteria. I. Influence of temperature and pH on requirements of *Pseudomonas fluorescens. Journal of Bacteriology* **79**, 572-578.
WODZINSKI, R. J. & FRAZIER, W. C. (1961a). Moisture requirements of bacteria. II. Influence of temperature, pH, and malate concentration on requirements of *Aerobacter aerogenes. Journal of Bacteriology* **81**, 353-358.
WODZINSKI, R. J. & FRAZIER, W. C. (1961b). Moisture requirements of bacteria. III. Influence of temperature, pH, and malate and thiamine concentration on requirements of *Lactobacillus viridescens. Journal of Bacteriology* **81**, 359-365.
WODZINSKI, R. J. & FRAZIER, W. C. (1961c). Moisture requirements of bacteria. IV. Influence of temperature and increased partial pressure of carbon dioxide on requirements of three species of bacteria. *Journal of Bacteriology* **81**, 401-408.
WODZINSKI, R. J. & FRAZIER, W. C. (1961d). Moisture requirements of bacteria. V.

Influence of temperature and decreased partial pressure of oxygen on requirements of three species of bacteria. *Journal of Bacteriology* **81**, 409-415.

ZAWADSKI, Z. (1973). Variability of superficial microflora of frozen meat. *Medycyna weterynaryjna* **29**, 171-173.

Thermal Injury and Inactivation in Vegetative Bacteria

R. I. TOMLINS

Department of Applied Biology & Food Science, Polytechnic of the South Bank, London, England

AND

Z. J. ORDAL

Departments of Food Science & Microbiology, University of Illinois, Urbana, Illinois, U.S.A.

CONTENTS

1. Thermal Resistance 153
   (a) Introduction 153
   (b) Measurement of heat resistance 153
   (c) Factors affecting heat resistance 157
2. Thermal injury and recovery 168
   (a) Introduction 168
   (b) Estimation of thermal injury by differential plating 169
   (c) Ribosome and rRNA degradation and resynthesis during thermal injury and recovery. 173
   (d) Cytoplasmic membrane alterations during thermal injury and recovery. . . 179
   (e) Metabolic damage during thermal injury and recovery 181
   (f) DNA damage and repair during thermal injury and recovery 182
3. References 184

1. Thermal Resistance

(a) *Introduction*

THE LITERATURE on various aspects of thermal injury and death in bacteria is considerable and includes several excellent reviews (Allwood & Russell, 1970; Brown & Melling, 1971; Ingram, 1971; Corry, 1973). This paper reviews the recent published work on thermal inactivation, and thermal injury and recovery in both non-spore forming bacteria and vegetative cells of spore forming bacteria. Attention has been centred on some of the important cultural and heating conditions which affect thermal resistance, and our current knowledge of mechanisms involved in thermal injury and recovery.

(b) *Measurement of heat resistance*

Interest regarding the order of thermal death of bacteria has stimulated many reports, advancing explanations for the experimental shapes of survival curves (Schmidt, 1957; Vas & Proszt, 1957; Hansen & Rieman, 1963; Moats, 1971).

The determination of thermal resistance of bacterial cells is usually achieved by exposing the cells to a variety of temperatures and determining, by viable plate counts, the rate of survival during heating. With bacteria the death rate is often logarithmic over a substantial portion of the curve, and this has been interpreted to show that death is a first-order reaction, caused by the inactivation of one critical site per cell (Charm, 1958; Stumbo, 1965). However, in practice, survival curves are of four main types, containing shoulders and tails as well as the logarithmic portions (Fig. 1). Curve A is a common type of survival curve in which there is an initial lag in death rate, producing a shoulder, followed by a logarithmic portion. Curve C is similar to curve A but with a tail. Curve D, which is concave, generally indicates a population of cells with a distribution of heat resistances (Moats, 1971).

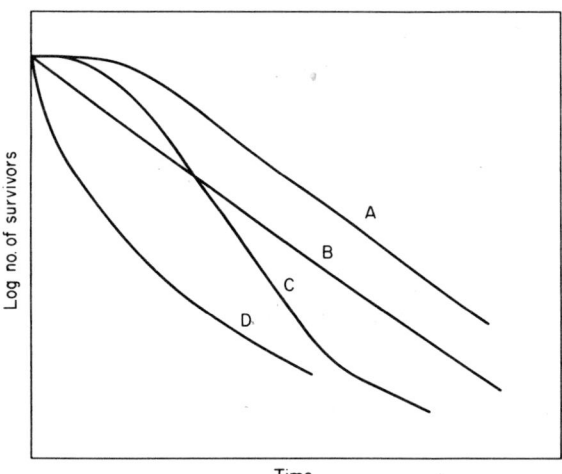

Fig. 1. Non-exponential survivor curves. Curve A, convex survivor curve — initial lag in death rate followed by an approximately logarithmic death rate, commonly observed; curve B, logarithmic death rate as described in textbooks; curve C, similar to curve A but with tails; curve D, commonly observed with cells in logarithmic growth phase and considered to indicate a heterogeneous population. From Moats (1971).

The two parameters which are used to describe the thermal resistance of bacterial cells are obtained from the logarithmic portion on the survival curve. They are called the D value and z value. The D value or decimal reduction time represents the time required, at a particular temperature, to reduce the surviving fraction of organisms 10-fold. At a constant D value the time for a given probability of inactivation is directly proportional to the initial numbers of organisms present. The z value is the number of degrees required to alter the D value 10-fold. Whereas the D value is susceptible to variation depending on the test conditions (Table 2), the z value is reasonably constant for a given organism over a range of test conditions.

The observation of non-logarithmic survival curves with heated bacteria has been reported for many different non-sporing bacteria, (White, 1953; Greenburg & Silliker, 1961; Dabbah et al., 1971a,b) (Fig. 2). This area has been reviewed by Moats et al. (1971a), who concluded that the complex survivor curves obtained experimentally represented populations of different heat resistances in a single culture of bacteria. The variation in heat resistance observed was reported to be a physiological rather than a genetic effect, since subcultures of survivors were no more resistant to heat than parent cultures. This observation is at variance with the work of other authors (Nagy, 1969; Corry & Roberts, 1970; Duitschaever & Jordan, 1974) who have reported that subcultured survivors of heating treatments showed increased heat resistances.

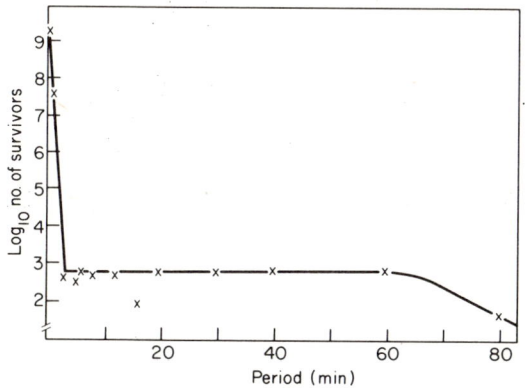

Fig. 2. An extreme example of tailing. *Escherichia coli* grown 24 h at 35° in trypticase-soy broth, heated at 62.5° in 0.1 N pH 7.0 phosphate buffer in 1 ml sealed ampoules, and plated on trypticase-soy agar (TSA). After an initial reduction of count of 6 log cycles in 2.7 min, no further kill occurred until after 60 min heating and survivors were still found after 80 min when the experiment was terminated. From Moats et al. (1971a).

The initial shoulder observed on survival curves can be explained by the gradual accumulated effects of the heating, resulting in injury but not loss in viability. Clumping of cells also was reported by Stumbo (1965) to explain such shoulders, and this was shown to be so for cells of *Streptococcus faecalis* heated in milk (Hansen & Rieman, 1963). Tailing of the survivor curves has been frequently noted by Moats and his co-workers, especially when curves were carried through 6-9 log cycles, and they have criticized the validity of D values obtained from cultures with such non-logarithmic death rates as misleading (Dabbah et al., 1971a,b; Moats et al., 1971a). Some authors (Rahn, 1945; Schmidt, 1957), have argued that assuming thermal death to be logarithmic was convenient in assessing survival of organisms under various conditions. However, Moats et al. (1971a) agreed with Ott et al. (1961) that thermal death time

Table 1

Heat resistance of psychrophilic/psychrotrophic bacteria

Organism	Heating medium	pH	Temp (°)	D value (min)	z value (°)	Reference
Pseudomonas fragi	Skim milk	—	52.0	2.0 – 3.0	—	Luedecke & Harmon (1966)
Ps. fragi	Cheese		54.5	2.0	11	Collins (1961)
Ps. fluorescens	Nutrient agar	—	55	1 – 2	—	Heather & Vanderzant (1957a,b)
Ps. fluorescens	Phosphate buffer (0.067M)	7.0	53	5 – 6	—	Calhoun & Frazier (1966)
Ps. fluorescens	Glutamate medium	7.0	36	60	—	Gray (1972)
Ps. fluorescens	Menstruum alone	—	60	3.2	7.5	Sevostyanova et al. (1971)
	Menstruum + 0.05% sorbic acid	—	60	1.6	4.5	
Microbacterium thermosphactum	Skim milk (10%)	—	50	2.5	—	Davidson et al. (1968)
Vibrio marinus	Seawater broth		31	11	9.5	Haight & Morita (1966)
Serratia spp.	3 day culture		30	6	9.5	Hagen et al. (1964)

measurements or F values, were more valid than D values as they did not make assumptions about the logarithmic nature of death.

The data reported in Table 1 illustrate the heat resistance of psychrotrophic and psychrophilic organisms reported in the literature. On the evidence available strict psychrophiles have much lower heat resistances than psychrotrophs, but generally there is a dearth of information on the heat resistances of these organisms.

The data reported in Table 2 illustrate typical heat resistances observed for non-sporing mesophilic bacteria. There is more information available for this group of bacteria and the information presented demonstrates the variability in observed D values with the experimental conditions used. Most of these organisms are important from a food spoilage and/or public health viewpoint and should normally be killed during the pasteurizing heat treatment given to foodstuffs. As Ingram (1971) has pointed out, D values are of the order of 1 min at 60° for most mesophiles at high water activities although *Salmonella senftenburg* 775W, *Microbacterium* spp. and many of the faecal streptococci have larger D values, of the order of 10 min at 60°. Likewise, the z value for many mesophilic bacteria is $c.$ 5° at high water activities, but at intermediate water activities (a_w 0.80-0.99) the z values of *Salmonella* spp. and *Escherichia coli* were increased as water activity decreased (Dega *et al.*, 1972; Gibson, 1973). The faecal streptococci are again unique, since their reported z values are high, even up to 20° in some cases (Zivanovic *et al.*, 1965).

(c) *Factors affecting heat resistance*

The important parameters which are known to affect the heat resistance of bacterial cells have been reviewed by Hansen & Rieman (1963) and more recently by Brown & Melling (1971).

(i) *Species variation and growth conditions*

The variation in the thermal resistance of different species of bacteria with their maximum growth temperature demonstrates that organisms with high growth temperatures have high heat resistances and vice versa (Table 3). A heat treatment of 10 min at 40° would eliminate true psychrophiles, heating at 50° would eliminate psychrotrophs and sensitive mesophiles, whilst heating at 60° would destroy all except the most resistant mesophiles, such as faecal streptococci which require 70° for their destruction (Ingram, 1971).

There is sufficient information to indicate that the heat resistance of an organism is in some way dependent on the composition of the growth medium. Gyllenberg & Sederholm (1961) reported that with a strain of *Pseudomonas* grown in a minimal medium, the presence of an organic nitrogen source in the medium upon subculture produced a population of cells with a greatly increased

Table 2

Heat resistance of vegetative mesophilic bacteria

Organism	Heating medium		a_w	pH	Temp (°)	D value (min)	z value (°)	Reference
Escherichia coli	Ringer soln.		–	7.0	55	4	–	Lemcke & White (1959)
E. coli	Nutrient broth		–	–	56	4.5	4.9	Chambers, Tabak & Kabler (1957)
E. coli	Tryptic soy broth		–	6.9	52	25.6	–	Stiles *et al.* (1973)
			0.99	6.9	57.2	1.2	–	
			0.96	6.9	57.2	10.1	–	
E. coli	Aqueous sucrose		0.93	6.9	57.2	33.9	–	Goepfert *et al.* (1970)
			0.90	6.9	57.2	46.5	–	
			0.87	6.9	57.2	43.7	–	
		10%	–	–	58	1.4	4.6	
E. coli	Milk solutions	30%	–	–	58	2.4	4.9	Dega *et al.* (1972)
		42%	–	–	58	7.3	6.3	
		51%	–	–	58	13.5	7.9	
E. coli	Milk		–	–	82.2	5.7×10^{-4}	6.5	Evans *et al.* (1970)
Pseudomonas aeruginosa	Nutrient agar		–	–	55	1.9	–	Nelson (1943)
Salmonella typhimurium	Phosphate buffer (0.1M)		–	6.5	65	0.0562	–	Corry (1974)
	same plus glucose/sucrose/		0.95	6.5	65	0.14 – 6.95	–	
	glycerol mixtures		0.85	6.5	65	0.26 – 31.5	–	
Salmonella (6 spp.)	Whole milk		–	–	62.8	4.5	5.2	Read *et al.* (1968)
			0.995	–	65.5	0.05	17.05	
Salm. typhimurium	Aqueous sucrose/glucose		0.98	–	65.5	0.066	14.49	Gibson (1973)
	mixtures		0.94	–	65.5	0.833	15.34	
			0.90	–	65.5	2.66	20.08	
			0.80	–	65.5	3.5	31.67	
		10%	–	–	57.2	1.4	4.0	
Salm. typhimurium	Milk solutions	30%	–	–	57.2	4.9	4.6	Dega *et al.* (1972)
		42%	–	–	57.2	9.8	6.0	
		51%	–	–	57.2	26.6	6.8	

Organism	Medium	a_w	pH	Temp (°C)	D (min)	z (°C)	Reference
Salmonella (7 spp.)	Sucrose soln. 15.4%	0.99	6.9	57.2	00.8 – 1.2	—	Goepfert et al. (1970)
	39.6%	0.96	6.9	57.2	5.3 – 35.5	—	
	51.3%	0.93	6.9	57.2	14.3 – 68.7	—	
	58.6%	0.90	6.9	57.2	21.1 – 80.0	—	
	63.7%	0.87	9.2	57.2	26.6 – 95.0	—	
Salm. typhimurium	Egg albumin	—	—	59.0	0.15	4.2	Corry & Barnes (1968)
Salm. typhimurium	Egg products	—	—	52–60	—	4.2 – 5.3	Garibaldi et al. (1969)
Salm. typhosa	Milk	—	—	82.2	1.1×10^{-5}	5.6	Evans et al. (1970)
Salm. anatum	Milk chocolate	—	—	71	1200	—	Barrile & Cone (1970)
	Milk chocolate + 2% H$_2$O	—	—	71	240	—	
Salm. typhimurium	Milk chocolate	—	—	70	816	19.0	Goepfert & Biggie (1968)
Salm. oranienburg	Egg yolk + 10% NaCl (no storage)	—	—	57	19	8.3	Cotterill & Glauert (1971)
	Egg yolk + 10% NaCl	—	—	57	45	6.8	
	Egg albumin	—	9.2	59.0	1.25	4.5	Corry & Barnes (1968)
	Skim milk	—	—	60	10.8	6.0	Thomas et al. (1966)
	Egg albumin	—	9.0	56.7	3.1	—	Kohl (1971)
	Egg albumin + 0.75% HMP	—	9.5	56.7	0.34	—	
	Phosphate buffer (0.1M)	—	6.5	65	0.290	—	
	ditto plus sucrose (% w/v) 30 : 70	—	6.5	65	1.4 : 43	—	Corry (1974)
	glucose (% w/v) 30 : 70	—	6.5	65	2.0 : 17.0	—	
	glycerol (% w/v) 30 : 70	—	6.5	65	0.95 : 0.7	—	
Salm. senftenberg (775W)	Aqueous sucrose/glucose mixtures	0.99	—	60	7.2	12.27	Gibson (1973)
		0.98	—	60	7.2	10.87	
		0.94	—	60	14.5	13.46	
		0.90	—	60	10.4	16.48	
		0.85	—	60	11.4	19.70	
	Aqueous sucrose 15.4%	0.99	6.9	57.2	14.5	—	Goepfert et al. (1970)
	39.6%	0.96	6.9	57.2	48.3	—	
	51.3%	0.93	6.9	57.2	55.0	—	
	58.6%	0.90	6.9	57.2	62.0	—	

Table 2–*cont.*

Organism	Heating medium	a_w	pH	Temp (°)	D value (min)	z value (°)	Reference
Salm. senftenberg (775W)	Milk chocolate	–	–	70	440	18	Goepfert & Biggie (1968)
Salm. (221 strains)	Heart infusion broth	–	7.4	60	0.2 – 6.5	–	
Salm. senftenberg (775W)	Heart infusion broth	0.99	7.4	60	6.1	6.8	Baird-Parker *et al.* (1970)
	+ NaCl	0.90	7.4	60	2.7	13	
	+ glycerol	0.90	7.4	60	2.5	7.2	
	+ sucrose	0.90	7.4	60	75.2	8.9	
Staphylococcus aureus	Phosphate buffer (0.1M)	–	7.2	52	3 – 32	–	Hurst *et al.* (1974)
Staph. aureus	Custard or pea soup	–	–	60	7.8	4.5	Thomas *et al.* (1966)
Staph. aureus	Phosphate buffer (0.067M)	–	7.0	58	1.8	–	Walker & Harmon (1966)
Staph. aureus	Milk	–	–	82.2	1.2×10^{-3}	5.1	Evans *et al.* (1970)
Lactobacillus brevis (3 strains)	TSB + NaCl	0.95		65	3.9 – 9.4	–	Vrchlabsky & Leistner (1971)
L. viridescens	TSB + salts mixture	0.95		65	4.4 – 9.7	–	
L. plantarum	TSB + sucrose	0.95		65	4.7 – 8.1	–	
Lactobacillus (2 spp.)	Tomato juice broth		7.0	62.8	3 & 12	–	Niven *et al.* (1954)
Lactobacillus (2 spp.)	Tomato serum	0.958	4.9	61.0	3.3	–	Casolari & Campanini (1973)
Leuconostoc dextranicum	Tomato serum	0.958	4.9	61.0	1.4	–	
Lactobacillus casei	Tomato juice	–	4.5	70.0	4.0	11.5	Hernandez & Feria (1971)
L. plantarum	Tomato juice	–	4.5	70.0	11.0	12.5	
L. delbrueckii	Tomato juice	–	4.5	70.0	5.0	10.0	
Streptococcus faecalis	Pork (3% fat + curing salts)	–	6.0	60	8.0	9.0	Zakula (1969)
	Pork (3% fat + curing salts + 0.4% poly PO$_4$ + 40% lard)	–	6.0	60	10.0	10.0	
Strep. faecalis	Chicken-a-la-king	–	–	60	13.5	6.8	Ott *et al.* (1961)
Strep. faecalis	Citrate–phosphate buffer (1.0M)	–	6.6	60	12.2	–	White (1963)

Organism	Medium	pH	Temp (°C)			Reference
Strep. faecalis	Phosphate buffer (0.1M)	—	—	—	—	Clark *et al.* (1968)
Strep. faecalis	Ringer soln.	—	—	—	—	White (1963)
	Skim milk	—	—	—	—	
Strep. faecalis var. *liquefaciens* (7 strains)	Broth	—	—	—	—	
Strep. faecalis var. *zymogenes* (3 strains)	Broth	—	—	—	—	
Strep. faecium	Broth	—	—	—	—	
Group I 2 str		6.0	60	20.0 – 1.0	19.0	Zivanovic *et al.* (1965)
II 2 str		6.0	60	0.85 – 1.0	17.0	
III 4 str		6.0	60	3.3 – 10		
IV 4 str		6.0	60	5.0	3	
V 4 str		6.0	60	11.0	3.5	
VI 4 str		6.0	60	15	3	
		6.0	60	11	2.5	
			60	10	—	
			60	8	—	
			60	15		
			60	7		
Streptococcus (Grp D) (milk isolates)	Skim milk	—	62.8	2.56	—	Lenistea *et al.* (1970)
Strep. faecium	Skim milk	—	62.8	10.3	—	
Strep. durans	Skim milk	—	62.8	7.5	—	
Strep. faecalis	Skim milk	—	62.8	3.5	—	
Strep. bovis	Skim milk	—	62.8	2.6	—	
Microbacterium lacticum	Skim milk	—	70	4.0	—	Davidson *et al.* (1968)
Micro. flavum	Skim milk	—	65	2.0	—	

heat resistance. The addition of glucose or NaCl, at the same water activities, to a trypticase-soy broth growth medium increased the heat resistance of *E. coli* and *P. fluorescens,* but not that of *Staphylococcus aureus* (Calhoun & Frazier, 1966). The authors concluded that the water activity and the permeability of

Table 3
Summary of approximate heat resistances of some non-sporing mesophiles, psychrotrophs and psychrophiles

Organism	D value (min) at		
	60°	50°	40°
Mesophiles with typical heat resistance			
Salmonella spp.			
E. coli			
Lactobacillus	1		
Staph. aureus			
Mesophiles with atypically high heat resistance			
Microbacterium	10		
Salm. senftenberg	10		
faecal streptococci	5 – 20		
Psychrotrophs		1 – 5	
Psychrophiles			1

Adapted from Ingram (1971)

these solutes to the cells was responsible for the observed variations in heat resistance. The addition of fat, up to 20%, to a skim-milk growth medium, reduced the heat resistance of *P. fragi*. The heat resistance also varied with the incubation temperature, being higher with cells grown at 7° than at 25°. (Luedecke & Harmon, 1966). This observation agreed with the work of Sherman & Cameron (1934) on *E. coli*. Dega *et al.* (1972) have also reported that the growth temperature of *Salm. typhimurium* greatly affected its heat resistance. However, as heat resistance is affected by the stage of growth, in order to make valid comparisons the stage of growth as well as the temperature must be known (White, 1953).

It has been variously reported that lag phase cells were more heat resistant than exponentially growing cultures, and it has been proposed that for *E. coli* this could be attributed to variations in the cell wall during the growth cycle (Hoffman *et al.,* 1966). Strange & Shon (1964) reported that in cultures of *Enterobacter aerogenes* grown in chemically defined medium, the heat resistance of exponential phase cells compared to stationary phase cells was greater after washing in distilled water, but less after washing in phosphate buffer. They concluded from these and other experiments, that the higher magnesium ion concentrations of exponential phase cells after washing in water exerted a protective effect which increased the heat resistance. Goepfert et al. (1970) reported that the heat resistance of salmonellae washed in 0.1% peptone water had D values 25-75% less than those of unwashed cells, heated under identical conditions.

Beuchat & Lechowich (1968) demonstrated that maximum heat resistance in *Strept. faecalis* was obtained with cells in the stationary phase of growth. Similarly, stationary phase cells of *Vibrio marinus* are more heat resistant than log phase cells, (Kenis & Morita, 1968; Griffiths & Haight, 1973). Recently, Hurst et al. (1974) have reported on the heat resistance of *Staph. aureus* of different physiological ages. They observed D_{55} values of 3.0 min for log phase cells and 9.0 min for late stationary phase cells, but strikingly, they observed a 10-fold increase in heat resistance, D_{55} equals 32 min, at the beginning of the stationary phase (Fig. 3). Interestingly also, the ability of cells to form colonies on a high salt medium after heating did not vary much as the culture aged. They speculated from their results that the significant changes reported in the membrane phospholipids observed during the growth cycle could contribute to their observed variations in heat resistance.

(ii) *Effect of heating menstruum composition*

It is well documented that bacterial cells are more resistant to dry heat than moist heat. Early work on the heat resistance of vegetative bacterial cells established that a reduction of water activity due to the presence of sugars, and especially sucrose, in the heating menstruum increased survival (Fay, 1934; Baumgartner, 1938). More recently, reports have been published which have demonstrated that the mechanism of this protective effect is nore complex than was thought originally (Calhoun & Frazier, 1966; Baird-Parker et al., 1970; Goepfert et al., 1970; Vrchlabsky & Leistner, 1970, 1971; Gibson, 1973; Corry, 1974; Horner & Anagnostopoulos, 1975). The effects of water activity on the thermal resistance of micro-organisms has recently been reviewed by Corry (1973).

Calhoun & Frazier (1966) demonstrated that sodium chloride in the heating menstruum protected *E. coli* and *Staph. aureus* but sensitized *Ps. fluorescens;* whereas glucose protected *E. coli* and *Ps. fluorescens* but hastened the thermal destruction of *Staph. aureus.*

The effect of sodium chloride at various water activities on the heat resistance of salmonellae has been reported by Baird-Parker et al. (1970). They showed that over the limited a_w range 0.85-0.98, NaCl in broth increased the heat resistance of heat-sensitive strains. This effect was also demonstrated on the addition of glycerol. Sucrose increased the heat resistance of all strains although

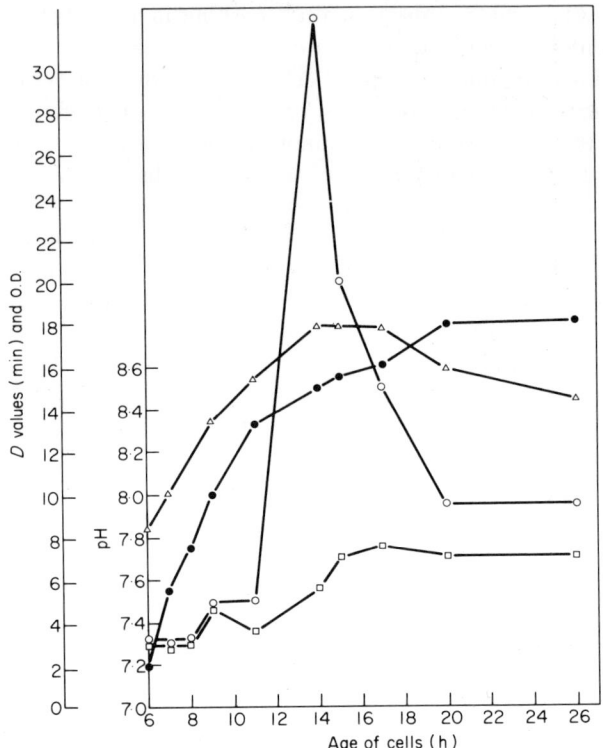

Fig. 3. Response of *Staphylococcus aureus* of different ages to sub-lethal heating. ●, pH; △, OD; ○, decimal reduction times (D_{52}) of cultures counted on TSA; □, pseudo-decimal reduction times (D'_{52}) of cultures counted on TSAS (high salt content). From Hurst et al. (1974).

the heat resistance of heat-sensitive strains was increased proportionately more than heat-resistant strains. The authors concluded that there was no direct relationship between heat resistance and the water activity of the medium. At the same time Goepfert et al. (1970) produced a report on the effect of water activity on the heat resistance of eight strains of salmonellae. Their results concurred with those of Baird-Parker et al. (1970), as sucrose was shown to give the greatest degree of protection compared with fructose, glycerol and sorbitol, (Table 2).

The effects of sucrose, glucose, sorbitol, fructose and glycerol on the heat resistance of salmonellae has recently been investigated by Corry (1974). The heat resistances of the three strains of *Salmonella* tested were increased as the solute concentration increased, and as reported by Goepfert et al. (1970) and Baird-Parker et al. (1970), no linear relationship between water activity and heat resistance was observed. However, except for glycerol, all solutes showed a linear relationship between log D_{65} and solute concentration (%w/w).

At any given water activity, sucrose exerted the most, and glycerol the least protective effect of all the solutes tested (Table 2). With sucrose–glycerol and sucrose–glucose mixtures, the heat resistance depended on both the concentration (%w/w) and the a_w (Fig. 4). It was concluded that the different protective effects observed were probably due to the extent to which the solute was able to penetrate the cell membrane. Glycerol caused no plasmolysis and sucrose, severe plasmolysis; caused by the other solutes being intermediate. The replacement of cytoplasmic water by glycerol was therefore less protective to the cell than the dehydration caused by sucrose.

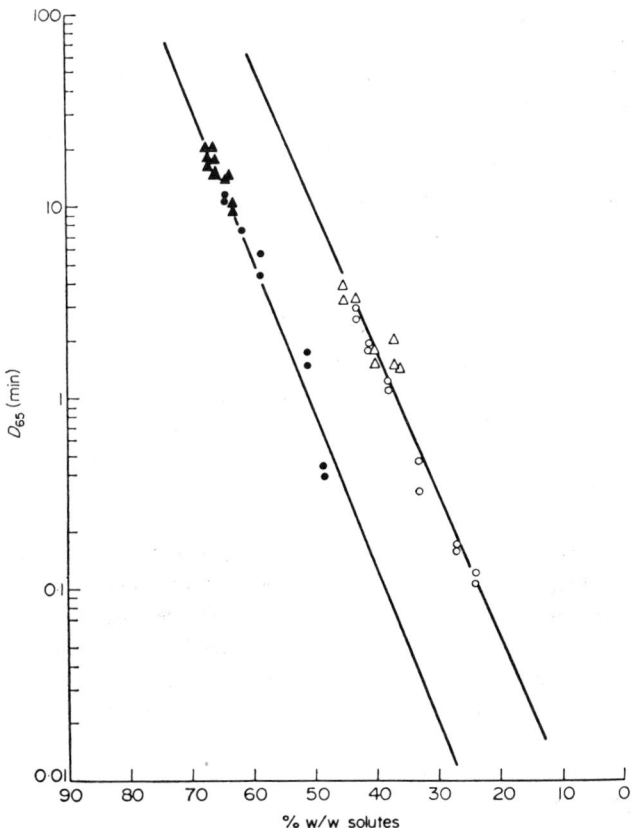

Fig. 4. The effect of total solute concentration on the heat resistance of *Salmonella typhimurium* 7M 4087 in sucrose–glycerol mixtures of: ●, a_w = 0.85; ○ a_w = 0.95, or in sucrose–glucose mixtures of: ▲, a_w = 0.85; △, a_w = 0.95. From Corry (1974).

Gibson (1973) has also investigated the effects of sucrose–glucose mixtures on the heat resistance of two strains of *Salmonella* and also some yeasts. The D and z values reported for various water activities (Table 2) did not correlate closely with the values reported by Corry (1974), and may have reflected the different experimental conditions used. The addition of 1% glucose to a 40% sucrose solution was shown to reduce the heat resistance of *Salm. montevideo* by c. 50% (Foster *et al.*, 1970).

Besides sugars, polyols and NaCl other chemicals added to the heating menstruum have been shown to have a significant effect on the heat resistance of bacteria. Moats *et al* (1971a,b) have reported on the survival of *Salm. anatum* heated at 55° for 30 min in menstrua containing various chemical constituents found in foods. Compared to 0.1 M phosphate buffer (pH 7.0) most protection against heat was given by commercially sterilized whole milk, on the other hand raw milk gave substantially less protection. Trypticase-soy broth and other peptide mixtures protected against heat. Results from carbohydrates were variable since sucrose increased, but glucose decreased, heat resistance. Cysteine, glutathione and sodium citrate also decreased heat resistance and pure proteins had little effect. The water activity of the menstruum also had little effect under the test conditions used. The authors concluded from their results that protection from heat probably resulted from complexing of the added substances with the heat sensitive proteins of the cells.

The influence of sodium nitrite and ascorbic acid on the death rate of stationary phase *Strep. faecium* and *Strep. faecalis* was investigated by Greenburg & Silliker (1961). At 148°F death rates were independent of added nitrite, but at 155°F and 158°F death rates were increased by the presence of nitrite. However, this effect was abolished by the addition of ascorbic acid.

Phosphate and polyphosphates have been shown to decrease the thermal resistance of bacteria when present in the heating menstruum (Oluski, 1963; Kelch & Buehlmann, 1958). Garibaldi *et al.* (1969) have reported on the effect of a commercial polyphosphate mixture on the heat resistance of *Salm. typhimurium* and *Salm. senftenburg* in egg white, The addition of 1% of polyphosphate was shown to reduce the $D_{52.5}$ value at any pH value by a factor of 2–6. Similar results were obtained by Kohl (1971) who showed that hexametaphosphate added to egg white increased the thermal destruction of the four *Salmonella* spp. tested, and on these results proposed a new pasteurization process for egg white. Zakula (1969) has reported on the effect of three commercial polyphosphate samples in homogenized meat systems on the heat resistance of *Micrococcus candidus* and *Strep. faecalis*. The reduction in thermal resistance due to the added polyphosphate was reversed in the presence of tallow or lard, which had a protective effect and increased heat resistance (Table 2). The antimicrobial effects of phosphates in foods has recently been reviewed by Hargreaves *et al.* (1972).

Recently, Kniewallner & Prandl (1974) have reported on the effect of 147 different chemcials on the heat resistance of *Staph. aureus* and *Strep. faecium,* as well as two spore-forming bacteria. The temperatures used for the non-sporing organisms was 50-75° and the heating menstrua were dextrose broth, emulsions, and meat–fat–water mixtures. Many of the compounds that decreased heat resistance depended on their ability to decrease the pH of the system. Of those chemcials that did not fall into this group 5-diazo-uracil was most effective, since 5 p/m in a meat–fat–water mixture reduced the D value of *Strep. faecalis* by 50-60%. Dihydronorguaiaretic acid, lauryl gallate, propyl gallate, lysozyme, sodium pyrophosphate and sodium lauryl sulphate were also reported to be effective against *Strep faecium* in glucose broth and to some extent against *Staph. aureus*. Unfortunately no actual D values were reported so comparison with other work was not possible. Sevostyanova *et al.* (1971) have reported on the isolation of a *Ps. fluorescens* strain from canned cucumbers, which had an unusually high heat resistance. The heat resistance of the organisms in the presence of sorbic acid was studied. At sorbic acid concentrations of 0.025 and 0.05% the D_{50} value was not affected, however, at 55° and 60° both the D values and z values were reduced significantly (Table 1). A sorbic acid concentration of 0.01% gave approximately the same results irrespective of temperature.

The effect of the water activity of a range of foods on the heat resistance of vegetative bacteria has also been investigated. Cotterill & Glauert (1969, 1971) investigated the effect of NaCl, and sucrose in whole egg or egg yolk, on the heat resistance of *Salm. oranienburg*. Maximum heat resistance was found with 10% of NaCl and 10% of sucrose in both products, the maximum concentrations tested. Maximum heat resistance was achieved by 4% NaCl and 10% sucrose in trypticase-soy broth. Storage of the salmonella in salted egg yolk for 8 h prior to heating gave maximum heat resistance but on prolonged storage the heat resistance was decreased (Table 2). It was therefore suggested that salted yolk should be pasteurized as soon as possible after the addition of salt.

The protective effect of sucrose in raw skimmed milk, on the heat resistance of *Staph. aureus* was reported by Kadan *et al.* (1963). Thirty minutes at 60° eliminated survivors if the sucrose concentration was $< 14\%$. When the sucrose concentration was $> 14\%$ the number of survivors increased as the concentration of added sucrose was increased further. Dega *et al.* (1972) extended the observation of McDonough & Hargrove (1968) and reported on the heat resistance of *E. coli, Salm. typhimurium* and *Salm. alachura* in concentrated skimmed milk solutions (10-51% w/w). Their results showed that *E. coli* was relatively more heat resistant at 10% of solids, but less heat resistant at 42 and 51% of solids, than the two salmonella strains. Heating the cells under reduced pressure greatly diminished the D value of *E. coli* and *Salm. typhimurium* and in addition virtually eliminated the protective effect of the higher levels of milk solids. Shannon *et al.* (1970) reported that the addition of extra milk solids or fat to skimmed milk did not result in the protection of the streptococcus species tested. Lenistea *et al.* (1970) have also investigated the heat resistance in skimmed milk, of 48 group D streptococci, originally isolated from commercially pasteurized milk. At any given test temperature the D values obtained from these strains was markedly variable. The 11 most heat-resistant strains had D_{95} values which ranged from 0.62 to 0.89 min. The plating medium used after heating, for the enumeration of survivors, was shown to affect significantly the counts obtained.

The heat resistance of salmonellae in chocolate is much greater than that observed in environments with high water activities (Goepfert & Biggie, 1968; Barrile *et al.*, 1970; Goepfert, 1972). Barrile & Cone (1970) have shown that the D_{71} value for *Salm. anatum* in milk chocolate was reduced from 20 h to 4.0 h by the addition of 2% of water (Table 2). The D values of *Salm. anatum* in milk chocolate was also shown to depend on the level of contamination. With an inoculum of 30 salmonella/g the D_{71} value was 24 h, but with an inoculum of 10^4-10^5 salmonella/g the D_{70} value was 60-90 min.

Zakula *et al.* (1970) have investigated the thermal resistance of *Stept. faecalis* and *M. candidus* as well as some spore-formimg bacteria in blended pork and beef systems. The D and z values obtained are presented in Table 2. Generally, heat resistance was greater in pork containing 3.0% of fat than in beef with 1.5% of fat. Zaleski *et al.*, (1971) have demonstrated that the heat resistance of *Staph. aureus* in hydrated soy bean oil was markedly reduced on the addition of 1-3 ml of water/1.

Hernandez & Feria (1971) investigated the thermal resistance of a range of micro-organisms isolated from spoiled canned tomatoes. The non-sporing bacteria identified were *Lactobacillus casei, L. plantarum, L. delbrueckii* and an *Enterobacter* sp., and their thermal death times ranged from 1.5 to 11 min at 70°.

The effect of pH of the heating menstruum on the heat resistance of bacteria has been discussed by Hansen & Rieman (1963) and Ingram (1971), but there seems to be few recently published reports on this subject. Generally, bacteria show maximum heat resistance over the range pH 6.0-8.0 although this can be altered by varying the growth and testing conditions. White (1963) demonstrated that when *Strep. faecalis* was heated at 60° in phosphate and citrate buffers at various pH values, maximum survival was obtained at pH 6.8, but on either side of this value there was a sharp decrease in survival. Similar results have also been reported for *Enterobacter aerogenes* (Strange & Shon, 1964).

2. Thermal Injury and Recovery

(a) *Introduction*

Information relating to the heat resistance of a wide range of vegetative bacteria, and how heat resistance is affected by environmental changes, is of practical importance to the food processor although it tells us little about the nature of heat resistance or the identity of important molecules and biochemical processes involved during thermal death. To answer some of these questions several groups of workers have investigated the effects of sublethal thermal damage and recovery from it in vegetative bacteria. The extent of injury produced during sublethal heating is dependent on factors already mentioned in the discussion on thermal death e.g. the time of exposure, physiological age, the cell density during injury, the composition of the injury menstruum and recovery medium, the diluent and the plating medium used for enumeration, (Nelson, 1943, 1956; Heather & Vanderzant, 1957a,b; Busta & Jezeski, 1963; Hansen & Rieman, 1963; White, 1963; Strang & Shon, 1964; Hurst et al., 1973; Smolka et al., 1974). Allwood & Russell (1970) have reviewed causes of thermal injury in non-sporing bacteria.

Damage to the cytoplasmic membrane, with a loss in the selective permeability character of the cells, such as leakage of proteins, amino acids, potassium ions, 260 nm-absorbing material and increased sensitivity to salt and selective media, have been used to demonstrate thermal injury (Califano, 1952; Hancock, 1958; Strange & Shon, 1964; Haight & Morita, 1966; Iandolo & Ordal, 1966; Allwood & Russell, 1967a,b, 1968, 1969a,b; Russell & Harries, 1967; Sogin & Odal, 1967; Beuchat & Lechowich, 1968; Clark et al., 1968; Clark & Ordal, 1969; Alsobrook et al., 1972; Erwin & Haight, 1973; Gray et al., 1973; Hurst et al., 1973; Stiles et al., 1973; Pierson et al., 1974).

Heat-injured cells have also shown altered metabolic and biosynthetic capabilities such as an increased lag time (Jackson & Woodbine, 1963; Iandolo & Ordal, 1966; Clark et al., 1968; Allwood & Russell, 1969a,b; Clark & Ordal, 1969); an increased nutritional demand for recovery (Nelson, 1943; Chambers, 1957; Heather & Vanderzant, 1957a; Baird-Parker & Davenport, 1965; Allwood & Russell, 1968, 1969b); minimal medium recovery (Gomez et al., 1973); alteration in enzyme activity (Langridge & Morita, 1966; Bluhm & Ordal, 1969; Tomlins et al. 1971); reduction in the efficiency of glucose transport (Pierson & Ordal, 1971); degradation of ribosomal RNA (Sogin & Ordal, 1967; Rosenthal & Iandolo, 1970; Tomlins & Ordal, 1971a,b; Miller & Ordal, 1972; Rosenthal et al., 1972; Gray et al., 1973); and damage to the DNA (Bridges et al., 1969a,b; Sedgewick & Bridges, 1972; Woodcock & Grigg, 1972; Gomez et al., 1973; Gomez & Sinskey, 1973).

(b) *Estimation of thermal injury by differential plating*

Loss in the selective permeability characteristic of cells has been used for the quantitative enumeration of the degree of injury in a heat-injured preparation, since injured cells show an increased sensitivity to salt and other compounds. The growth and injury conditions used for the production and estimation of

Table 4
Production and estimation of sublethal thermal injury

Organism	Growth medium & temperature	Heating menstruum composition	Heating temperature and time	Differential plating media	Reference
Salmonella typhimurium	Trypticase-soy broth (TSB) 37°	100mM potassium phosphate buffer pH 6.0	48°/30 min	Trypticase-soy agar (TSA)/ eosin methylene blue agar + 2.0% NaCl (EMBS)	Clark & Ordal (1969)
Salm. typhimurium	Citrate salts medium 37°	100mM potassium phosphate buffer pH 6.0	48°/30 min	TSA + 0.25% Na citrate EMBS	Tomlins & Ordal (1971a)
Staphylococcus aureus	TSB 37°	100mM potassium phosphate buffer pH 7.2	55°/15 min	TSA/TSA + 7.5% NaCl	Iandolo & Ordal (1966)
Streptococcus faecalis	TSB 37°	100mM potassium phosphate buffer pH 6.8	60°/15 min	TSA/TSA + 6.0% NaCl	Clark et al. (1968)
Pseudomonas fluorescens	Glutamate-salts medium 25°	Glutamate-salts medium, pH 7.0	36°/120 min	Direct microscopic count/ TSA	Gray et al. (1973)
Bacillus subtilis (vegetative cells)	TSB 37°	100mM potassium phosphate buffer pH 6.0	47°/30 min	TSA/modified sulphite polymyxin sulphadiazine agar	Miller & Ordal (1973)
Clostridium botulinum (E) (vegetative cells)	Peptone-yeast extract broth (PY) 30°	P.Y. broth pH 7.0	40.5°/120 min	Peptone-yeast extract agar (PYA)/PYA + 1.5% NaCl PYA + 0.07% bile salts	Pierson et al. (1974)
Cl. perfringens (vegetative cells)	Peptone-yeast extract (PY) broth 30°	Peptone-yeast extract (PY) broth	51°/30–90 min	PY agar/PY agar + 1.5% NaCl	Ades (1973)
Vibrio marinus	Artificial seawater broth 15°	Tris buffer pH 7.3	31°/20 min	Bacto peptone modified Agar/Russell double sugar modified agar	Griffiths & Haight (1973)

thermal injury in a range of vegetative bacterial cells are summarized in Table 4. The injury and differential plating conditions adopted for any particular organism were chosen for the following characteristics: (i) identical productivity on the two media with uninjured cells; (ii) large difference in the two viable counts after thermal injury; (iii) composition of the menstruum, heating time and temperature to give the minimum amount of death.

The sensitivity of heat-injured cells, to the plating diluent is also of critical importance. Gray (1972) demonstrated that 0.1% peptone, distilled water, or Butterfield's buffer diluents were lethal to heat stressed *Ps. fluorescens*. The greatest survival was achieved using the defined growth medium as a diluent. The effects of plating diluents on bacterial viability has been discussed by Farwell & Brown (1971).

Figure 5a illustrates the differential plating curve for *Salm. typhimurium*,

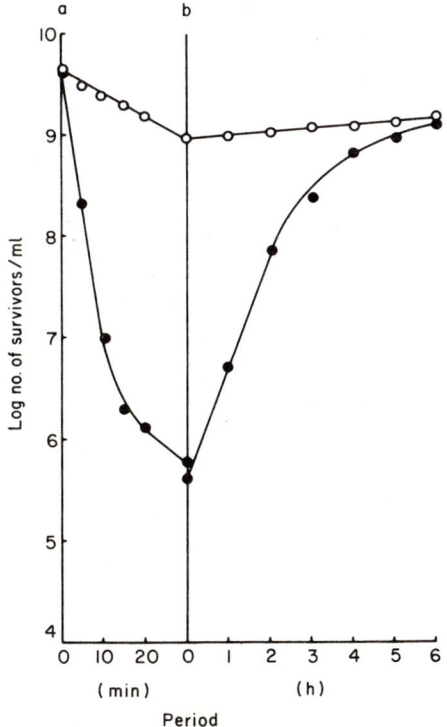

Fig. 5. a. Survival curve of CM-grown *Salmonella typhimurium* 7136 heated in 100 mM phosphate buffer (pH 6.0) at 48°. ○, samples plated on TSA-citrate: ●, samples plated on EMB-NaCl. b. Recovery curve of CM-grown *Salm. typhimurium* 7136 heat-injured at 48° for 30 min, inoculated into fresh CM medium, and recovered at 37 . ○, samples plated on TSA–citrate; ●, samples plated on EMB–NaCl. From Tomlins & Ordal (1971a).

heat injured at 48°. The difference in the plate counts is an estimation of the injured population since heat-injured cells will not grow on the EMBS medium. The inoculation of heat-injured cells into fresh growth medium allowed the repair of the heat-induced lesions and the cells rapidly regained their competence on the EMBS medium, (Fig. 5b). The injury and recovery curves illustrated in this figure are similar to those produced in the other systems given in Table 4. Such curves amply illustrate why preincubation is necessary when attempting to enumerate stressed populations of cells from food materials, on selective media (Clark & Ordal, 1969; Gray et al., 1974).

Hurst et al. (1974) have studied the effect of sublethal heating on *Staph. aureus* of different physiological ages. Using a differential plating system of Iandolo & Ordal (1966), (Table 4), they showed that the maximum difference in the counts occurred in the late log-early stationary phase because cells had a greatly increased heat resistance. The ability of cells to form colonies on high salt agar after sublethal heating varied little during the growth cycle (Fig. 3).

To give a preliminary indication as to the mechanisms involved during

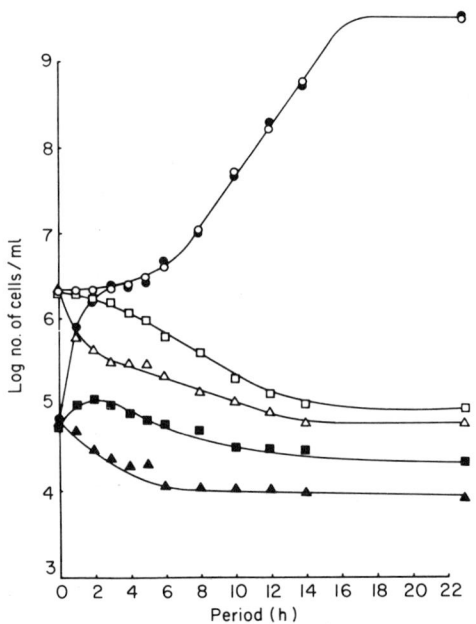

Fig. 6. Recovery and growth of heat-injured *Salmonella typhimurium* 7136 incubated in CM medium plus 5 μg of rifamycin or 100 μg of 5-fluorouracil/ml. The cells were heat-injured in 100 mM potassium phosphate buffer, pH 6.0 at 48° for 30 min. Cells incubated in CM medium: ○, plated on TSA-citrate; ●, plated on EMB–NaCl. Cells incubated in CM plus 5 μg of rifamycin/ml: △, plated on TSA-citrate; ▲, plated on EMB–NaCl. Cells incubated in CM plus 100 μg of 5-fluorouracil/ml: □, plated on TSA-citrate; ■, plated on EMB–NaCl. From Tomlins & Ordal (1971a).

recovery selective metabolic inhibitors have been added to the recovery medium, since an identification of the metabolic processes occurring during repair should reasonably indicate those sites most affected during thermal injury. The typical data presented in Fig. 6, yielded presumptive evidence that RNA synthesis was necessary for recovery of *Salm. typhimurium* from thermal injury. Addition of other metabolic inhibitors to the recovery medium has demonstrated whether DNA, protein, ATP and cell wall synthesis were necessary requirements for the recovery of cells from thermal injury (Table 5). All the organisms reported required RNA synthesis, but only *Salm. typhimurium* and *Clostridium perfringens* required protein synthesis for recovery.

(c) *Ribosome and rRNA degradation and resynthesis during thermal injury and recovery*

In order to understand better the high temperature growth of thermophilic organisms, numerous workers have studied the *in vitro* thermal denaturation of ribosomes from both mesophilic and thermophilic bacteria, with particular reference to the destabilization due to phosphate ions and EDTA and stabilization due to magnesium ions, (Natori *et al.*, 1966; Saunders & Campbell, 1966; Haight & Ordal, 1969; Miller *et al.*, 1969), the stimulation of ribonuclease activity (Stenesh & Yang, 1967) and stabilization due to ribosomal proteins (Friedman *et al.*, 1967; Nomura *et al.*, 1968; Moller *et al.*, 1969; Tal, 1969).

Degradation of ribosomes and rRNA *in vivo* during thermal injury has been studied in *Staph. aureus*, (Sogin & Ordal, 1967; Rosenthal & Iandolo, 1970; Rosenthal *et al.*, 1972); *B. subtilis* (Miller & Ordal, 1972) and *Ps. fluorescens* (Gray *et al.*, 1973). Tomlins & Ordal (1971*b*) compared the sedimentation rates of ribosomes, from unheated and heat-injured *Salm. typhimurium*, on sucrose gradients (Fig. 7). The heat-injured cells contained only one type of particle with a sedimentation coefficient of 47S and no 30S particles. The total loss of the 30S ribosomal particle has been also reported after sublethal thermal injury of *Staph. aureus* (Sogin & Ordal, 1967; Rosenthal & Iandolo, 1970; Rosenthal *et al.*, 1972). However, during the mild thermal stress to *Ps. fluorescens* 47S particles were produced without the total loss of 30S particles (Gray, 1972).

When rRNA was extracted from unheated and heat-injured cells of *Salm. typhimurium* and separated by polyacrylamide gel electrophoresis the profile (Fig. 8) showed the complete degradation of the 16S RNA species with only partial degradation of the 23S RNA species (Tomlins & Ordal, 1971*b*). Profiles similar to these have been reported for heat-injured *Staph. aureus* (Rosenthal *et al.*, 1972) and vegetative cells of *B. subtilis* (Miller & Ordal, 1972). Ribosome degradation in vegetative cells of *B. cereus* heated at 65° has been also observed by electron microscopy of thin sections (Silva & Sousa, 1972).

It has been reported that phosphate and magnesium ions exert considerable

Table 5
Requirements for the recovery from thermal injury

Organism	Protein synthesis	RNA synthesis	DNA repair	ATP synthesis	Cell wall synthesis	Reference
Salmonella typhimurium	+	+	+	+	0	Tomlins & Ordal (1971a) Gomez *et al.* (1973)
Staphylococcus aureus	−	+	0	−	−	Iandolo & Ordal (1966)
Streptococcus faecalis	−	+	0	0	−	Clark & Ordal (1969−)
Pseudomonas fluorescens	−	+	0	0	0	Gray *et al.* (1973)
Bacillus subtilis (vegetative cells)	−	+	0	0	−	Miller & Ordal (1973)
Clostridium perfringens (vegetative cells)	+	+	0	0	−	Ades (1973)
Escherichia coli	0	0	+	0	0	Woodcock & Grigg (1972)

+, required for recovery; −, not required fro recovery;)
+, required for recovery; −, not required for recovery; 0, not assayed.

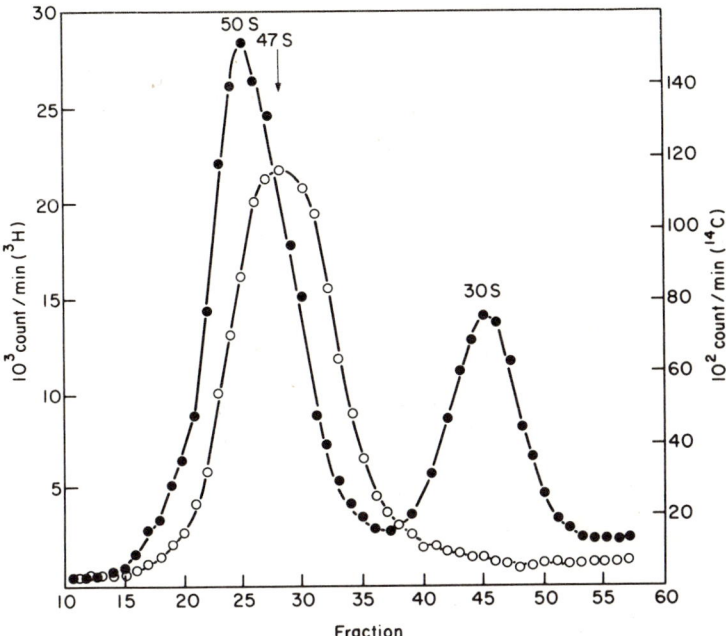

Fig. 7. Effect of thermal injury on the 50S and 30S ribosomal particles of *Salm. typhimurium* 7136. The cells were steady-state-labelled with uracil-6-^{3}H and uracil-2-^{14}C for 5 h before harvesting. The ^{14}C-labelled cells were heat injured at 48° for 30 min. Both control and injured cells were lysed into 0.01 M Tris-hydrochloride buffer (pH 7.2) containing 5×10^{-4} M Mg^{2+}. Sedimentation analysis of the crude extract was carried out on a 20 to 5% linear sucrose gradient prepared in the same buffer for 6.5 h at 26,500 rev/min. Six-drop fractions were collected directly into scintillation vials and counted. ●, uracil-6-^{3}H counts; ○, uracil-2-^{14}C counts. From Tomlins & Ordal (1971b).

influence on the stability of ribosomes *in vitro*, (Stenesh & Yang, 1967; Haight & Ordal, 1969). Natori *et al.* (1966) observed that when *E. coli* ribosomes were incubated *in vitro* for short periods in various phosphate concentrations the ribosomes unfolded with significant loss of protein. These phosphate treated ribosomes were also shown to be more sensitive to ribonuclease than untreated ribosomes. When cells of *Salm. typhimurium* were thermally injured at 48° in phosphate buffer containing 100mM magnesium, or in the absence of phosphate, i.e. in distilled water, after 30 min heating no selective degradation of the 30S ribosomal particle or 16S rRNA occurred, and the ribosome and rRNA profiles were identical to unheated controls (Figs 7, 8). Ribosome degradation again did not occur when calls of *Salm. typhimurium* were incubated for extended periods in phosphate buffer at 37° (Tomlins, 1971). This was in agreement with Strange & Shon (1964) who concluded that the increased heat resistance observed with

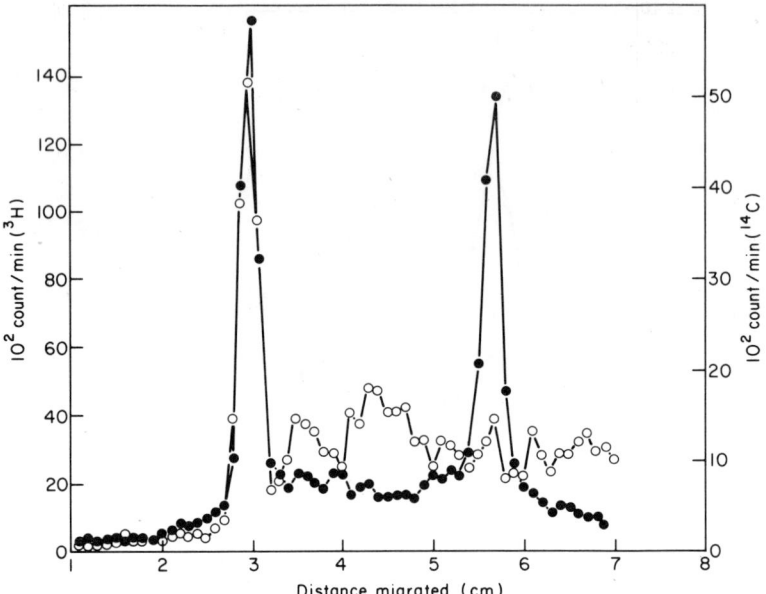

Fig. 8. Effect of thermal injury on the 23S and 16S rRNA species of *Salm. typhimurium* 7136. The cells were steady-state-labelled with uracil-6-^{3}H and uracil-2-^{14}C for 5 h before harvesting. The ^{14}C-labelled cells were heat injured at 48° for 30 min. The RNA was extracted from both control and heat injured cells and co-electrophoresed on a 3.0% polyacrylamide gel for 6 h at 5 ma/gel. The gel was frozen, sliced into 1 mm slices, digested and counted. ●, uracil-6-^{3}G counts; ○, uracil-2-^{14}C counts. From Tomlins & Ordal (1971b).

Enterobacter aerogenes was due to increased magnesium ion content of the cell, or the presence of magnesium ions in the heating menstruum. Bacterial cells heated in the presence of the phosphate ion produced conditions which were conducive to RNA degradation by the stimulation of ribonuclease activity (Chakraburtty & Burma, 1968), and by perturbation of ribosome stability due to chelation of intracellular magnesium ions by the phosphate ions. The latter condition was indicated by the lack of ribosome degradation in cells heated in distilled water. The presence of excess magnesium ions in the phosphate buffer heating menstruum protected RNA against degradation by inhibition of ribonuclease activity (Chakraburtty & Burma, 1968) and by stabilizing the ribosomal subunits, thereby made the rRNA substrate less susceptable to attack. Consequently this selective degradation of bacterial ribosomes was due to the combination of the presence of phosphate ions and heat, both of which made the ribosome unstable and stimulated ribonuclease activity.

Evidence presented in Table 5 has indicated the need for heat-injured bacteria to resynthesize RNA as an obligatory requirement for recovery. Allwood &

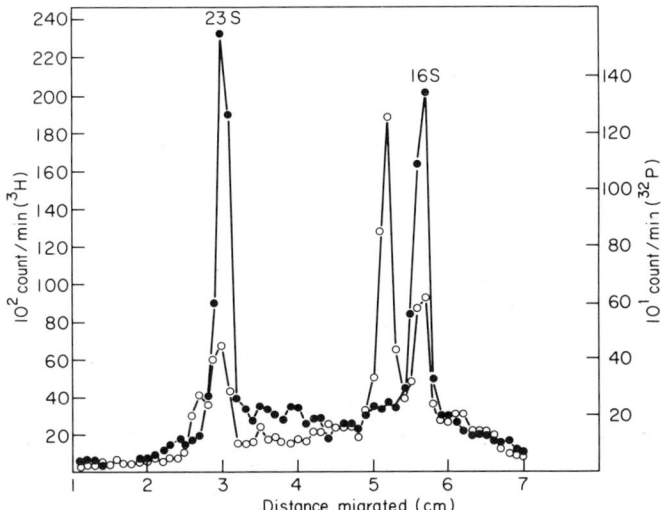

Fig. 9. Profiles of rRNA from normal and recovering cells of *Salm. typhimurium* 7136. Control cells were steady-state-labelled with uracil-6-^{3}H for 5 h before harvesting, and the RNA was extracted. Test cells (unlabelled) were heat-injured and recovered for 4 h in CMP containing 2 Ci of Na$_2$H^{32}PO$_4$/ml, and the RNA was extracted. The two samples of RNA were mixed and co-electrophoresed on a 3.0% polyacrylamide gel for 6 h at 5 ma/gel. The gel was then frozen, sliced into 1 mm slices, digested and counted. ●, uracil-6-^{3}H counts; ○, ^{32}P counts. From Tomlins & Ordal (1971*b*).

Russell (1969) reported that RNA was synthesized during the early stages of recovery of heat-injured *Staph. aureus* but at a rate much less than that of unheated cells in the log phase of growth. In *Salm. typhimurium* however, it has been demonstrated that during recovery the rate of RNA synthesis/viable cell was slightly greater than that observed with unheated cells in the same medium (Tomlins, 1971). Tomlins & Ordal (1971*a*) verified by incorporation experiments that both RNA and protein synthesis occurred during the recovery of *Salm. typhimurium* from thermal injury. Since RNA synthesis ceased before recovery was completed, RNA synthesis was not the rate-limiting step for recovery. Protein was synthesized during the latter stages of recovery and included new ribosomal protein.

Ribosomal RNA profiles from control cells of *Salm. typhimurium* and cells recovering from thermal injury, obtained after co-electrophoresis in polyacrylamide gels are presented in Fig. 9. In comparison to the control, rRNA synthesized during recovery was fractionated into four peaks; 23S and 16S RNA and their respective precursor molecules 24S and 17S RNA. When injured cells were similarly incubated, but in the presence of chloramphenicol, the maturation of 17S RNA to 16S RNA was completely abolished, (Tomlins & Ordal.

1971b). Similar RNA profiles (Fig. 9) have been reported for vegetative cells of *B. subtilis* during recovery, (Miller & Ordal, 1972). Also, recovering *Ps. fluorescens* accumulated 17S RNA but not 24S RNA (Gray et al., 1973). However, during the recovery of *Stap. aureus* from thermal injury only 23S and 16S rRNA were synthesized (Rosenthal & Iandolo, 1970).

The ribosomal particles generated during the recovery of *Salm. typhimurium* from thermal injury are given in Fig. 10. The partially recovered cells had particles at 50S and 48S, and 31S, 28S, 26S and 22S when co-sedimented with the 50S and 30S control ribosomal particles. When injured cells were allowed to recover for a longer period of time, there was a shift in the relative amounts of the 30S precursor particles towards the 30S region. The addition of chloramphenicol to the recovery medium altered this ribonucleoprotein profile. However, the 50S and 48S particles were observed and those contributing to the major peak sedimented at 22S and caused a shoulder in the curve at 32S. This demonstrated that the inhibition of protein synthesis had resulted in the inhibition of 30S ribosomal particle maturation, without affecting the formation of mature or precursor 50S particles (Tomlins & Ordal, 1971b). A similar profile of ribosomal precursor particles (Fig. 10) has also been identified during the recovery of *Ps. fluorescens* following mild thermal shock (Gray, 1972). However,

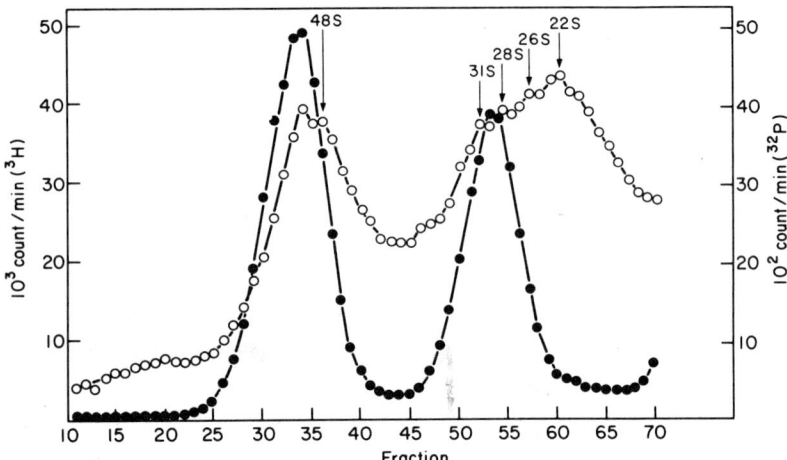

Fig. 10. Ribosomal particle profile from normal and recovering cells of *Salm. typhimurium* 7136. Control cells were steady-state-labelled with uracil-6-^{3}H for 5 h before harvesting. Injured cells (unlabelled) were recovered for 4 h in CMP containing 2 Ci of $Na_2H^{32}PO_4$/ml. Both control and recovering cells were lysed into 0.01 M Tris-hydrochloride buffer (pH 7.2) containing 5×10^{-4} M Mg^{2+}. Sedimentation analysis of the crude extracts was carried out on a 20 to 5% linear sucrose gradient, prepared in the same buffer for 6.5 h at 26,500 rev/min. Six-drop fractions were collected directly into scintillation vials and counted. ●, uracil-6-^{3}H counts; ○, ^{32}P counts. From Tomlins & Ordal (1971b).

Rosenthal *et al.* (1972), have reported that ribosomal precursor particles are not accumulated during recovery of *Staph. aureus* from thermal injury.

It has been reported that during other conditions of physiological stress, such as magnesium ion starvation, which produce ribosome degradation in bacteria, the original ribosomal proteins were conserved and re-used for the production of new ribonucleoprotein particles in any subsequent recovery period. This has also been reported to be the case during the recovery of *Staph. aureus* from thermal injury (Rosenthal *et al.*, 1972). However, Tomlins & Ordal (1971*b*) have shown that in *Salm. typhimurium* the synthesis of some ribosomal proteins required for the formation of mature 30S ribosomal particles was an obligatory requirement for recovery from thermal injury.

The accumulated ribosomal precursor particles identified during various stages of recovery from thermal injury in *Salm. typhimurium* and *Ps. fluorescens* have all been identified by numerous other workers, using a variety of techniques or mutant bacteria. The 47S-48S particle identified during both thermal injury and recovery had not been previously identified, but Nomura (1972) has reported on the formation of a 46S-48S particle during the *in vitro* reconstitution of *B. stearothermophilus* 50S ribosomes.

It has been shown, using a range of taxonomically unrelated bacteria that ribosome and rRNA degradation occurred during thermal injury. The extent to which the 50S and 30S subunits were damaged depended largely on the severity of the heating and the heating menstruum composition, however, the 50S ribosomal particle and its constituent RNA and proteins seem slightly more heat stable than the 30S ribosomal particle. Generally the ribosomal proteins from degraded ribosomes are re-used during recovery for the synthesis of new particles. However, during the recovery of *Salm. typhimurium* from thermal injury the synthesis of new 30S ribosomal proteins was required (Tomlins & Ordal, 1971*b*).

(d) *Cytoplasmic membrane alterations during thermal injury and recovery*

It has already been established by indirect methods that during heat injury substantial alterations to the functional properties of the bacterial cell membrane occur, so that there is reduced control over what passes into the cell and out of it.

Damage which has decreased the normal tolerance of the cell to salt and dyes, and allowed leakage of intracellular components has been used to quantify the degree of injury in a heat-damaged population of cells (Table 4). Pierson *et al.* (1971) have demonstrated using a cell fractionation technique that during the recovery of *Salm. typhimurium* from thermal injury lipid material was formed rapidly during the early stages of recovery. This indicated that the re-establishment of a competent membrane was needed for the re-construction of intracellular pools, required for the repair of other heat-induced lesions. This has also been verified indirectly by the observation that *Salm. typhimurium* recovered normal tolerance to Levine's eosin–methylene blue medium containing 2% of NaCl, and that *Staph. aureus* recovered tolerance to NaCl, several hours before any cell division occurred (Tomlins & Ordal, 1971a; Hurst *et al.*, 1973).

The effects of heat injury on the transport of methyl-α-D-glucopyranoside (αMG) in *Salm. typhimurium* has been reported by Pierson & Ordal (1971). In the absence of an energy source the rate of αMG accumulation was less for heat-injured than for unheated cells. This decreased level of incorporation, was accounted for by the lower rate of endogenous metabolism in injured cells. However, in the presence of an exogenous energy source, the rate of αMG transport and its level of accumulation was greater for heat-injured cells in comparison to unheated cells. It was postulated that this increased accumulation could be due to an alteration of the cells normal αMG exit reaction caused during heat injury.

In order to demonstrate the physical state of the membrane of *Salm. typhimurium* Tomlins *et al.* (1972) investigated the degree of destruction of lipid species during injury at 48° and the extent of their re-synthesis during recovery. An isotope dilution assay demonstrated that 26.6% of the cellular lipid of recovered cells was synthesized during the recovery period. However, as the heating menstruum yielded almost no lipid at all, disruption of the membrane during injury occurred without any significant release of lipid. The fatty acid profiles from both normal and heat-injured *Salm. typhimurium* were similar, quantitatively and qualitatively, which indicated that there was no selective destruction of the fatty acids during injury. The quantities of the neutral lipid and phospholipid species produced during recovery was similar to those from unheated controls. However, the amounts of fatty acids synthesized during recovery were quite different from the controls. The most striking difference was the large increase in unsaturated fatty acids, ($C_{16:1}$, $C_{18:1}$) with a concomitant decrease in both short chain saturated fatty acids up to and including $C_{14:0}$, and cyclopropane fatty acids ($C_{17:0}$ cyclo and $C_{19:0}$ cyclo).

Hurst *et al.* (1973a) investigated the return of salt tolerance in *Staph. aureus* during recovery from thermal injury at 52°. They related the return of tolerance to 7.5% NaCl to membrane fatty acid composition, aspartate transport and the K/Na ratio; all of which are membrane-related functional properties. Injury resulted in a 30% loss of lipid from the membrane. During recovery, amounts of C_{15} and C_{17} branched chain fatty acids returned to normal, and an increase in C_{16} and C_{18} unsaturated fatty acids occurred. Penicillin abolished this increase in unsaturated fatty acids without affecting recovery. After injured cells had regained their tolerance to 7.5% NaCl during recovery, the K/Na ratio was still the same as for injured cells and aspartate uptake was only 35% of the unheated control. It was demonstrated that cells of *Staph. aureus* recovered salt tolerance while various membrane functions were still impaired.

Recently, Duitschaever & Jordan (1974) have investigated the altered fatty acid profile of *Strep. faecium* produced during recovery from thermal injury at 55°. Cells undergoing repair from heat injury showed a significant increase in their saturated fatty acid concentrations, especially $C_{12:0}$, $C_{14:0}$, $C_{16:0}$ and $C_{18:0}$, from 18.5% in control cells to 38.0% after recovery. These results are at variance with those of Hurst *et al.* (1973*a*). Repeated subculture of these recovered cells did not alter their new fatty acid profile. Concomitant with this increased level of saturated fatty acids in the membrane, the cells also showed an increased resistance to both NaCl and heat, which was not lost after 12 subcultures.

Corry & Roberts (1970) have also reported on the production of increased heat resistance in *Salm. typhimurium* after repeated thermal injury at 55°, which was not lost after 14 subcultures. Investigations into membrane fatty acid composition were not undertaken but it was noted there was no apparent change in pathogenicity or common biochemical characteristics in the heat-resistant strain, compared with the heat-sensitive parent.

Clearly there is mounting evidence to suggest that the loss of membrane integrity during thermal injury, and subsequent lipid synthesis during recovery, result from a lesion of major importance. Based on both experimental and calculated values for *Strep faecalis,* Moats' (1971) kinetic explanation of thermal injury and death suggested that there were >100 possible sites for thermal inactivation, a concept which is in agreement with multiple injury to the membrane.

The changes in lipid composition and related increases in heat resistance should be investigated thoroughly so as to reveal the extent to which micro-organisms important in the processing and handling of foods may be progressively selected for increased heat resistance.

(e) *Metabolic damage during thermal injury and recovery*

There have been several published reports concerning the loss of viability and metabolic activity in bacteria submitted to sublethal heat damage (Hershey, 1939; Robeson & Morita, 1966; Bluhm & Ordal, 1969). Several heat-sensitive enzymes have been implicated in the loss of metabolic activity. These enzymes include: aldolase and lactate dehydrogenase (Bluhm & Ordal, 1969); NADH oxidase and cytochrome *c* reductase (Purohit & Stokes, 1967); pantothenate synthesizing enzyme (Maas, 1952); pyrophosphatase (Blumenthal, 1967) formate dehydrogenlyase (Upadhyay & Stokes, 1963) and malate dehydrogenase (Burton & Morita, 1963; Langridge & Morita, 1966).

Pierson & Ordal (1973) reported on the metabolic activity of heat-injured *Salm. typhimurium*. They concluded from their results using Warburg manometry, radiorespirometry and metabolite transport experiments, that heat injury produced a stimulated metabolic activity, which was possibly directed towards the repair of heat-induced lesions. The effects of thermal injury on the TCA cycle enzymes of both *Staph. aureus* and *Salm. typhimurium* has been reported by Tomlins et al. (1971). In extracts from *Salm. typhimurium* there was only a minor loss in specific activity with fumarate hydratase, glutamate dehydrogenase, fructose diphosphate aldolase, lactic dehydrogenase and the NAD(P) oxidases after thermal injury at 48°. Clearly, metabolic damage is not an important mechanism contributing to the thermal injury of this bacterium. In extracts of *Staph. aureus* oxoglutarate dehydrogenase, malate dehydrogenase and lactate dehydrogenase were inactivated strongly after thermal injury at 52°. Substantial renaturation of lactate dehydrogenase and malate dehydrogenase occurred during recovery both in the absence and presence of chloramphenicol but no renaturation of oxoglutarate dehydrogenase was found under the same conditions. It was concluded that the irreversible denaturation of this enzyme may be responsible for increased nutritional requirement for recovery in *Staph. aureus* (Iandolo & Ordal, 1966).

There is little evidence to suggest that enzyme inactivation is a significant mechanism in the thermal injury of vegetative bacteria. Although it has been reported that the maximum temperature for growth of psychrophiles is limited by thermolabile enzymes (Hagen & Rose, 1962; Purohit & Stokes, 1967) recent evidence has shown that enzyme inhibition is not the only mechanism responsible for the inhibition of their growth (Gray et al., 1973). Towards the latter stages of recovery from thermal injury, protein synthesis may occur, so the re-synthesis of any heat-inactivated enzymes is a possibility, (Tomlins & Ordal, 1971a). However, if enzymes or structural proteins associated with ribosome biosynthesis are also destroyed during thermal injury, recovery may be further inhibited owing to the lack of competent ribosomal particles required for protein synthesis, (Tomlins & Ordal, 1971b). The correlation of enzyme inactivation with the production of thermal injury seems to be a tenuous one, particularly in *Salm typhimurium*, although at higher temperatures than those used for thermal injury, but still within the range of pasteurization processes, damage to enzyme systems may well become an added factor in decreasing the level of bacterial survival.

(f) *DNA damage and repair during thermal injury and recovery*

Until recently there has been scant attention paid to the effect of mild thermal stress on the *in vivo* stability of bacterial DNA. The relative heat stability of *in vitro* DNA has been reported by Eigner et al. (1961) and Baldwin (1964). Haight

& Morita (1966) have demonstrated that DNA along with other intracellular components, leaked from cells of *Vibrio marinus,* held at an elevated temperature, but the DNA of cells of *Staph. aureus* stored at 60° (Allwood & Russell, 1968) and *B. psychrophilus* stored at 40° (Alsobrook *et al.,* 1972), remained unchanged.

Bridges *et al.* (1969a,b) reported that *E. coli* incubated at 52° suffered damage to the DNA similar to that induced by γ-radiation. They concluded that bacteria may use the same repair processes to recover from both types of damage. In further studies Sedgwick & Bridges (1972) concluded that the DNA strand degradation observed in several *E. coli* strains differing in their DNA polymerase I activity was consistant with attack by native nucleases which were released or activated by mild heating.

Woodcock & Grigg (1972) have reported that after heating in a defined medium, also at 52°, *E. coli* DNA had *c.* 100 heat-induced double-strand breaks. Subsequent incubation in phosphate buffer plus thymidine, at 37°, gave an increase in DNA molecular weight, returning it to its former value in 30 min. Viable numbers also increased during this period. Inhibition of DNA synthesis during recovery reduced both the viability and DNA mol. wt. Little or no loss of DNA was observed during injury and recovery, neither did heating the DNA *in vitro* at 52° result in strand breakage. They concluded that repair of the DNA strand breaks, rather than replication, occurred during the recovery period, indicating a direct involvement of enzymic DNA breakage in the thermal injury of *E. coli.*

When cells of *Salm. typhimurium, Salm. thompson* and *Salm. heidelberg,* grown in a glucose–salts medium, were heated at 50° a higher survival on a glucose–salts agar than on a nutritionally complete medium was observed (Gomez *et al.,* 1973). They discussed the similarity of this phenomenon to the minimal medium recovery observed with irradiated bacteria, and questioned the practice of pre-enrichment for food pathogens in complex medium, without consideration of the cells prior nutritional status. Gomez & Sinskey (1973) went on to show that with *Salm. typhimurium* heated in water at 50°, followed by a short incubation at 37° in a glucose–salts medium, the DNA sedimentation patterns in alkaline sucrose gradients did not change. However, if the heat-damaged bacteria were incubated instead in trypticase-soy broth plus 0.5% of yeast extract (TSY), a marked increase in DNA single-strand breakage, accompanied by a loss in viability, was observed. If heated cells were incubated in glucose–salts broth prior to incubation in TSY broth there was a decrease in the single-strand breakage that occurred in the TSY broth. They concluded that the minimal medium recovery observed after heating was due to DNA damage caused by sequential exposure to heat and TSY broth.

The recently published reports which demonstrated an important alteration to the DNA of bacteria after thermal shock, have contributed to our understanding of the important processes involved during thermal injury and recovery. However, further investigation should be undertaken perhaps with a Gram positive bacterium, to resolve whether thermally induced DNA breakage is an isolated or general phenomenon in bacteria.

The detection of minimal medium recovery, after heat damage, and its important implications regarding pre-enrichment procedures for the isolation of food-borne pathogens should be critically assessed by further experimentation.

3. References

ADES, G. L. (1973). Heat injury of the vegetative cells of *Clostridium perfringens*. Ph.D Thesis, Virginia P. I. and S. U., U.S.A.
ALLWOOD, M. C & RUSSELL, A. D. (1967a). Mechanisms of thermal injury in *Staphylococcus aureus*. 1. Relationship between viability and leakage. *Applied Microbiology* 15, 1266-1269.
ALLWOOD, M. C. & RUSSELL, A. D. (1967b). The leakage of intracellular constituents from heated suspensions of *Staphylococcus aureus*. *Experientia* 23, 878-885.
ALLWOOD, M. C. & RUSSELL, A. D. (1968). Thermally induced ribonucleic acid degradation and leakage of substances from the metabolic pool in *Staphylococcus aureus*. *Journal of Bacteriology* 95, 345-349.
ALLWOOD, M. C. & RUSSELL, A. D. (1969a). Thermally induced changes in the physical properties of *Staphylococcus aureus*. *Journal of Applied Bacteriology* 32, 68-78.
ALLWOOD, M. C. & RUSSELL, A. D. (1969b). Growth and metabolic activities of heat treated *Staphylococcus aureus*. *Journal of Applied Bacteriology* 32, 79-85.
ALLWOOD, M. C. & RUSSELL, A. D. (1970). Mechanisms of thermal injury in non-sporing bacteria. *Advances in Applied Microbiology* 12, 89-119.
ALSOBROOK, D., LARKIN, J. M. & SEGA, M. W. (1972). Effects of temperature on the cellular integrity of *Bacillus psychrophilus*. *Canadian Journal of Microbiology* 18, 1671-1678.
BAIRD-PARKER, A. C. & DAVENPORT, E. (1965). The effect of recovery medium on the isolation of *Staphylococcus aureus* after heat treatment and storage of frozen or dried cells. *Journal of Applied Bacteriology* 28, 390-402.
BAIRD-PARKER, A. C., BOOTHROYD, M. & JONES, E. (1970). The effect of water-activity on the heat resistance of heat sensitive and heat resistant strains of salmonellae. *Journal of Applied Bacteriology* 33, 515-522.
BALDWIN, R. L. (1964). Molecular aspects of the gene: replication mechanisms. In *The Bacteria* Vol. V, eds Gunsalus, I. C. & Stanier, R. Y. New York: Academic Press.
BARRILE, J. C. & CONE, J. F. (1970). Effect of added moisture on the heat resistance of *Salmonella anatum* in milk chocolate. *Applied Microbiology* 19, 177-178.
BARRILE, J. C., CONE, J. F. & KEENEY, P. G. (1970). A study of salmonella survival in milk chocolate. *Manufacturing Confectioner* 50, 34-39.
BAUMGARTNER, J. G. (1938). Heat sterilized reducing sugars and their effects on the thermal resistance of bacteria. *Journal of Bacteriology* 36, 369-375.
BEUCHAT, L. R. & LECHOWICH, R. V. (1968). Effect of salt concentration in the recovery medium on heat-injured *Streptococcus faecalis*. *Applied Microbiology* 16, 772-776.
BLUHM, L. & ORDAL, Z. J. (1969). Effect of sub-lethal heat on the metabolic activity of *Staphylococcus aureus*. *Journal of Bacteriology* 97, 140-150.
BLUMENTHAL, B. I., JOHNSON, M. K. & JOHNSON, E. J. (1967). Distribution of heat-labile and heat-stable inorganic pyrophosphatases among some bacteria. *Canadian Journal of Microbiology* 13, 1695-1699.
BRIDGES, B. A., ASHWOOD-SMITH, M. J. & MUNSON, R. J. (1969a). Susceptability of

mild thermal and of ionising radiation damage to the same recovery mechanisms in *Escherichia coli. Biochemical and Biophysical Research Communications* **35**, 193-196.

BRIDGES, B. A., ASHWOOD-SMITH, M. J. & MUNSON, R. J (1969b). Correlation of bacterial sensitivities to ionising radiation and mild heating. *Journal of General Microbiology* **58**, 115-124.

BROWN, M. R. W. & MELLING, J. (1971). Inhibition and destruction of microorganisms by heat. In *Inhibition and Destruction of the Microbial Cell,* ed. Hugo, W. B. London: Academic Press.

BURTON, S. D. & MORITA, R. Y. (1963). Denaturation and renaturation of malic dehydrogenase in cell-free extract from a marine psychrophile. *Journal of Bacteriology* **86**, 1019-1024.

BUSTA, F. F. & JEZESKI, J. J. (1963). Effect of sodium chloride concentration in an agar medium on growth of heat shocked *Staphylococcus aureus. Applied Microbiology* **11**, 404-407.

CALHOUN, C. L. & FRAZIER, W. C. (1966). Effect of available water on thermal resistance of three non-sporeforming species of bacteria. *Applied Microbiology* **14**, 416-420.

CALIFANO, L. (1952). Liberation d'acide nucleique par les cellules bacteriennes sous l'action de la chaleur. *Bulletin of the World Health Organization* **6**, 19-34.

CASOLARI, A. & CAMPANINI, M. (1973). Resistenza termica in Lactobacillaceae. *Industria Conserve* **48**, 140-143.

CHAKRABURTTY, K. & BURMA, D. P. (1968). The purification and properties of a ribonuclease from *Salmonella typhimurium* extract. *Journal of Biological Chemistry* **243**, 1133-1139.

CHAMBERS, C. W., TABAK, H. H. & KABLER, P. W. (1957). Effect of Krebs cycle metabolites on the viability of *Escherichia coli* treated with heat and chlorine. *Journal of Bacteriology* **73**, 77-84.

CHARM, S. E. (1958). The kinetics of bacterial inactivation by heat. *Food Technology* **12**, 4-8.

CLARK, C. W. & ORDAL, Z. J. (1969). Thermal injury and recovery of *Salmonella typhimurium* and its effect on enumeration procedures. *Applied Microbiology* **18**, 332-336.

CLARK, C. W. WITTER, L. D. & ORDAL, Z. J. (1968). Thermal injury and recovery of *Streptococcus faecalis. Applied Microbiology* **16**, 1764-1769.

COLLINS, E. B, (1961). Resistance of certain bacteria to cottage cheese cooking procedures. *Journal of Dairy Science* **44**, 1989-1996.

CORRY, J. E. L. (1973). The water relations and heat resistance of microorganisms. *Progress in Industrial Microbiology* **12**, 73-108.

CORRY, J. E. L. (1974). The effect of sugars and polyols on the heat resistance of salmonellae. *Journal of Applied Bacteriology* **37**, 31-43.

CORRY, J. E. L. & BARNES, E. M. (1968). The heat resistance of salmonellae in egg albumin. *British Poultry Science* **9**, 253-256.

CORRY, J. E. L. & ROBERTS, T. A. (1970). A note on the development of resistance to heat and gamma radiation in *Salmonella. Journal of Applied Bacteriology* **33**, 733-737.

COTTERILL, O. J. & GLAUERT, J. (1969). Thermal resistance of salmonella in egg yolk products containing sugar or salt. *Poultry Science* **48**, 1156-1166.

COTTERILL, O. J. & GLAUERT, J. (1971). Thermal resistance of salmonella in egg yolk containing 10% sugar or salt after storage at various temperatures. *Poultry Science* **50**, 109-115.

DABBAH, R., MOATS, W. A. & EDWARDS, V. M. (1971a). Heat survivor curves of food-borne bacteria suspended in commercially sterilized whole milk. I. Salmonellae. *Journal of Dairy Science* **54**, 1583-1588.

DABBAH, R., MOATS, W. A. & EDWARDS, V. M. (1971b). Heat survivor curves of

food-borne bacteria suspended in commercially sterilized whole milk. II. Bacteria other than salmonellae. *Journal of Dairy Science* 54, 1772-1779.
DAVIDSON, C. M., MOBBS, P. & STUBBS, J. M. (1968). Some morphological and physiological properties of *Microbacterium thermosphactum*. *Journal of Applied Bacteriology* 31, 551-559.
DEGA, C. A., GOEPFERT, J. M. & AMUNDSON, C. H. (1972). Heat resistance of salmonellae in concentrated milk. *Applied Microbiology* 23, 415-420.
DUITSCHAEVER, C. L. & JORDAN, D. C. (1974). Development of resistance to heat and sodium chloride in *Streptococcus faecium* recovering from thermal injury. *Journal of Milk and Food Technology* 37, 382-386.
EIGNER, J., BOEDTKER, H. & MICHAELS, G. (1961). The thermal degradation of nucleic acids. *Biochemica et biophysica acta* 51, 165-168.
ERWIN, D. G. & HAIGHT, R. D. (1973). Lethal and inhibitory effects of sodium chloride on thermally stressed *Staphylococcus aureus*. *Journal of Bacteriology* 116, 337-340.
EVANS, D. A., HANKINSON, D. J. & KITSKY, W. (1970). Heat resistance of certain pathogenic bacteria in milk using a commercial plate heat exchanger. *Journal of Dairy Science* 53, 1659-1665.
FARWELL, J. A. & BROWN, M. R. W. (1971). The influence of inoculum history on the response of microorganisms to inhibitory and destructive agents. In *Inhibition and Destruction of the Microbial Cell,* ed. Hugo, W. B. London: Academic Press.
FAY, A. C. (1934). The effect of hypertonic sugar solutions on the thermal resistance of bacteria. *Journal of Agricultural Research* 48, 453-455.
FOSTER, E. M., GOEPFERT, J. M. & DIEBEL, R. H. (1970). Detecting presence of salmonella: rapid methods. *Manufacturing Confectioner* 50, 57-60.
FRIEDMAN, S. M., AXEL, R. & WEINSTEIN, I. B. (1967). Stability of ribosomes and ribonucleic acid from *Bacillus stearothermophilus*. *Journal of Bacteriology* 93, 1521-1526.
GARIBALDI, J. A., IJICHI, K. & BAYNE, H. G. (1969). Effect of pH and chelating agents on the heat resistance and viability of *Salmonella typhimurium* Tm-1 and *Salmonella senftenburg* 775-W in egg white. *Applied Microbiology* 18, 318-322.
GIBSON, B. (1973). The effect of high sugar concentrations on the heat resistance of vegetative microorganisms. *Journal of Applied Bacteriology* 36, 365-376.
GOEPFERT, J. M. (1972). Salmonellae study focuses on variables affecting heat resistance. *Candy and Snack Industry* 137, 42, 44, 86.
GOEPFERT, J. M. & BIGGIE, R. A. (1968). Heat resistance of *Salmonella typhimurium* and *Salmonella senftenburg* 775-W in milk chocolate. *Applied Microbiology* 16, 1939-1940.
GOEPFERT, J. M., ISKANDER, I. K. & AMUNDSON, C. H. (1970). Relation of the heat resistance of salmonellae to the water activity of the environment. *Applied Microbiology* 19, 429-438.
GOMEZ, R. F. & SINSKEY, A. J. (1973). DNA breaks in heated *Salmonella typhimurium* LT-2 after exposure to nutritionally complex media. *Journal of Bacteriology* 115, 522-528.
GOMEZ, R. F., DAVIES, R., SINSKEY, A. J. & LABUZA, T. P. (1973). Minimal medium recovery of heated *Salmonella typhimurium* LT2. *Journal of General Microbiology* 74, 267-274.
GRAY, R. J. H. (1972). Molecular and cellular events initiated by thermal stress in *Pseudomonas fluorescens* P7. Ph.D Thesis, University of Illinois, U.S.A.
GRAY, R. J. H., GASKE, M. A. & ORDAL, Z. J. (1974). Enumeration of thermally stressed *Staphylococcus aureus* MF-31. *Journal of Food Science* 39, 884-886.
GRAY, R. J. H., WITTER, L. D. & ORDAL, Z. J. (1973). Characterization of mild thermal stress in *Pseudomonas fluorescens* and its repair. *Applied Microbiology* 26, 78-85.
GREENBERG, R. A. & SKILLIKER, J. H. (1961). Evidence of heat injury in enterococci. *Food Research* 26, 622-625.
GRIFFITHS, R. P. & HAIGHT, R. D. (1973). Reversible heat injury in the marine

psychrophilic bacterium *Vibrio marinus* MP-1. *Canadian Journal of Microbiology* 19, 557-561.
GYLLENBERG, H. & SEDERHOLM, H. (1961). Interdependent changes in growth requirements and heat resistance by a strain of *Pseudomonas*. *Acta Agriculturae scandinavica* 11, 3-12.
HAGEN, P. O. & ROSE, A. H. (1962). Studies on the biochemical basis of the low maximum temperature in a psychrophilic *Cryptococcus*. *Journal of General Microbiology* 27, 89-99.
HAGEN, P. O., KUSHNER, D. J. & GIBBONS, N. E. (1964). Temperature induced death and lysis of a psychrophilic bacterium. *Canadian Journal of Microbiology* 10, 813-822.
HAIGHT, R. D. & MORITA, R. Y. (1966). Thermally induced leakage from *Vibrio marinus*, an obligately psychrophilic marine bacterium. *Journal of Bacteriology* 92, 1388-1393.
HAIGHT, R. D. & ORDAL, Z. J. (1969). Thermally induced degradation of staphylococcal ribosomes. *Canadian Journal of Microbiology* 15, 15-19.
HANCOCK, R. (1958). The intracellular amino acids of *Staphylococcus aureus:* release and analysis. *Biochimica et biophysica acta* 28, 402-412.
HANSEN, N. H. & RIEMAN, H. (1963). Factors affecting the heat resistance of non-sporing organisms. *Journal of Applied Bacteriology* 26, 314-333.
HARGREAVES, L. L., WOOD, J. M. & JARVIS, B. (1972). The antimicrobial effect of phosphates with particular reference to food products. *B.F.M.I.R.A. Scientific and Technical Survey* No. 76.
HEATHER, C. D. & VANDERZANT, W. C. (1957a). Effect of the plating medium on the survival of heat-treated cells of *Pseudomonas fluorescens*. *Journal of Food Research* 22, 164-169.
HEATHER, C. D. & VANDERZANT, W. C. (1957b). Effects of temperature and time of incubation and pH of the plating medium on enumerating heat treated psychrophilic bacteria. *Journal of Dairy Science* 40, 1079-1085.
HERNANDEZ, E. & FERIA, M. A. (1971). Typical flora of spoiled canned tomatoes. Thermal death time. *Revista de Agroquimica y Tecnologia de Alimentos* 11, 126-131.
HERSHEY, A. D. (1939). Factors limiting bacterial growth. VII. Respiration and growth properties of *Escherichia coli* surviving sub-lethal temperatures. *Journal of Bacteriology* 38, 563-578.
HOFFMAN, H., VALDINA, J. & FRANK, M. E. (1966). Effects of high incubation temperature upon the cell wall of *Escherichia coli*. *Journal of Bacteriology* 91, 1635-1637.
HORNER, K. J. & ANAGNOSTOPOULOS, G. D. (1975). Effect of water activity on heat survival of *Staphylococcus aureus, Salmonella typhimurium* and *Salmonella senftenburg*. *Journal of Applied Bacteriology* 38, 9-17.
HURST, A., COLLINS-THOMPSON, D. L. & KRUSE, H. (1973b). Effect of glucose and pH on growth and enterotoxin B synthesis by *Staphylococcus aureus* strain S6, after heat injury in sodium or potassium phosphate buffer. *Canadian Journal of Microbiology* 19, 823-829.
HURST, A., HUGHES, A., BEARE-ROGERS, J. L. & COLLINS-THOMPSON, D. L. (1973a). Physiological studies on the recovery of salt tolerance by *Staphylococcus aureus* after sub-lethal heating. *Journal of Bacteriology* 116, 901-907.
HURST, A., HUGHES, A. & COLLINS-THOMPSON, D. L. (1974). The effect of sub-lethal heating on *Staphylococcus aureus* at different physiological ages. *Canadian Journal of Microbiology* 20, 765-768.
IANDOLO, J. J. & ORDAL, Z. J. (1966). Repair of thermal injury of *Staphylococcus aureus*. *Journal of Bacteriology* 91, 137-142.
INGRAM, M. (1971). The microbiology of food pasteurization. *Pastorisering av Livsmedel, S.I.K. Rapport* No. 292, pp. A1-A43.
JACKSON, H. & WOODBINE, M. (1963). The effect of sub-lethal heat treatment on the

growth of *Staphylococcus aureus. Journal of Applied Bacteriology* **26**, 152-158.

KADAN, R. S., MARTIN, W. H. & MICKELSEN, R. (1963). Effects of ingredients used in condensed and frozen dairy products on thermal resistance of potentially pathogenic staphylococci. *Applied Microbiology* **11**, 45-49.

KELCH, F. & BUEHLMANN, X. (1958). Effect of commercial phosphates on the growth of microorganisms. *Fleischwirtschaft* **10**, 325-328.

KENIS, P. R. & MORITA, R. Y. (1968). Thermally induced leakage of cellular material and viability in *Vibrio marinus*, a psychrophilic marine bacterium. *Canadian Journal of Microbiology* **14**, 1239-1244.

KNIEWALLNER, K. & PRANDL, O. (1974). Experiments in lowering the heat resistance of microorganisms. *Fleischwirtschaft* **54**, 1189-1192, 1195-1200.

KOHL, W. F. (1971). A new process for pasteurising egg whites. *Food Technology, Champaign* **25**, 102-110.

LANGRIDGE, P. & MORITA, R. Y. (1966). Thermostability of malic dehydrogenase from the obligate psychrophile *Vibrio marinus. Journal of Bacteriology* **92**, 418-423.

LEMCKE, R. M. & WHITE, H. R. (1959). The heat resistance of *Escherichia coli* from cultures of different ages. *Journal of Applied Bacteriology* **22**, 193-201.

LENISTEA, C., CHITU, M. & ROMAN, A. (1970). Heat resistance in milk of some strains of group D streptococci from pasteurised milk and the influence exerted on their growth by selective media. *Zentralblatt für Bakteriologie, Parasitenkunde, Infektionskrankheiten und Hygiene Abt. I. Originale* **215**, 173-181.

LUEDECKE, L. O. & HARMON, L. G. (1966). Thermal resistance of *Pseudomonas fragi* in milk containing various amounts of fat. *Applied Microbiology* **14**, 716-719.

MAAS, W. K. (1952). Production of an altered pantothenate synthesizing enzyme as a result of mutation. *Federation Proceedings* **11**, 475.

McDONOUGH, F. E. & HARGROVE, R. E. (1968). Heat resistance of salmonella in dried milk. *Journal of Diary Science* **51**, 1587-1591.

MILLER, L. L. & ORDAL, Z. J. (1972). Thermal injury and recovery of *Bacillus subtilis. Applied Microbiology* **24**, 878-884.

MOATS, W. A. (1971). Kinetics of thermal death of bacteria. *Journal of Bacteriology* **105**, 165-171.

MOATS, W. A., DABBAH, R. & EDWARDS, V. M. (1971a). Interpretation of non-logarithmic survivor curves of heated bacteria. *Journal of Food Science* **36**, 523-526.

MOATS, W. A., DABBAH, R. & EDWARDS, V. M. (1971b). Survival of *Salmonella anatum* heated in various media. *Applied Microbiology* **21**, 476-481.

MOLLER, W., AMONS, R., GREENE, J. C. L., GARRETT, R. A. & TERHORST, C. P. (1969). Protein-ribonucleic acid interactions in ribosomes. *Biochimica et biophysica acta* **190**, 381-390.

NAGY, E. (1969). In vitro studies on the thermal resistance of certain saprophytic bacteria and organisms of the Salmonella group. *Wiener tierärztliche Monatasschrift* **56**, 410-418.

NATORI, S., NOZAWA, R. & MIZUNO, D. (1966). The turnover of ribosomal RNA of *Escherichia coli* in a magnesium deficient stage. *Biochimica et biophysica acta* **114**, 245-253.

NELSON, F. E. (1943). Factors which influence the growth of heat treated bacteria. *Journal of Bacteriology* **45**, 395-403.

NELSON, F. E. (1956). The influence of the pH of the plating medium upon enumeration of heat-treated bacteria. *Bacteriological Proceedings* p. 40.

NIVEN, C. F., BUETTNER, L. G. & EVANS, J. B. (1954). Thermal tolerance studies on the heterofermentative lactobacilli that cause greening of cured meat products. *Applied Microbiology* **2**, 26-29.

NOMURA, M. (1972). Assembly of bacterial ribosomes. *Federation Proceedings* **31**, 18-20.

NOMURA, M., TRAUB, P. & BECHMAN, H. (1968). Hybrid 30S ribosomal particles reconstituted from components of different bacterial origins. *Nature, London* **219**, 793-799.

OLUSKI, A. (1963). Contribution to knowledge on thermoresistance of some micrococcus strains. (II). *Technologija Mesa.* **5**, 131-133.

OTT, T. M., EL-BISI, H. M. and ESSELEN, W. B. (1961). Thermal destruction of *Streptococcus faecalis* in prepared frozen foods. *Journal of Food Science* **26**, 1-9.

PIERSON, M. D. & ORDAL, Z. J. (1971). The transport of methyl-α-D-gluco-pyranoside by thermally stressed *Salmonella typhimurium*. *Biochemical Biophysical Research Communications* **43**, 378-383.

PIERSON, M. D. & ORDAL, Z. J. (1973). The metabolic activity of heat-injured *Salmonella typhimurium*. *Bacteriological Proceedings* (Abstract) p. 18.

PIERSON, M. D., TOMLINS, R. I. & ORDAL, Z. J. (1971). Biosynthesis during recovery of heat-injured *Salmonella typhimurium*. *Journal of Bacteriology* **105**, 1234-1236.

PIERSON, M. D., PAYNE, S. L. & ADES, G. L. (1974). Heat injury and recovery of vegetative cells of *Clostridium botulinum* type E. *Applied Microbiology* **27**, 425-426.

PUROHIT, K. & STOKES, J. L. (1967). Heat-labile enzymes in a psychrophilic bacterium. *Journal of Bacteriology* **93**, 199-206.

RAHN, O. (1945). Physical methods of sterilization of microorganisms. *Bacteriological Reviews* **9**, 1-47.

READ, R. B., BRADSHAW, J. G., DICKERSON, R. W. & PEELER, J. T. (1968). Thermal resistance of salmonellae isolated from dry milk. *Applied Microbiology* **16**, 998-1001.

ROBESON, S. M. & MORITA, R. Y. (1966). The effect of moderate temperature on the respiration and viability of *Vibrio marinus*. *Zeitschrift für Allgemeine Mikrobiologie* **6**, 181-187.

ROSENTHAL, L. J. & IANDOLO, J. J. (1970) Thermally induced intracellular alteration of ribosomal ribonucleic acid. *Journal of Bacteriology* **103**, 833-835.

ROSENTHAL, L. J., MARTIN, S. E., PARIZA, M. W. & IANDOLO, J. J. (1972). Ribosome synthesis in thermally shocked cells of *Staphylococcus aureus*. *Journal of Bacteriology* **109**, 243-249.

RUSSELL, A. D. & HARRIES, D. (1967). Some aspects of thermal injury in *Escherichia coli*. *Applied Microbiology* **15**, 407-410.

SAUNDERS, G. F. & CAMPBELL, L. L. (1966). Ribonucleic acid and ribosomes of *Bacillus stearothermophilus*. *Journal of Bacteriology* **91**, 332-339.

SCHMIDT, C. F. (1957). Thermal resistance of microorganisms. In *Antiseptics, Disinfectants, Fungicides and Sterilization*, 2nd edn, ed. Reddish, G. F. London: Henry Kimpton.

SEDGWICK, S. G. & BRIDGES, B. A. (1972). Evidence for indirect production of DNA strand scissions during mild heating of *Escherichia coli*. *Journal of General Microbiology* **71**, 191-193.

SEVOSTYANOVA, N. A., BOGDANOVA, N. V. & KHETSURIANI, K. G. (1971). The effect of sorbic acid and heating on *Pseudomonas fluorescens*. *Konservnaya i ovoshchesushil'naya promyshlennost'* **1971**, 30-31.

SHANNON, E. L., REINBOLD, G. W. & CLARK, W. E. Jr. (1970). Heat resistance of enterococci. *Journal of Milk and Food Technology* **33**, 192-196.

SHERMAN, J. M. & CAMERON, G. M. (1934). Rate of growth and viability in *Bacterium coli*. *Journal of Bacteriology* **27**, 23-28.

SILVA, M. T. & SOUSA, J. C. F. (1972). Ultra-structural alterations induced by moist heat in *Bacillus cereus*. *Applied Microbiology* **24**, 463-476.

SMOLKA, L. R., NELSON, F. E. & KELLET, L. M. (1974). Interaction of pH and NaCl on enumeration of heat-stressed *Staphylococcus aureus*. *Applied Microbiology* **27**, 443-447.

SOGIN, S. J. & ORDAL, Z. J. (1967). Regeneration of ribosomes and ribosomal ribonucleic acid during repair of thermal injury in *Staphylococcus aureus*. *Journal of Bacteriology* **94**, 1082-1087.

STENESH, J. & YANG, C. (1967). Characterization and stability of ribosomes from mesophilic and thermophilic bacteria. *Journal of Bacteriology* **93**, 930-936.

STILES, M. E., ROTH, L. A. & CLEGG, L. F. L. (1973). Heat injury and resuscitation of

Escherichia coli. Canadian Institute of Food Science and Technology Journal **6**, 226-229.

STRANGE, R. E. & SHON, M. (1964). Effects of thermal stress on viability and ribonucleic acid of *Aerobacter aerogenes* in aqueous suspensions. *Journal of General Microbiology* **34**, 99-114.

STUMBO, C. R. (1965). *Thermobacteriology in Food Processing.* New York: Academic Press.

TAL, M. (1969). Thermal denaturation of ribosomes. *Biochemistry* **8**, 424-435.

THOMAS, C. T., WHITE, J. C. & LONGREE, K. (1966). Thermal resistance of salmonellae and staphylococci in foods. *Applied Microbiology* **14**, 815-820.

TOMLINS, R. I. (1971). Thermal injury and recovery in *Salmonella typhimurium* 7136. The *in vivo* degradation and resynthesis of ribosomes and ribosomal ribonucleic acid. Ph.D. Thesis, University of Illinois. U.S.A.

TOMLINS, R. I. & ORDAL, Z. J. (1971a). Requirements of *Salmonella typhimurium* for recovery from thermal injury. *Journal of Bacteriology* **105**, 512-518.

TOMLINS, R. I. & ORDAL, Z. J. (1971b). Precursor ribosomal ribonucleic acid and ribosome accumulation in vivo during the recovery of *Salmonella typhimurium* from thermal injury. *Journal of Bacteriology* **107**, 134-142.

TOMLINS, R. I., PIERSON, M. D. & ORDAL, Z. J. (1971). Effect of thermal injury on the TCA cycle enzymes of *Staphylococcus aureus* MF31 and *Salmonella typhimurium* 7136. *Canadian Journal of Microbiology* **17**, 759-765.

TOMLINS, R. I., VAALER, G. L. & ORDAL, Z. J. (1972). Lipid biosynthesis during the recovery of *Salmonella typhimurium* from thermal injury. *Canadian Journal of Microbiology* **18**, 1015-1021.

UPADHYAY, J. & STOKES, J. O. (1963). Temperature sensitive formic hydrogenlyase in a psychrophilic bacterium. *Journal of Bacteriology* **85**, 177-185.

VAS, K. & PROSZT, G. (1957). Observation on the heat destruction of spores of *Bacillus cereus. Journal of Applied Bacteriology* **20**, 431-441.

VRCHLABSKY, J. & LEISTNER, L. (1970). Hitzeresistenz der Enterokokken bei unterschiedlichen a_W-Werten. *Fleischwirtschaft* **50**, 1237-1238.

VRCHLABSKY, J. & LEISTNER, L. (1971). Hitzeresistenz von Laktobazillen bei unterschiedliechen a_W-Werten *Fleischwirtschaft* **51**, 1368-1370.

WALKER, G. C. & HARMON, L. G. (1966). Thermal resistance of *Staphylococcus aureus* in milk, whey and phosphate buffer. *Applied Microbiology* **14**, 584-590.

WHITE, H. R. (1953). The resistance of *Streptococcus faecalis. Journal of General Microbiology* **8**, 27-37.

WHITE, H. R. (1963). The effect of variation in pH on the heat resistance of cultures of *Streptococcus faecalis. Journal of Applied Bacteriology* **26**, 91-99.

WOODCOCK, E. & GRIGG, G. W. (1972). Repair of thermally induced DNA breakage in *Escherichia coli. Nature-New Biology* **237**, 76-79.

ZAKULA, R. (1969). Results on investigation of thermoresistance of some bacteria suspended in meat, lard and tallow. *Fifteenth European Meeting of Meat Research Workers* b.7 p. 157-163.

ZAKULA, R., TODOROVIC, M. & TOMCOR, D. (1970). Heat resistance of bacilli, micrococci, streptococci and clostridia introduced into minced pork and beef. *Zbornik Radora, Tehnoloski Fakultet i Jugoslavenski Institut za Prehrambeau Industriju, Novi Sad* **2**, 171-182.

ZALESKI, S., SOBOLEWSKA-CERONIK, K. & CERONIK, E. (1971). The effect of hydrated soybean oil on the heat resistance of an enterotoxigenic *Staphylococcus aureus* strain. *Annales de l'Institut Pasteur de Lille* **22**, 263-267.

ZIVANOVIC, R., OLUSKI, A. & TADIC, Z. (1965). Contribution to knowledge of the thermoresistance of streptococci of Lancefield group D. *Technologija Mesa.* **6**, 198-205.

Effects of Temperature on Filamentous Fungi

J. G. ANDERSON AND J. E. SMITH

*Department of Applied Microbiology, Strathclyde University
Glasgow, Scotland*

CONTENTS

1. Introduction 191
2. Effects of temperature on growth 192
3. Heat resistance 194
4. Effects of temperature on cell components and cell structure 196
   (a) Effects on enzymes and other proteins 197
   (b) Effects on lipids 198
   (c) Effects on membranes 199
5. Effects of temperature on morphology 201
   (a) Induction of budding yeast-like cells 202
   (b) Induction of non-budding spherical cells and unusual hyphal forms . . . 207
   (c) Induction of microcycle conidiation 211
6. References 212

1. Introduction

TEMPERATURE is clearly a major factor influencing the growth, development and metabolic activities of filamentous fungi. As such it provides one of the most basic means of controlling both the biosynthetic and biodegradative activities of these organisms which are of great economic importance particularly in the food and fermentation industries. The filamentous fungi are also of considerable importance as agents of human and animal infection and in at least some instances temperature appears to play an important role in influencing the *in vivo* morphology of the organisms.

Most filamentous fungi have a life cycle comprised of several well-defined stages and there is a large literature, summarized in various reviews, dealing with the effects of temperature on spore germination (Sussman, 1966; Sussman & Halvorson, 1966), hyphal growth (Deverall, 1965) and sporulation (Hawker, 1966). The influence of temperature on fungal survival (Mazur, 1968) and the topics of fungal psychrophilism (Deverall, 1968) and fungal thermophilism (Emerson, 1968; Crisan, 1973) have also received considerable attention. While a great deal has been written, most of this is observational and there is little information on the physiological and biochemical basis of the induction or inhibition of fungal activities by temperature.

More precise information on the effects of temperature on metabolic processes has been obtained for the bacteria and to a lesser extent also the yeasts. While the underlying physiological effects of temperature are likely to be similar in various micro-organisms (see Crisan, 1973, for an excellent comparative review), nevertheless the filamentous fungi possess characteristics not shared by the unicellular organisms. This relates particularly to growth mechanism, the filamentous colonial structure and the production of complex reproductive structures. Clearly then, there will be temperature effects which are unique to this group of organisms.

It is the intention of this review to deal only with the effects of medium to elevated temperatures on filamentous fungi and no reference will be made to low or freezing temperature effects. In most of the studies which are referred to in this review it has been assumed that temperature is the major environmental factor affecting the organism. It should be borne in mind however that temperature affects most of the physicochemical characteristics of the environment and that consequently a complexity of factors, influenced by, but distinct from temperature, may also be involved in the resultant alteration to cell structure and metabolism.

2. Effects of Temperature on Growth

The cardinal growth temperatures have been determined for a wide variety of filamentous fungi and these have been summarized in various reviews (Wolf & Wolf, 1947; Hawker, 1950; Cochrane, 1958; Deverall, 1965). In general it is known that most fungi make at least some growth over a 25 or 30° range and that most fungi are unable to grow at 35-40°. There are however now known to be > 50 fungi that appear to be thermophilic (fungi with growth temperature optima of 40° or higher) and these have been listed together with their cardinal temperatures by Crisan (1973). They include representatives from the ascomycetes, basidiomycetes, deuteromycetes, zygomycetes and mycelia sterilia. The most heat tolerant of these organisms have a maximum temperature for growth of 61° and this may well represent the highest temperature at which filamentous fungi can be found in an active growing state.

Temperature however interacts with other environmental factors and it is clear that temperature optima and ranges reported are valid only under specified conditions of time, medium and method of measurement. It is known, for example, that the nutritional requirements of some organisms change during growth at supraoptimal temperatures. Fries (1953) found that a strain of *Coprinus fimetarius* Fr. which had a temperature optimum for growth between 35 and 40° could continue to grow at 44° if supplemented with methionine or to a lesser extent with homocysteine. Fries & Källströmer (1965) observed a requirement for biotin in *Aspergillus niger* when grown on a rhamnose medium at high temperature. Other examples of this phenomenon have been discussed by Langridge (1963) and Deverall (1965).

Water activity is another factor which appears to interact strongly with temperature to affect microbial growth. A detailed study showing the interacting effects of three factors, temperature, water activity and pH, on the growth and spoilage potential of filamentous fungi has been carried out by Horner & Anagnostopoulos (1973). Their results (Table 1) showed that fungal growth could be markedly decreased by a simultaneous decrease in a_w (water activity) and temperature. While this could be expected the results were particularly interesting since the extent of growth inhibition by this treatment was greater than could have been predicted by considering the individual effects of a_w and temperature which were obtained in separate experiments. The results also showed that there was little interaction of pH with temperature or a_w.

Table 1

Interaction between water activity a_w and temperature on the growth of three filamentous fungi

Organism	Temperature (°)	Water activity (a_w)	Time for formation of a colony of 2 mm diam (h)
Aspergillus niger	15	0.94	104
	15	0.997	57
	25	0.94	44
	25	0.997	26
Penicillium sp.	15	0.94	98
	15	0.997	46
	25	0.94	42
	25	0.997	21
Rhizopus sp.	15	0.94	51
	15	0.997	22
	25	0.94	28
	25	0.997	11

Horner & Anagnostopoulos (1973).

Many other factors, in addition to those already mentioned, could be expected to affect the temperature response of organisms although there is little quantitative information available. Temperature, for instance, clearly affects gas solubility and it has been suggested that at least in some instances growth limitation or metabolic changes in filamentous fungi at supraoptimal temperatures may be related to a decreased amount of dissolved oxygen in the culture medium (Sumner *et al.*, 1969).

The effect of temperature on fungal growth rates has been discussed in previous reviews (Cochrane, 1958; Deverall, 1965; Waid, 1968). It is pertinent however to mention the recent detailed study of Alberghina (1973) on the effect of temperature on growth regulation in *Neurospora crassa*. When the growth rates of *N. crassa* at different temperatures were expressed as an Arrhenius plot (Fig. 1) a sharp change in the slope was observed at 20°. The slope of the straight line for the temperature range 15-20° was much greater than that for the 20-35° range. An activation energy of 46 Kcal/mole was calculated for the average slope of the lower temperature range and one of 8 Kcal/mole for the higher temperature range. This marked deviation of the Arrhenius plot of the constant of the rate of growth at 20° led to the suggestion that either denaturation of a rate-limiting enzyme or ribosome damage may be the molecular basis of the reduction in growth rate which occurred below 20°.

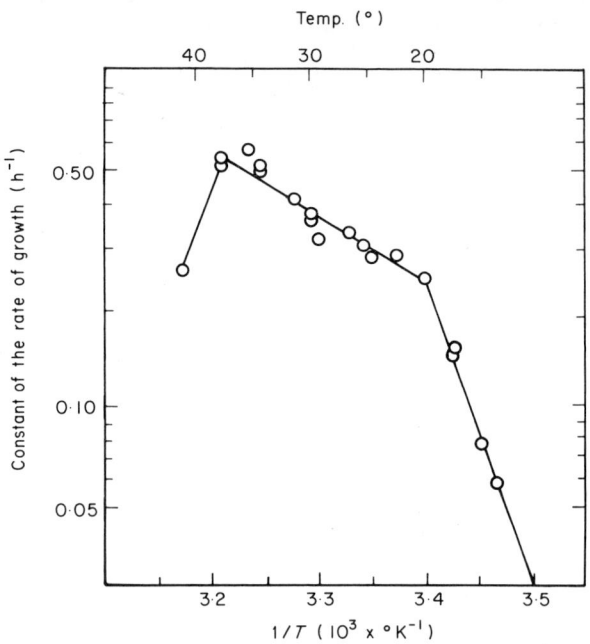

Fig. 1. Arrhenius plot of the rate of growth of *Neurospora crassa* (Alberghina, 1973).

3. Heat Resistance

The majority of fungi, with the exception of some thermophiles, have thermal death times similar to those recorded for the vegetative cells of mesophilic bacteria. There are however no known fungal spores which possess the degree of heat resistance as shown by their bacterial counterparts. Consequently studies on both the theoretical and practical aspects of metabolic inactivation and sterilization by heat have been carried out predominantly with bacteria and in only a few instances have the filamentous fungi received attention.

One situation where filamentous fungi can cause a spoilage problem after heat treatment concerns the processing of various fruits and juices where the heat treatment is kept relatively mild to avoid flavour damage. In these cases the fungus involved usually belong to the genus *Byssochlamys* which encompasses two species, *B. fulva* and *B. nivea*. The spoilage potential of *Byssochlamys* was first noticed in England in the early 1930s and these fungi have since caused spoilage outbreaks in many other countries (Ruyle *et al.*, 1946; Put & Kruiswijk, 1964; Richardson, 1965; Yates *et al.*, 1968).

The *Byssochlamys* fungi are ascomycetes which produce spherical eight-spored asci and it is the ascospore which is believed to be the heat resistant stage in the life cycle (Brown & Smith, 1957). Conidia are produced in large numbers but they lack heat resistance (Hull, 1939). *Paecilomyces varioti,* which is considered to be the imperfect form of *B. fulva* (Brown & Smith, 1957) has also been isolated as a heat resistant fungus (Lüthi & Vetsch, 1955; Put, 1964: King *et al.*, 1969). This suggests that these are strains which produce ascospores under natural conditions but not in artificial culture.

In addition to *Byssochlamys* and *Paecilomyces* a number of other heat resistant fungi have been isolated from fruit juices and processing plants. These include *Thermoascus aurantiacum,* certain *Aspergillus* and *Penicillium* spp., *Monascus purpurea, Aureobasidium pullulans* and *Phialophora* sp. (Jensen, 1960; King *et al.*, 1969). In experiments on the thermal resistance of a number of their isolates King *et al.,* (1969) found that *B. nivea, T. aurantiaceum* and *Penicillium* sp. were the most heat resistant, surviving several heat treatments, the most severe of which was at 88° for 60 min (Table 2).

Table 2
Heat-resistant moulds isolated from grape material

Species	No. of isolates surviving successive heat treatments				
	80° for 30 min	followed by:	88° for 45 min	followed by:	88° for 60 min
Byssochlamys nivea	18		15		4
Paecilomyces varioti	7		3		0
Thermoascus aurentiacum	4		3		3
Aspergillus sp.	20		9		0
Penicillium sp.	1		1		1
Others	2		0		
Total surviving strains	52		31		8

King *et al.* (1969).

Byssochlamys is therefore not unique in its heat resistant properties; nevertheless, because of its commercial importance as a spoilage organism it has received most attention. The growth range of *B. fulva* extends from *c.* 6-53°

with an optimum at 35°. It is able to grow in extremely low levels of oxygen (in a gas mixture containing only 0.27% of O_2) and this ability together with its heat resistance probably contributes to the spoilage potential of the organism. A considerable amount of experimental work has been carried out on the ascospores, and conditions which give good ascospore production to facilitate these studies have been reported by Splittstoesser et al. (1969). Details of the ultrastructure of the *Byssochlamys* asci and ascospores have been given by Partsch et al. (1969). In heat resistance studies King et al. (1969) have shown that the asci have a D value (time, at a given temperature, after which 10% of the population survives) of c. 10 min at 88° when heated in fruit juice. In the presence of 90 μl of SO_2/l the D value could be reduced by half.

A problem which has been encountered in heat resistance studies on *Byssochlamys* was the difficulty of obtaining accurate counts of survivors. The viable counting procedure was complicated by the fact that some of the asci break up into their constituent ascospores whereas others remain intact. Recently, Michener & King (1974) have described a method of preparing a uniform suspension of free ascospores for heat resistance studies based on a pressure treatment which ruptures the asci but does not affect the heat resistance of the ascospores.

Another finding which emerged during heat resistance studies was that germination of the ascospores of *Byssochlamys* could be activated by heat shock treatment. Optimum germination was obtained by a heat shock of 75° for 10 min with *B. fulva* (Hull, 1939) and 75° for 5 min with *B. nivea* (Yates et al., 1968). The possibility that heat processing might provide a heat shock to stimulate ascospore germination has been discussed by Yates et al. (1968). This activation of germination by heat shock has also been observed with several other fungi. Optimum conditions for heat activation of sporangiospores of *Phycomyces blakesleeanus* involves exposure at 50° for 30 min (Sommer & Halbsguth, 1957). For basidiospores of *Coprinus radiatus* optimum conditions are 44-46° for 4 h and for ascospores of *N. crassa*, 50-60° for 10-20 min (Sussman, 1966).

4. Effects of Temperature on Cell Components and Cell Structure

Considering the multiplicity of reactions and the ultrastructural complexities which occur in living cells it is not surprising that temperature should exert such a profound effect on all aspects of metabolism. A considerable number of studies have been successful in identifying the precise effect of temperature on a particular metabolic process. It has however been less easy to speculate as to how such individual metabolic changes have a causal or contributory role in the overall response of the cell to the altered thermal environment. Most of the work has been done with bacteria, and to a lesser extent also with yeasts, and those studies which have been done ont the filamentous fungi have tended to be of a comparative nature. In general, similar concepts have been developed for these various micro-organisms concerning the metabolic and ultrastructural effects of temperature. These concepts have been reviewed by Crisan (1973) for the thermophilic fungi and the yeasts have been treated in similar detail by Stokes (1971).

(a) *Effects on enzymes and other proteins*

There is considerable evidence that enzyme destruction is involved to some extent in setting the upper limits for the growth of bacteria (Stokes, 1967; Farrell & Rose, 1967a,b) and yeasts (Stokes, 1971). Enzymes of the filamentous fungi, which have been examined for thermostability, include ribonucleases (Craveri & Colla, 1966), β-glucosidase (Longinova & Tashpulatov, 1967), carboxypeptidase (Zuber, 1967), acid phosphatase (Crisan, 1969), lipase (Somkuti & Somkuti, 1969), aminopeptidase (Chapuis & Zuber, 1970), glucose-6-phosphate dehydrogenase (Broad & Shepherd, 1971) and RNA polymerases (Boguslawski *et al.*, 1974). In cases where enzyme thermostability was compared in mesophilic and thermophilic strains, it was generally found that although the enzymes from thermophiles showed higher optimum and maximum temperatures there was no evidence that the thermophilic fungi possessed enzymes of unusually high thermostability.

In studies on *Penicillium dupontii* and *Humicola lanuginosa,* Crisan (1969, 1973) separated by electrophoresis a large number of cellular proteins from the organisms after growth at different temperatures and then tested these for thermostability. The results indicated that the fungi when grown at higher temperatures produced proteins which showed some but not a great deal of increased heat resistance. Crisan (1973) has concluded from these results that the absence of proteins of unusually high thermostability is in contrast to the situation in the thermophilic bacteria and that the explanation for the existence of thermophilic fungi is not simply by the possession of very thermostable enzymes.

A word of caution however, must be sounded when tests on enzymes and proteins are carried out on cell extracts. The difficulty in biochemical studies, of extrapolating from *in vitro* enzyme studies to the *in vivo* situation is now well recognized. It may well be also that thermal characteristics of proteins differ in the intact cell. For example, Yu *et al.* (1967) have shown that the trehalase of *N. crassa* is more stable at temperatures of 60 or $65°$ in intact ascospores than in extracts. More recently Hecker & Sussman (1973a,b) have speculated that it is either the enzyme's substrate, trehalose, present in large amounts in the ascospores, or an association of the enzyme with the cell wall, which protects the trehalase from heat inactivation at $65°$.

(b) *Effects on lipids*

Apart from the work on the thermostability of proteins the most intensively studied aspect of fungal metabolism in relation to temperature has been concerned with the quantity and composition of the cellular lipids. The impetus for such studies has undoubtedly arisen from the previous finding, with higher organisms and with bacteria and yeasts of an interesting temperature-lipid association. In general, for most poikilothermic organisms examined, it has been found that cellular lipids become increasingly saturated as the growth temperature is increased. Because of the importance of lipids in membrane structure and function it has been postulated that the degree of saturation of such lipids may be an important factor governing the temperature range for growth.

Although there have been some conflicting reports in the literature regarding the effect of temperature on the composition of the lipids of filamentous fungi, the most recent detailed studies have shown that the typical temperature-lipid relationship is also found in the filamentous fungi. In studies on the zygomycetes *Mucor* and *Rhizopus,* Sumner *et al.* (1969) found that the lipids of psychrophilic, thermotolerant and thermophilic fungi were generally more unsaturated when grown at a lower temperature. With the deuteromycete *Humicola* it has been found that in general the thermophilic species contained more lipid than the mesophilic species and that the lipids in the thermophilic species were more saturated (Mumma *et al.,* 1970, 1971). Also working with a *Humicola* spp. Crisan (1973) found that this thermophile produced almost nine times more lipid at 37 than at 52°. It was suggested that the unusually high lipid content at 37° might represent the production of a larger amount of storage lipid at the suboptimal temperature.

Although the effect of temperature on lipid composition is now well established the underlying biochemical mechanism is not yet clear although a number of proposals have been made. Kates & Baxter (1962) have suggested that at lower temperatures the synthesis of unsaturated fatty acids may be less retarded than their degradation. Sumner *et al.* (1969) have suggested that the temperature effects observed with the Mucorales fungi may be indirect in as much as the temperature affects the solubility of oxygen which in turn affects fatty acid synthesis. They further speculated that the temperature-oxygen effect may be mediated through the desaturase enzymes which require oxygen as a co-factor.

Any interpretation of the effects of temperature upon the metabolism and cellular composition of micro-organisms is undoubtedly complicated by the fact that other environmental factors may be interacting with temperature. This is particularly true in any batch-growth system where the environment is constantly changing as growth progresses. Also it must be established whether the changes in composition are a direct result of a change in temperature or whether they are partly or entirely due to the change in growth rate that results from a change in temperature. Both of these problems can be greatly minimized by the use of the chemostat culture technique and although this does not yet appear to have been used in temperature studies on filamentous fungi some excellent studies have been done with yeasts (Brown & Rose, 1969a,b; Jones & Hough, 1970).

(c) *Effects on membranes*

There have been several studies on the effects of supraoptimal temperatures on the ultrastructure of filamentous fungi and these have invariably cited membrane damage as one of the most obvious cytological features. Langvad (1972) has carried out a detailed study on the wood rotting fungus *Serpula lacrymans* which is very sensitive to heat. The optimum growth temperature is 23° and yet growth is inhibited completely at 28°. After exposure to 37.5° for 20 min the cristae of mitochondria and the nuclear membrane began to break down and material from the nucleolus leaked out. After exposure for 1 h normal mitochondria and nuclei could no longer be observed, and after 4 h the hyphae were completely disorganized and all organelles and membrane systems were disrupted.

Somewhat similar ultrastructural changes have been observed in fungal spores held at supraoptimal temperatures. Ojha & Turian (1968) compared the ultrastructure of conidia of *N. crassa* produced at different temperatures. In 37° cultures the mitochondria were swollen with altered cristae and the endoplasmic reticulum (ER) was irregular and more prominent than in 25° cultures. By contrast, at 40° the mitochondria in conidia were shrunken with dispersed cristae and the ER was broken and occasionally formed vesicles. Baker & Smith (1970) have examined the ultrastructural changes in germinating spores of *Rhizopus stolonifer* and *Monilinia fructicola* when exposed to elevated temperatures. With *R. stolonifer* spores, after 2.5 min exposure at 52°, the nuclei and mitochondria were either disrupted or disorganized and the ribosomes tended to aggregate. With *M. fructicola* the effects were qualitatively similar although not as extreme. Sargent & Payne (1974) have reported ultrastructural changes in the spores of the plant pathogen *Bremia lactucae* when held at the supraoptimal temperature of 28°. Conidia maintained at this temperature became highly vacuolate and their membranes were disrupted. From observations on the movement and properties of intracellular lipid droplets, Sargent & Payne (1974) suggested that the elevated temperature might be acting to prevent the conversion of lipid reserves to intermediates necessary for the supply of energy and the maintenance and synthesis of membrane systems required for germination.

Other studies on filamentous fungi also point to the significant effect which elevated temperatures have on the cell membranes. In studies on the germination of rust urediospores, Maheshwari & Sussman (1971) have postulated that reversible changes in membrane structure and permeability are caused by cycles of extreme cold and subsequent warming to 40°. In relation to germination of ascospores of *N. crassa,* Hecker & Sussman (1973b) have pointed out that the ascospores contain large amounts of the disaccharide trehalose together with the enzyme trehalase yet the trehalose is not metabolized until activation of the ascospore by heat shock or treatment with furfural. Their results indicate that trehalase may be associated with the innermost wall of the ascospore whereas trehalose is located in the cytoplasm; thus the two are physically separated by the plasma membrane. They have suggested that heat or various chemicals which can activate the spores may act by causing an increase in the permeability of the plasma membrane thus allowing trehalose to diffuse to the vicinity of its hydrolase, thereby providing the energy and intermediates for germination. Significantly perhaps, Mandels *et al.* (1965) have also found that treatment of the spores of *Myrothecium verrucaria* with either heat (60° for 20 min) or toluene results in the release of internal stores of trehalose.

A number of studies have been carried out with yeasts which also illustrate the susceptibility of the cellular membranes to damage by supraoptimal temperatures (Stokes, 1971). While the present review has been restricted to the filamentous fungi it is nevertheless relevant to discuss the recent significant finding by Hagler & Lewis (1974) of a temperature–membrane effect in yeasts which may also be applicable to the filamentous fungi. In studies with a number of yeast strains it was found that the yeast cells could be heated above their maximum growth temperature (T_{max}) in water suspension without membrane damage and lethal effect. However injury occurred readily when glucose was present during heating or added after heat stress. Such yeast lost the ability to establish or maintain a concentration gradient of sorbose and, simultaneously, ATPase activity could be measured; both of these features are characteristic of yeast cells with a ruptured cytoplasmic membrane. The net effect was a substantial, irreversible loss of yeast contents and ultimate cell death. It was concluded from these results that no growth of yeasts could occur above their T_{max} because the utilizable sugar of the growth medium could damage the cytoplasmic membrane. An important observation from these studies was that thermal characteristics of micro-organisms might differ substantially depending on whether the cells are suspended in water or a growth medium containing utilizable sugar.

It is clear then from the results of both ultrastructural and biochemical studies that temperature can have marked effects on the cellular membranes of fungi. This indeed is consistent with much current thinking on the importance of the membranes and ultrastructural organization in general, in determining the response of microbial and other cells to temperature. The merits of a hypothesis on ultrastructural thermostability in relation to other theories on the effects of temperature on living cells has been presented in detail by Crisan (1973). Within the context of this theory on ultrastructural thermostability Crisan considers that the optimum temperature of an organism is that temperature which is optimal for achieving an integration of all essential metabolic activities. At the minimum temperature for growth the fluidity and flexibility of the membranes is reduced to such an extent that essential membrane functions are impaired. At the maximum temperature for growth excess fluidity and flexibility act to affect membrane function and at higher temperatures disorganization reaches irreparable proportions and thermal death results. Crisan considers that this idea of ultrastructural thermostability is the only hypothesis at present which can explain the existence of the thermophilic fungi.

5. Effects of Temperature on Morphology

The life cycle of most filamentous fungi comprises a series of quite marked changes in morphology extending from the conversion of a more or less spherical cell to a hyphal structure during germination and then the subsequent elaboration of quite complex reproductive structures from the filamentous mycelium. Thus, there clearly lies within these organisms an inherent ability to undergo extensive changes in form. It is now well realized that such differentiation processes may require the operation of unique metabolic processes and undoubtedly considerable membrane synthesis and ultrastructural reorganization must be involved. It is hardly surprising, considering what has already been discussed in Section 4, that temperature can exert a powerful influence on differentiation processes and consequently on the resultant morphology of the fungal organism.

In this review attention will be drawn to very dramatic effects which temperature can have on the most characteristic morphological feature of filamentous fungi, namely their hyphal form. The hyphal structure results from the operation of the well-known apical growth process. The importance of understanding this process is now well recognized since it is at the hyphal tip that most morphological form is programmed by interaction with the environment. It is relevant therefore, before proceeding to consider temperature effects, to summarize briefly, what is known about hyphal tip growth since it will then be apparent how interruption or malfunction of this process can lead to altered morphology.

The fungal hypha can never be considered as a homogeneous system but rather as a complex gradient of cytochemical changes. Ultrastructural studies have shown that a young hypha consists of three relatively distinct zones namely an apical zone, a subapical zone and a zone of vacuolation. The apical zone is characterized by an accumulation of cytoplasmic vesicles and is nearly devoid of other cell components. The vesicles, produced by the endomembrane system of the cell, are believed to contain wall precursors and both lytic and synthetic enzymes. The vesicles accumulate at the site of growth and by fusing with the plasmalemma they contribute to the expanding apex and thus mediate cell extension (Grove et al., 1970). This restriction of wall synthesis to the apex results in the characteristic cylindrical shape of the hyphae. Consequently, factors which interfere with this process either directly or indirectly as a result of an altered cellular metabolism, can generate alternative morphologies in normally filamentous fungi.

With numerous fungi there is evidence that the processes involved in the generation and maintenance of the filamentous structure are particularly susceptible to elevated temperatures. In a few cases, temperature seems to be the sole determining factor but in most instances there appears to be an interaction between temperature and a complexity of other environmental factors. With some fungi, particularly those which cause systemic mycoses, there is a conversion from filamentous growth to a budding yeast-like growth. With others, filamentous growth is suppressed and alternative morphologies, particularly spherical non-budding cells, are produced. The term dimorphism is applied universally to the former situation but has also been used with particular examples of the latter form of vegetative development. Dimorphism as defined by Romano (1966) is an environmentally controlled reversible interconversion of mycelial and yeast forms, denoted as $M \rightleftharpoons Y$. While many mycologists apply the term yeast-like growth to the situation where cells arise and reproduce by budding, others have used this to describe the production of cells which have a yeast-like appearance but which are incapable of budding, and this can lead to some confusion in the literature. Consequently, in this review a basic distinction will be made between the influence of temperature on the induction, in filamentous fungi, of a budding cellular growth phase and the induction of a variety of other more or less spherical but non-budding cell forms. In relation to the topic of this symposium, namely, the inhibition and inactivation of vegetative microbes, such morphological changes in fungi can be considered in many cases to result directly from conditions which are inhibitory to the filamentous form or as may be the case with certain pathogenic fungi, that the alterations to the filamentous morphology may represent changes which have evolved to allow survival and growth in the otherwise inhibitory *in vivo* environment. An unusual microcycle conidiation process will also be discussed since this arises as a direct consequence of inhibition of the hyphal growth form by exposure to supraoptimal temperature.

(a) *Induction of budding yeast-like cells*

The ability, shown by a number of filamentous fungi, to grow in either a filamentous or a budding yeast-like form is a phenomenon which has fascinated

Table 3

Influence of temperature and other factors on the conversion from mycelial (M) to budding yeast-like (Y) cells in various fungi

Species	Morphological change	Major inducing conditions	Reference
Blastomyces dermatitidis	M→Y	Temp. (37°)	See Mariat (1964), Gilardi (1965) and Romano (1966)
B. [Paracoccidioides] brasiliensis	M→Y	Temp. (37°)	
Histoplasma capsulatum	M→Y	Temp. (37°), nutrients, —SH	
Histoplasma farciminosum	M→Y	Temp. (37°), CO_2	
Sporotrichum [Sporothrix] schenckii	M→Y	Temp. (37°), CO_2	
Fusarium moniliforme	M→Y	Temp. (37°), agitation	Kidd & Wolf (1973)
Penicillium lilacinum	M→Y	Temp. (37°), low E_h	Rippon et al. (1965)
Aspergillus syndowi	M→Y	Temp. (37°), low E_h	Rippon et al. (1965)
Mycotypha sp.	M→Y	Low O_2, CO_2, nutrients, temp. (up to 45°), pH	Schulz et al. (1974) and Hall & Kolankaya (1974).

M→Y indicates conversion from mycelial to yeast-like form.

mycologists for many years. Many different environmental factors may be involved in the expression of M⇌Y dimorphism but in a wide range of fungi an elevated temperature is one of the most if not the most powerful inducer of the Y phase (Table 3).

It is useful to consider firstly the three genera *Histoplasma, Sporotrichum* and *Blastomyces* which all contain organisms causing deep mycoses and which produce septate mycelia *in vitro* at room temperature and a budding yeast-like parasitic phase. Because of the medical importance of these organisms there have been extensive studies on the conditions which influence the M⇌Y transformations and these have been dealt with in detail in previous reviews (Mariat, 1964; Gilardi, 1965; Romano, 1966; Austwick, 1968). With *Blastomyces dermatitidis* and *B. (Paracoccidioides) brasiliensis*, elevated temperature ($37°$) is the factor inducing the Y phase. With *Sporotrichum (Sporothrix) schenckii, Histoplasma farciminosum* and *H. capsulatum*, elevated temperature ($37°$) interacts with the nutritional and physicochemical status of the medium to induce the Y phase. The interaction of CO_2 and sulphydryl containing compounds with temperature appears to be of particular significance in these organisms.

There has been considerable interest in understanding the physiological basis of the temperature and temperature-nutritionally determined forms of dimorphism. With *H. capsulatum* the physiological and chemical characteristics of the M and Y phases have been compared (Mahvi, 1965; Kobayashi & Guiliacci, 1967; Gupta & Howard, 1971; Cino & Tewari, 1972; San-Blas & Carbonell, 1974). Studies have also been carried out to characterize the biochemical events associated with RNA metabolism during the Y→M and the M→Y transformations and these appear to have special significance. It has been observed that just after the shift from 37 to $23°$ which induces the Y→M conversion, net RNA synthesis stops for some time (Cheung *et al.*, 1974). The study of this phenomenon as well as the analysis of the RNA polymerase patterns in the yeast phase (Boguslawski *et al.*,1974) have suggested that the conversion process may require some change in the RNA synthesizing components of the cell. Subsequent experiments (Boguslawski *et al.*, 1975) have revealed that in contrast to extracts from the yeast phase, those from mycelia contained an inhibitor of RNA polymerase. This inhibitor, called histin, has been partially purified and found to be a heat stable, acidic protein, mol. wt 24,000. There is as yet no information on the role played by histin in *H. capsulatum*; however, Boguslawski *et al.* (1975) have speculated that a temperature-dependent critical level of histin in the cell could conceivably be a determinant in the phase transitions of this organism.

Most of the work which has been done on dimorphism in *Sporotrichum (Sporothrix) schenckii* has involved studies on the ultrastructure of the fungus. Details of the fine structure of both the Y form and the M form have been reported (Kitamura, 1965; Lane et al., 1969; Lurie & Still, 1969). Several studies have also been carried out to determine the ultrastructural changes which are involved in the Y→M and the M→Y transformations. During the Y→M conversion discrete intracytoplasmic membrane systems were formed shortly after the 37-25° temperature change (Lane & Garrison, 1970). These structures appeared to be similar to those observed during early Y→M conversion of both *Blastomyces dermatitidis* and *H. capsulatum* (Garrison et al., 1970). According to Lane & Garrison (1970) the formation of these systems appears to be the earliest detectable change in ultrastructural reorganization after the initiation of phase conversion as a result of a change in thermal environment. Further comparison of the ultrastructural changes which occur during the Y→M conversion in these fungi together with the results of Carbonell (1969) with *B. (Paracoccidioides) brasiliensis* has shown that the Y→M conversion in these dimorphic pathogenic fungi proceeds by a very similar sequence of ultrastructural events thus suggesting a common conversion mechanism (Lane & Garrison, 1970).

Ultrastructural changes involved in the M→Y transformation of *Sporotrichum (Sporothrix) schenckii* have also been studied (Garrison et al., 1975). The results of this study suggest that the M→Y transition of *S. schenckii* may be regulated by at least two mechanisms involving alterations of the biochemical and/or biophysical nature of the cell wall of the M phase cell in response to conversion stimuli. The results also indicated that direct budding of the yeast from the hyphae of *S. schenckii* by blastic action resembles that in *H. farciminosum* and *Phialophora dermatitidis* (Oujezdsky et al., 1973). With *P. dermatitidis* Grove et al. (1973), have also suggested that the budding process may be mediated by cytoplasmic vesicles derived from rudimentary Golgi apparatus.

In early studies on the physiology of dimorphism in *B. dermatitidis* and *P. brasiliensis*, Nickerson & Edwards (1949) proposed an interesting theory which has yet neither been proved nor disproved. They considered that the change in morphology upon conversion to the M form resulted from the selective inhibition of cell division, without simultaneous inhibition of growth, with selective inhibition being dependent only on temperature. They further considered that there is competition for a common substrate between the enzymes responsible for M form development and the enzymes responsible for Y form development. They assumed that a normally higher affinity of the mould–enzyme system for the common substrate would be offset at higher temperatures by an increasing rate in its reversible thermal inactivation thus explaining the observed dependence of the cell division mechanism on the maintenance of an elevated temperature. Taylor (1961) compared the nucleic acid and protein contents of both phases of *B. dermatitidis*. While DNA content was essentially the same, differences were found in protein and RNA levels and

Taylor considered that variations in structure of the fungus would result from the operation of many metabolic processes rather than by an effect on any individual enzyme.

More recent studies on dimorphism in *B. (Paracoccidioides) brasiliensis* have been concerned with a comparison of the ultrastructural and chemical characteristics of the walls of the mycelial and yeast phases. The advantages of this approach have been discussed in detail by Bartnicki-Garcia (1969) since whatever the underlying biochemical mechanism of dimorphism the change in morphology must be reflected in the cell wall since this is where cell shape is determined. Most of the work on this aspect of dimorphism in *P. brasiliensis* has been carried out by Carbonell and co-workers (Carbonell, 1967; Carbonell *et al.*, 1970; Kanetsuna *et al.*, 1969; Kanetsuna & Carbonell, 1970, 1971; Kanetsuna *et al.*, 1972). As a result of these studies it has been found that the main polysaccharide of hexoses of the Y form cell wall is α-glucan, whereas the polysaccharides of the M form cell wall are β-glucans and galactomannan. Cell free extracts of whole cells of the Y form had five times more protein disulphide reductase activity than the M form, whereas extracts of the M form contained five to eight times more β-glucanase activity than the Y form. It has been concluded from these studies that changes of cell wall glucans may play an important role in the dimorphism of *P. brasiliensis* and also that the information necessary for the synthesis of those enzymes regulating alterations of cell wall configuration during M→Y conversion may not be evenly distributed throughout the fungus.

On the basis of their findings and speculations Kanetsuna *et al.* (1972) have proposed a detailed structural model to explain temperature dimorphism in *B. (Paracoccidioides) brasiliensis.* It was suggested that the Y form cell wall is composed of two layers with α-glucan as the main hexose polysaccharide in the outer layer and chitin restricted to the inner layer. The cell wall contained small areas of β-glucan apparently localized as 'islets'. To explain budding at 37° it was suggested that there is a loss of rigidity around an islet of β-glucan and that this is blown out as a bud. At 37° the α-glucan and chitin layers are synthesized to form the Y form wall structure. To explain filamentous growth at 20° it was considered that at a budding site there was a predominance of β-glucan synthesis. Chitin is also found in the wall of the filamentous form but in this case it was suggested to be interwoven with other fibres (not in a distinct layer as in the Y form cell wall) and thus making the wall more rigid. An important aspect of the model is that the spherical cell growth, characteristic of the Y form, results from a dispersion of wall synthesis around the periphery of the cell whereas in the filamentous form, wall synthesis is restricted to the apical region. The importance of localized as opposed to dispersed wall synthesis as a controlling factor in dimorphism has been discussed in detail by Bartnicki-Garcia (1973).

The pathogenic fungi causing deep mycoses have been the most intensively studied examples of filamentous fungi in which an elevated temperature can promote a budding yeast-like growth phase. The phenomenon has however been observed with several other fungi which in some cases may be opportunist pathogens (Table 3). Rippon et al. (1965) have pointed out that among the major deterrents to *in vivo* growth of fungi are temperature and the reduced oxidation–reduction environment of living tissues and that organisms which can adapt to these conditions may invade the deep tissues of the body. These authors were able to induce dimorphism in *Aspergillus sydowi* and *Penicillium lilacinum* by growth of the organisms on increasing concentrations of cysteine on gradient tilt plates at 37°. Accompanying the conversion to a thick-walled budding yeast-like stage was an increase in metabolic rate, measured respirometrically and an increase in pathogenicity as manifested by invasion of the deep organs of mice.

In recent years there has been considerable interest in the pathogenicity of some *Fusarium* spp., particularly in relation to corneal ulcers and mycotic keratitis. It has been shown (Kidd & Wolf, 1973) that *Fusarium moniliforme*, which has been involved in such infections, can exhibit dimorphism. In *in vitro* studies it was concluded that temperature (37°) and agitation were the factors of greatest importance in the production of yeast phase growth in this fungus (Kidd & Wolf, 1973).

There are then clear examples where it has been established that temperature (37°) strongly influences the *in vivo* yeast-like morphology of certain fungal pathogens. A relationship between yeast-like development and pathogenicity has in fact been argued (Ainsworth, 1958) but the issues involved are complex and no definitive statements can be made (Romano, 1966). For example, the Mucorales fungi are essentially non-pathogenic and yet an elevated temperature is one of a number of environmental factors which have been shown to enhance a budding yeast-like phase in a number of *Mycotypha* spp. (Schulz et al., 1974; Hall & Kolankaya, 1974). It was speculated that the primary effects of environmental influences such as temperature may be to alter membrane structure and permeability and that regulation of metabolic activites at the cell membrane level plays a decisive role in the phenotypic determination of Y→M dimorphism in *Mycotypha*. One aspect of the effects on cellular membranes could be the alteration of the fermentation–respiration equilibrium which was implicated in the above studies on the dimorphism of the *Mycotypha* organisms.

(b) *Induction of non-budding spherical cells and unusual hyphal forms*

With a considerable number of fungi elevated temperatures can influence a reduction from the filamentous state to a more or less spherical cell form which does not display cellular multiplication by budding (Table 4). Such spherical cell forms range from quite complex cells which can develop into a multicellular structure to much simpler forms which remain as non-dividing unicells. With a number of these fungi the spherical cell form is characteristic of an *in vivo* parasitic phase and consequently they are of considerable medical importance. Among these are the dematiaceous fungi which cause chromomycosis and which produce sclerotic cells *in vivo* and also a number of other pathogenic fungi in which the *in vivo* form is described as a 'spherule'. Only the most relevant features of these fungi will be discussed here and for fuller details the reader should consult Mariat (1964).

The fungi causing chromomycosis, which include *Phialophora pedrosoi, P. verrucosa* and *Cladosporium carrionii,* form 'sclerotic cells' *in vivo* which are single, rounded, generally thick-walled cells, 6-12 μm or more in diameter. These do not bud but multiply by producing septa. These sclerotic cells can be produced *in vitro* on complex media at $37°$ (Silva, 1957).

Among the organisms with an *in vivo* spherule form are included *Coccidioides immitis, Emmonsia crescens* and *Rhinosporidium seeberi*. With *C. immitis* the spherules take the form of round cells 10-80 μm or more in diameter bounded by a thick-walled membrane and which when mature contain numerous endospores. These spherules have been produced on synthetic media at temperatures only above $26°$, with optimum production at $33-34°$ (Converse, 1957).

The spherule (or adiaspore) of *E. crescens,* observed in the lungs of rodents, appears as a large spore or multinucleate structure up to 450 μm or more in diameter and is surrounded by a very thick wall. The adiaspore can be induced *in vitro* on complex media at $37°$ (Emmons & Jellison, 1960). *Rhinosporidium seeberi* also produces an enormous spherical structure *in vivo*. The spherule of this fungus is up to 300 μm in diameter and contains an enormous number of endospores when mature (Ashworth, 1923).

Spherical or bulbous cell forms which are quite unrelated to the sclerotic cells or spherules described have occasionally been observed in infections with various other fungi. With *Penicillium marneffei* the *in vivo* invasive form is an arthrospore which multiplies by schizogenesis and these can be produced in culture at $37°$ (Segretain, 1959; Garrison & Boyd, 1973). With certain *Trichophyton* and *Microsporon* spp. which are normally mycelial dermatophytes it has been found that under conditions of elevated temperature and lowered oxidation-reduction potential the organisms will assume a spherical form and a concomitant invasive ability for animal tissue (Rippon & Scherr, 1959; Rippon, 1968). In the rare mycoses caused by *Absidia, Mucor* and *Aspergillus* mycelial elements in tissue sections have been observed to occur as spherical cells or as distorted or bulbous hyphae (Rippon *et al.,* 1965). A detailed description of the *in vivo* morphology of *A. fumigatus* has been given by Austwick (1965) and which may be characteristic for all pathogenic aspergilli. Wide globose, or oval, celled hyphae are found at the centre of acute lesions and these represent the

Table 4

Influence of temperature and other factors on the production of non-budding spherical cells from normally filamentous fungi

Species	Type of spherical cell produced	Major inducing conditions	Reference
Phialophora pedrosoi	sclerotic cell	Temp. (37°), nutrients, –SH	See Mariat (1964) and Gilardi (1965)
Phialophora verrucosa	sclerotic cell	Temp. (37°), nutrients, –SH	
Cladosporium carrionii	sclerotic cell	Temp. (37°), nutrients, –SH	
Coccidioides immitis	spherule	Temp. (40°), nutrients, CO_2	
Emmonsia crescens	spherule	Temp. (37°), nutrients	
Rhinosporidium seeberi	spherule	*In vivo* conditions	Segretain (1959)
Penicillium marneffei	arthrospore	Temp. (37°), nutrients	Anderson & Smith (1972)
Aspergillus niger	Spherical cells produced by abnormal conidial germination	Temp. (38°–44°), CO_2	Austwick (1965)
Aspergillus fumigatus		*In vivo* conditions	Anderson & Smith (unpublished)
Other *Aspergillus* sp.		Temp. (>35°)	Cortat & Turian (1974)
Neurospora crassa		Temp. (46°)	Sekiguchi *et al.* (1975a)
Penicillium urticae		Temp. (37°)	

germinated conidium and its primary hyphae. In subsequent stages of the infection various other morphological forms are produced.

These *in vivo* morphological forms of primarily saprophytic fungi are undoubtedly an expression of fungal response to an unfavourable environment. A complexity of factors is likely to be involved including elevated temperature and the host defence mechanism but in most cases no studies have been carried out to determine the importance of these factors in influencing the morphology of the fungi. A detailed study which was not concerned with pathogenicity has however been carried out with *Aspergillus niger* and this has demonstrated several clearly definable effects of temperature on the morphology of the germinating conidia and newly formed hyphae of this organism (Anderson & Smith, 1972). In this study on *A. niger* a gradation of changes was observed to occur when the conidia were incubated under submerged conditions at temperatures from 30 to 44°. Whereas all conidia produced germ tubes at 30° at temperatures from 38 to 43° the proportion of conidia which produced germ tubes decreased progressively and at 44° germ tube formation was completely inhibited. At this temperature, however, enlargement of the conidia continued over a prolonged period resulting in the formation of large spherical cells (20 μm mean diameter). At 45° no enlargement occurred, the conidia apparently remaining dormant. Thus at a critical temperature just below the maximum which allowed activation of metabolism the development of the normal filamentous form of the fungus was inhibited and growth occurred in a non-dividing spherical form. This was a true growth process with the increase in size being accompanied by an increase in dry weight and the rapid synthesis of macromolecules including nucleic acids (Kuboye et al., 1976). Ultrastructural studies (Smith et al., 1976) have shown that there is an increase in the typical cellular organelles and that extensive nuclear division occurs to give a large multinucleate structure. These ultrastructural studies have also shown that throughout cell enlargement there is a progressive thickening of the cell wall to produce a thick multilayered cell wall structure.

These striking effects of elevated temperatures on the morphology of *A. niger* during germination have been explained in terms of the localization of wall synthesis (Anderson & Smith, 1972; Smith et al., 1976). Most fungal spores undergo some degree of enlargement during germination, and germ tube emission normally occurs before maximum cell size has been attained. The production of the germ tube marks a critical change in germination, a change from a non-polarized (dispersed) form of wall synthesis or extension to a polarized (localized) form of wall growth required for germ tube emission and hyphal extension. With *A. niger* conidia at 44° the initial enlargement phase occurs but the polarization mechanism is inhibited and the dispersed form of wall synthesis continues to accompany growth thus leading to the formation of large spherical cells with thickened walls. It was also observed in these studies that temperature influenced the morphology of the germ tubes and hyphae of *A. niger*. At 30° hyphae were thin and relatively unbranched; at 41° hyphal swelling and excessive branch formation occurred; at 43° a thick multiseptate form was produced by a process resembling arthrospore formation and at 44° growth occurred in the unicellular form described. These changes were interpreted to result from an increasingly severe inhibition of the apical growth process in response to a progressive increase in temperature.

It has been found recently (Kuboye *et al.*, 1976) that the unicellular spherical cell form of *A. niger* can be produced at 41° in the presence of 5% CO_2. Under these conditions cell enlargement occurs very rapidly and after only 15 h there is complete and rapid lysis of the population. It is unlikely that the reason for lysis is due simply to a damaging effect of high temperature since cell bursting could be prevented by an increase in temperature from 41 to 44°. A possible explanation is that the increased rate of enlargement which occurs at 41° and which must involve rapid wall extension may lead to a weakening of the wall structure. This could occur by inhibition or restriction in some essential part of the wall synthesis mechanism or alternatively, by an increase in the lytic activity associated with well expansion. Lysis appears to originate from small areas in the wall where substantial thinning has occurred. These may well be incipient germ pores and lysis could occur by extrusion of the protoplast from the spore without concomitant wall synthesis.

It is interesting to note that a somewhat similar situation has been observed with the yeast *Cryptococcus*. The saprophytic cryptococci fail to grow at 37° apparently due to temperature-induced lysis. When the cells of these yeasts are transferred to 37° they undergo abnormal spherical growth and then lyse (Dabbagh *et al.*, 1974*b*). In studies to define the basis of this effect in *C. diffluens* Dabbagh *et al.* (1974*a*) found that low molecular weight compounds were excreted at non-permissive but not at permissive growth temperatures. The cells were apparently unable to maintain their intracellular pools presumably due to factors affecting membrane permeability. It was considered that lysis occurred either because of an imbalance of protoplasmic growth and the biosynthesis of wall probably at the site of bud formation or alternatively because of temperature-induced conformational changes in the membrane leading to malfunction of cell wall synthesis.

(c) *Induction of microcycle conidiation*

The effect of elevated temperature in allowing selectively the enlargement or spherical growth of conidia while inhibiting germ tube formation is not unique to *A. niger*. This has been observed in other *Aspergillus* spp. and *Paecilomyces varioti* (Anderson & Smith, unpublished material), *N. crassa* (Cortat & Turian, 1974) and *Penicillium urticae* (Sekiguchi *et al.*, 1975*a*). It may well be that this phenomenon is quite widespread among many types of fungi. Moreover it has been found that the enlarged cells of these fungi, produced by growth of conidia at elevated temperatures, possess quite remarkable properties since they can display a rapid 'microcycle' conidiation when the culture temperature is lowered. With *A. niger,* maintenance of the conidia at 44° for 48 h, which produces large spherical cells, followed by a temperature reduction to 30° leads to the direct production of a conidiophore (the asexual reproductive structure) from the enlarged cell in the complete absence of hyphal growth (Anderson & Smith, 1971). Ultrastructural studies (Smith *et al.,* 1976) have shown that conidiophore emergence from the enlarged cells occurs by a localization of wall synthesis at one area of the wall and that conidiophore extension involves the accumulation of cytoplasmic vesicles at the growing tip which is characteristic of the apical growth process in fungi.

Microcycle conidiation has been induced in *N. crassa* by incubation of conidia for 15 h at 46° followed by incubation at 25° (Cortat & Turian, 1974). After this treatment the swollen conidia produced germ tubes which instead of elongating produced proconidia by a process of basifugal budding and consequently they could be considered as simple conidiophores. Cortat & Turian (1974) have speculated that the heat treatment may inactivate the vesicular materials, aggregated in the tips of elongating hyphae, which are presumed to support glycolytic enzyme activities. This inhibition of glycolysis would allow the function of the oxidative and gluconeogenic metabolic pathways which are thought to sustain active conidiogenesis (Turian & Bianchi, 1972).

With *Penicillium urticae* microcycle conidiation has been achieved by incubating conidia in a complete medium at 37° for 24 h followed by replacement to a nitrogen-poor medium at 35°. At the lower temperature the germ tubes which are produced from the enlarged conidia undergo highly abbreviated growth and then sporulate (Sekiguchi *et al.,* 1975*a*). Detailed ultrastructural studies have been made on the conidial enlargement phase (Sekiguchi *et al.,* 1975*a*), the germ tube outgrowth phase (Sekiguchi *et al.,* 1975*b*) and on the process involved in the production of conidia from the abbreviated germ tubes (Sekiguchi *et al.,* 1975*c*). The temperature changes involved in the induction of microcycle conidiation in these various fungi induce a considerable degree of developmental synchrony thus making the microcycle system particularly useful for biochemical studies on certain differentiation processes in fungi.

6. References

AINSWORTH, G. C. (1958). Pathogenic yeasts. In *The Chemistry and Biology of Yeasts,* ed. Cook, A. H. New York & London: Academic Press.

ALBERGHINA, F. A. M. (1973). Growth regulation in *Neurospora crassa.* Effects of nutrients and of temperature. *Archiv für Mikrobiologie* **89**, 83-94.

ANDERSON, J. G. & SMITH, J. E. (1971). The production of conidiophores and conidia by newly germinated conidia of *Aspergillus niger* (microcycle conidiation). *Journal of General Microbiology* **69**, 185-197.

ANDERSON, J. G. & SMITH, J. E. (1972). The effects of elevated temperatures on spore swelling and germination in *Aspergillus niger*. *Canadian Journal of Microbiology* **18**, 289-297.

ASHWORTH, J. H. (1923). On *Rhinosporidium seeberi* with special reference to its sporulation and affinities. *Transactions of the Royal Society of Edinburgh b* **53**, 301.

AUSTWICK, P. K. C. (1965). Pathogenicity. In *The Genus Aspergillus*, eds Raper K. B. & Fennell, D. I. Baltimore: Williams and Wilkins.

AUSTWICK, P. K. C. (1968). Effects of adjustment to the environment on fungal form. In *The Fungi*, Vol. 3. eds Ainsworth G. C. & Sussman, A. S. New York & London: Academic Press.

BAKER, J. E. & SMITH, W. L. (1970). Heat-induced ultrastructural changes in germinating spores of *Rhizopus stolonifer* and *Monilinia fruticola*. *Phytopathology* **60**, 869-874.

BARTNICKI-GARCIA, S. (1969). Cell wall differentiation in the Phycomycetes. *Phytopathology* **59**, 1065-1071.

BARTNICKI-GARCIA, S. (1973). Fundamental aspects of hyphal morphogenesis. In *Microbial Differentiation*, eds Ashworth, J. M. & Smith, J. E. London: Cambridge University Press.

BOGUSLAWSKI, G., SCHLESSINGER, D., MEDOFF, G. & KOBAYASHI, G. (1974). Ribonucleic acid polymerases of the yeast phase of *Histoplasma capsulatum*. *Journal of Bacteriology* **118**, 480-485.

BOGUSLAWSKI, G., KOBAYASHI, G. S., SCHLESSINGER, D. & MEDOFF, G. (1975). Characterisation of an inhibitor or ribonucleic acid polymerase from the mycelial phase of *Histoplasma capsulatum*. *Journal of Bacteriology* **122**, 532-537.

BROAD, T. E. & SHEPHERD, M. G. (1971). Purification and properties of glucose-6-phosphate dehydrogenase from the thermophilic fungus *Penicillium dupontii*. *Biochimica et biophysica acta* **198**, 407-414.

BROWN, C. M. & ROSE, A. H. (1969a). Effects of temperature on composition and cell volume of *Candida utilis*. *Journal of Bacteriology* **97**, 261-272.

BROWN, C. M. & ROSE, A. H. (1969b). Fatty acid composition of *Candida utilis* as affected by growth temperature and dissolved oxygen tension. *Journal of Bacteriology* **99**, 371-378.

BROWN, A. H. S. & SMITH, G. (1957). The genus *Paecilomyces* Bainer and its perfect state *Byssochlamys* Westling. *Transactions of the British Mycological Society* **40**, 17-89.

CARBONELL, L. M. (1967). Cell wall changes during the budding process of *Paracoccidioides brasiliensis* and *Blastomyces dermatitidis*. *Journal of Bacteriology* **94**, 213-223.

CARBONELL, L. M. (1969). Ultrastructure of dimorphic transformation in *Paracoccidioides brasiliensis*. *Journal of Bacteriology* **100**, 1076-1082.

CARBONELL, L. M., KANETSUNA, F. & GIL, F. (1970). Chemical morphology of glucan and chitin in the cell wall of the yeast phase of *Paracoccidioides brasiliensis*. *Journal of Bacteriology* **101**, 636-642.

CHAPUIS, R. & ZUBER, N. (1970). Thermophilic aminopeptidases; AP 1 from *Talaromyces duponti*. In *Methods in Enzymology*, Vol. 19, eds Perlmann, G. E. & Lorand, L. New York & London: Academic Press.

CHEUNG, S. C., KOBAYASHI, G. S., SCHLESSINGER, D. & MEDOFF, G. (1974). RNA metabolism during morphogenesis in *Histoplasma capsulatum*. *Journal of General Microbiology* **82**, 301-307.

CINO, P. M. & TEWARI, R. P. (1972). Chemical composition of *Histoplasma capsulatum*. *Mycopathologia et mycologia applicata* **47**, 285-294.

COCHRANE, V. W. (1958). *Physiology of Fungi*. New York: John Wiley & Sons.

CONVERSE (1957). Effect of surface active agents on endosporulation of *Coccidioides immitis* in a chemically defined medium. *Journal of Bacteriology* **74**, 106-107.

CORTAT, M. & TURIAN, G. (1974). Conidiation of *Neurospora crassa* in submerged culture without mycelial phase. *Archiv für Mikrobiologie,* 95, 305-309.

CRAVERI, R., & COLLA, C. (1966). Ribonuclease activity of mesophilic and thermophilic molds. *Annals of Microbiology* 16, 97-99.

CRISAN, E. V. (1969). The proteins of thermophilic fungi. In *Current Topics in Plant Science,* ed. Gunckel, J. E. New York & London: Academic Press.

CRISAN, E. V. (1973). Current concepts of thermophilism and the thermophilic fungi. *Mycologia* 65, 1171-1198.

DABBAGH, R., CONANT, N. F. & BURNS, R. O. (1974a). Effect of temperature on saprophytic Cryptococci: Observations relating to wall biosynthesis at non-permissive growth temperatures. *Journal of General Microbiology* 85, 190-202.

DABBAGH, R. CONANT, N. F., NIELSEN, H. S. & BURNS, R. O. (1974b). Effect of temperature on saprophytic Cryptococci. Temperature-induced lysis and protoplast formation. *Journal of General Microbiology* 85, 177-189.

DEVERALL, B. J., (1965). The physical environment for fungal growth. 1. Temperature. In *The Fungi,* Vol. 1. eds Ainsworth, G. C. & Sussman, A. S. New York & London: Academic Press.

DEVERALL, B. J. (1968). Psychrophiles. In *The Fungi,* Vol. 3, eds Ainsworth, G. C. & Sussman, A. S. New York & London: Academic Press.

EMERSON, R. (1968). Thermophiles. In *The Fungi,* Vol. 3, eds Ainsworth, G. C. & Sussman, A. S. New York & London: Academic Press.

EMMONS, C. W. & JELLISON, W. L. (1960). *Emmonsia crescens* sp. n. and Adiaspiromycosis (Haplomycosis) in mammals. *Annals of the New York Academy of Science* 89, 91.

FARRELL, J. & ROSE, A. (1967a). Temperature relationships among micro-organisms. In *Thermobiology,* ed. Rose, A. H. New York and London: Academic Press.

FARRELL, J. & ROSE, A. (1967b). Temperature effects on microorganisms. *Annual Review of Microbiology* 21, 101-120.

FRIES, L. (1953). Factors promoting growth of *Coprinus fimetarius* under high temperature conditions. *Physiologia plantarum* 6, 551-563.

FRIES, N. & KALLSTROMER, L. (1965). A requirement for biotin in *Aspergillus niger* when grown on a rhamnose medium at high temperature. *Physiologia plantarum* 18, 191-200.

GARRISON, R. G. & BOYD, K. S. (1973). Dimorphism of *Penicillium marneffei* as observed by electron microscope. *Canadian Journal of Microbiology* 19, 1305-1309.

GARRISON, R. G., LANE, J. W. & FIELD, M. F. (1970). Ultrastructural changes during the yeast like to mycelial phase conversion of *Blastomyces dermatitidis* and *Histoplasma capsulatum. Journal of Bacteriology* 101, 628-635.

GARRISON, R. G., BOYD, K. S. & MARIAT, F. (1975). Ultrastructural studies of the mycelium—to yeast transformation of *Sporothrix schenckii. Journal of Bacteriology* 124, 959-968.

GILARDI, G. L. (1965). Nutrition of systemic and subcutaneous pathogenic fungi. *Bacteriological Reviews* 29, 406-424.

GROVE, S. N., BRACKER, C. E. & MORRÉ, D. J. (1970). An ultrastructural basis for hyphal tip growth in *Pithium ultimum. American Journal of Botany* 59, 245-266.

GROVE, S. N. OUJEZDSKY, K. B. & SZANISZLO, P. J. (1973). Budding in the dimorphic fungus *Phialophora dermatitidis. Journal of Bacteriology* 115, 323-329.

GUPTA, R. K. & HOWARD, D. H. (1971). Comparative physiological studies of the yeast and mycelial forms of *Histoplasma capsulatum:* Uptake and incorporation of L-leucine. *Journal of Bacteriology* 105, 690-700.

HAGLER, A. N. & LEWIS, M. J. (1974). Effect of glucose on thermal injury of yeast that may define the maximum temperature of growth. *Journal of General Microbiology* 80, 101-109.

HALL, M. J. & KOLANKAYA, N. (1974). The physiology of mould-yeast dimorphism in the genus *Mycotypha* (Mucorales). *Journal of General Microbiology* 82, 25-34.

HAWKER, L. E. (1950). *Physiology of Fungi.* London & New York: Oxford University Press.
HAWKER, L. E. (1966). Environmental influences on reproduction. In *The Fungi,* Vol. 2, eds Ainsworth, G. C. & Sussman, A. S. New York & London: Academic Press.
HECKER, L. I. & SUSSMAN, A. S. (1973a). Activity and heat stability of trehalase from the mycelium and ascospores of *Neurospora. Journal of Bacteriology* **115**, 582-591.
HECKER, L. I. & SUSSMAN, A. S. (1973b). Localization of trehalase in the ascospores of *Neurospora:* Relation to ascospore dormancy and germination. *Journal of Bacteriology* **115**, 592-599.
HORNER, K. J. & ANAGNOSTOPOULOS, G. D. (1973). Combined effects of water activity, pH and temperature on the growth and spoilage potential of fungi. *Journal of Applied Bacteriology* **36**, 427-436.
HULL, R. (1939). Study of *Byssochlamys fulva* and control measures in processed fruits. *Annals of Applied Biology* **26**, 800-822.
JENSEN, M. (1960). Experiments on the inhibition of some thermoresistant molds in fruit juices. *Annales de l'Institut Pasteur de Lille* **11**, 179-182.
JONES, R. C. & HOUGH, J. S. (1970). The effect of temperature on the metabolism of bakers yeast growing on continuous culture. *Journal of General Microbiology* **60**, 107-116.
KANETSUNA, F. & CARBONELL, L. M. (1970). Cell wall glucans of the yeast and mycelial forms of *Paracoccidioides brasiliensis. Journal of Bacteriology* **101**, 675-680.
KANETSUMA, F. & CARBONELL, L. M. (1971). Cell wall composition of the yeastlike and mycelial forms of *Blastomyces dermatitidis. Journal of Bacteriology* **106**, 946-948.
KANETSUNA, F., CARBONELL, L. M., MORENO, R. E. & RODRIGUEZ, J. (1969). Cell wall composition of the yeast and mycelial forms of *Paracoccidioides brasiliensis. Journal of Bacteriology* **97**, 1036-1041.
KANETSUNA, F., CARBONELL, L. M., AZUMA, I. & YAMAMURA, Y. (1972). Biochemical studies on the thermal dimorphism of *Paracoccidioides brasiliensis. Journal of Bacteriology* **110**, 208-218.
KATES, M. & BAXTER, R. M. (1962). Lipid composition of mesophilic and psychrophilic yeasts (*Candida* species) as influenced by environmental temperature. *Canadian Journal of Biochemical Physiology* **40**, 1213-1227.
KIDD, G. H. & WOLF, F. T. (1973). Dimorphism in a pathogenic fusarium. *Mycologia* **65**, 1371-1375.
KING, A. D. Jr, MICHENER, H. D. & ITO, K. A. (1969). Control of *Byssochlamys* and related heat resistant fungi in grape products. *Applied Microbiology* **18**, 166-173.
KITAMURA, K. (1965). *Sporotrichum schenckii.* An electron microscope study. *Japanese Journal of Dermatology* **75**, 285-304.
KOBAYASHI, G. S. & GIULIACCI, P. L. (1967). Cell wall studies of *Histoplasma capsulatum. Sabouraudia* **5**, 180-188.
KUBOYE, A. O., ANDERSON, J. G. & SMITH, J. E. (1976). Control and autolysis of a spherical cell form of *Aspergillus niger. Transactions of the British Mycological Society* (in press).
LANE, J. W. & GARRISON, R. G. (1970). Electron microscopy of the yeast to mycelial phase conversion of *Sporotrichum schenckii. Canadian Journal of Microbiology* **16**, 747-749.
LANE, J. W., GARRISON, R. G. & FIELD, M. F. (1969). Ultrastructural studies on the yeastlike and mycelial phases of *Sporotrichum schenckii. Journal of Bacteriology* **100**, 1010-1019.
LANGRIDGE, J. (1963). Biochemical aspects of temperature response. *Annual Review of Plant Physiology* **14**, 441-462.
LANGVAD, F. (1972). The effect of supraoptimal temperatures on the fine structure of *Merulius lacrymans* (Jacq.) Fr. *Journal of General Microbiology* **70**, 157-159.
LONGINOVA, L. G. & TASHPULATOV (1967). Multicomponent cellulolytic enzymes of

thermotolerant and mesophilic fungi, closely related to *Aspergillus fumigatus*. *Microbiology, Washington* **36**, 828-831.

LURIE, H. I. & STILL, W. J. S. (1969). The "capsule" of *Sporotrichum schenckii* and the evolution of the asteroid body. A light and electron microscope study. *Sabouraudia* **7**, 285-304.

LÜTHI, H. & VETSCH, V. (1955). Uber das vorkommen thermoresistenter Pilze in der Sussmosterei. *Schweizerische Zeitschrift für Obst- und and Wein-bau* **64**, 404-409.

MAHESHWARI, R. & SUSSMAN, A. S. (1971). The nature of cold-induced dormancy in urediospores of *Puccinia graminis tritici*. *Plant Physiology* **47**, 289-295.

MAHVI, T. A. (1965). A comparative study of the yeast and mycelial phases of *Histoplasma capsulatum*. 1. Pathways of carbohydrate dissimilation. *Journal of Infectious Diseases* **115**, 226-232.

MANDELS, G. R., VITOLS, R. & PARRISH, F. W. (1965). Trehalose as an endogenous reserve in spores of the fungus *Myrothecium verrucaria*. *Journal of Bacteriology* **90**, 1589-1598.

MARIAT, F. (1964). Saprophytic and parasitic morphology of pathogenic fungi. In *Microbial Behaviour, 'In vivo and In vitro'*, eds Smith, H. & Taylor, J. London: Cambridge University Press.

MAZUR, P. (1968). Survival of fungi after freezing and desiccation. In *The Fungi*, Vol. 3, eds Ainsworth, G. C. & Sussman, A. S. New York & London: Academic Press.

MICHENER, H. D. & KING, D. A. Jr (1974). Preparation of free heat-resistant ascospores from *Byssochlamys* asci. *Applied Microbiology* **27**, 671-673.

MUMMA, R. O., FERGUS, C. L. & SEKURA, R. D. (1970). The lipids of thermophilic fungi: Lipid composition comparisons between thermophilic and mesophilic fungi. *Lipids* **5**, 100-103.

MUMMA, R. O., SEKURA, R. D. & FERGUS, C. L. (1971). Thermophilic fungi 11. Fatty acid composition of polar and neutral lipids of thermophilic and mesophilic fungi. *Lipids* **6**, 584-588.

NICKERSON, W. J. & EDWARDS, G. A. (1949). Studies on the physiological bases of morphogenesis in fungi. 1. The respiratory metabolism of dimorphic pathogenic fungi. *Journal of General Physiology* **33**, 41-55.

OJHA, M. N. & TURIAN, G. (1968). Thermostimulation of conidiation and succinic oxidative metabolism of *Neurospora crassa*. *Archiv für Mikrobiologie* **63**, 232-241.

OUJEZDSKY, K. B., GROVE, S. N. & SZANISZLO, P. J. (1973). Morphological and structural changes during the yeast-to-mold conversion of *Phialophora dermatitidis*. *Journal of Bacteriology* **113**, 468-477.

PARTSCH, G., DRAZLER, H. & ALTMANN, H. (1969). The ultrastructure of spores of *Byssochlamys fulva*. *Mycopathologia et mycologia applicata* **39**, 305-313.

PUT, H. M. C. (1964). A selective method for cultivating heat resistant molds, particularly those of the genus *Byssochlamys* and their presence in Dutch soil. *Journal of Applied Bacteriology* **27**, 59-64.

PUT, H. M. C. & KRUISWIJK, J. T. (1964). Disintegration and organoleptic deterioration of processed strawberries caused by the mould *Byssochlamys nivea*. *Journal of Applied Bacteriology* **27**, 53-58.

RICHARDSON, K. C. (1965). Incidence of *Byssochlamys fulva* in Queensland grown canned strawberries. *Queensland Journal of Agricultural and Animal Science* **22**, 347-350.

RIPPON, J. (1968). Monitored environment system to control cell growth, morphology and metabolic rate in fungi by oxidation-reduction potentials. *Applied Microbiology* **16**, 114-121.

RIPPON, J. & SCHERR G. (1959). Induced dimorphism in dermatophytes *Mycologia* **51**, 902-914.

RIPPON, J., CONWAY, T. P. & DOMES, A. L. (1965). Pathogenic potential of *Aspergillus* and *Penicillium* species. *Journal of Infectious Diseases* **115**, 27-32.

ROMANO, A. H. (1966). Dimorphism. In *The Fungi*, Vol. 2, eds Ainsworth, G. C. & Sussman, A. S. New York & London: Academic Press.

RUYLE, E. H., PEARCE, W. E. & HAYES, G. L. (1946). Prevention of mold in kettled blueberries in No. 10 cans. *Food Research* **11**, 274-279.

SAN-BLAS, G. & CARBONELL, L. M. (1974). Chemical and ultrastructural studies on the cell walls of the yeastlike and mycelial forms of *Histoplasma farciminosum*. *Journal of Bacteriology* **119**, 602-611.

SARGENT, J. A. & PAYNE, H. L. (1974). Effect of temperature on germination, viability and fine structure of conidia of *Bremia lactucae*. *Transactions of the British Mycological Society* **63**, 509-518.

SCHULZ, B. E., KRAEPELIN, G. & HINKELMANN, W. (1974). Factors affecting dimorphism in *Mycotypha* (Mucorales): Correlation with the fermentation/respiration equilibrium. *Journal of General Microbiology* **82**, 1-13.

SEGRETAIN, G. (1959). *Penicillium marneffei* n. sp. Agent d'une mycose du système reticulo-endothélial. *Mycopctholgia et mycologia applicata* **11**, 327.

SEKIGUCHI, J., GAUCHER, G. M. & COSTERTON, J. W. (1975a). Microcycle conidiation in *Penicillium urticae:* an ultrastructural investigation of spherical spore growth. *Canadian Journal of Microbiology* **21**, 2048-2058.

SEKIGUCHI, J., GAUCHER, G. M. & COSTERTON, J. W. (1975b). Microcycle conidiation in *Penicillium urticae:* an ultrastructural investigation of conidial germination and outgrowth. *Canadian Journal of Microbiology* **21**, 2059-2068.

SEKIGUCHI, J., GAUCHER, G. M. & COSTERTON, J. W. (1975c). Microcycle conidiation in *Penicillium urticae:* an ultrastructural investigation of conidiogenesis. *Canadian Journal of Microbiology* **21**, 2069-2083.

SILVA, M. (1957). The parasitic phase of the fungi of chromoblastomycosis. Development of sclerotic cells *in vitro* and *in vivo*. *Mycologia* **49**, 318-331.

SMITH, J. E., GULL, K., ANDERSON, J. G. & DEANS, S. G. (1976). Organelle changes during fungal spore germination. In *The Fungal Spore*. New York: John Wiley & Sons.

SOMKUTI, G. A. & SOMKUTI, A. C. (1969). Lipase of *Mucor pusillus*. *Applied Microbiology* **17**, 606-610.

SOMMER, L. & HALBSGUTH, W. (1957). Forschungsber. Wirtsch. Verkehrministeriums Nordrhein-Westfalen No. 411. Westdoutscher, Köln.

SPLITTSTOESSER, D. F., CADWELL, M. C. & MARTIN, M. (1969). Ascospore production by *Byssochlamys fulva*. *Journal of Food Science* **34**, 248-250.

STOKES, J. L. (1967). In *Molecular Mechanisms of Temperature Adaptation*, ed. Prosser, C. L. Washington, D. C.: American Association for the Advancement of Science, pp. 311-323.

STOKES, J. L. (1971). Influence of temperature on the growth and metabolism of yeasts. In *The Yeasts*, Vol. 2, eds Rose, A. H. & Harrison, J. S. New York & London: Academic Press.

SUMNER, J. L., MORGAN, E. D. & EVANS, H. C. (1969). The effect of growth temperature on the fatty acid composition of fungi in the order Mucorales. *Canadian Journal of Microbiology* **15**, 515-520.

SUSSMAN, A. S. (1966). Dormancy and spore germination. In *The Fungi*, Vol. 2, eds Ainsworth, G. C. & Sussman, A. S. New York & London: Academic Press.

SUSSMAN, A. S. & HALVORSON, H. O. (1966). *Spores, their Dormancy and Germination*. New York & London: Harper & Row.

TAYLOR, J. J. (1961). Nucleic acids and dimorphism in *Blastomyces*. *Experimental Cell Research* **24**, 155-158.

TURIAN, G. & BIANCHI, D. E. (1972). Conidiation in *Neurospora*. *Botanical Reviews* **38**, 119-154.

WAID, J. S. (1968). Physiological and biochemical adjustment of fungi to their environment. In *The Fungi*, Vol. 3, eds Ainsworth, G. C. & Sussman, A. S. New York & London: Academic Press.

WOLF, F. A. & WOLF, F. T. (1947). *The Fungi*, Vol. 2. New York: John Wiley & Sons.

YATES, A. R., SEAMAN, A. & WOODBINE, M. (1968). Ascospore germination in *Byssochlamys nivea*. *Candian Journal of Microbiology* **14**, 319-325.

YU. S., SUSSMAN, A. S. & WOOLEY, S. (1967). Mechanisms of protection of trehalase

against heat inactivation in *Neurospora. Journal of Bacteriology* **94**, 1306-1312.
ZUBER, H. (1967). Sequence analysis of peptides with citrus carboxypeptidase and with a thermophilic carboxypeptidase from *Talaromyces duponti. Abstracts 7th International Congress of Biochemistry, Tokyo* **3**, 541-542.

Inhibition of Micro-organisms in Food by Water Activity

L. LEISTNER AND W. RÖDEL

Bundesanstalt für Fleischforschung, Institut für Bakteriologie und Histologie, 8650 Kulmbach, Blaich 4, West Germany

CONTENTS

1. Introduction 219
2. Tolerance of micro-organisms to a_w 219
3. Measurement of a_w of meats 223
4. Significance of a_w for meats 225
5. References 233

1. Introduction

FROM THE VIEWPOINT of the food microbiologist water activity (a_w) indicates the amount of water in a food which is available for micro-organisms. This is not the total water content of the food because a proportion of it may be bound by water soluble salts, proteins and carbohydrates; such bound water is not available to micro-organisms.

The a_w of a food influences the multiplication (Scott, 1953) and metabolic activity (including toxin production) of micro-organisms, also their survival and resistance. This is true not only for organisms that cause spoilage and food-poisoning, but also for those which are desirable for the fermentation of certain foods. Microbial spoilage, food-poisoning and fermentation take place if the a_w of the substrate is favourable for the multiplication and metabolic activity of the organisms involved. Most organisms occurring in foods proliferate at a high a_w, only a few require a low a_w for growth. Thus, if the a_w decreases then fewer genera of micro-organisms are able to multiply on or in a food. A decrease in a_w of a food and thereby an extension of its storage life can be accomplished by drying, salting, addition of sugar or freezing. Some of these processes are applied in combination. Furthermore, inhibition of micro-organisms in a food is frequently not caused solely by a decrease in a_w, but may be influenced by pH, E_h, temperature, preservatives or a competitive microflora.

2. Tolerance of Micro-organisms to a_w

In general, of the micro-organisms associated with foods, moulds are more tolerant of a decreased a_w than yeasts, and yeasts more tolerant than bacteria.

In high moisture foods ($a_w > 0.90$) bacteria are mainly responsible for spoilage, food-poisoning or fermentation. In intermediate moisture foods (a_w 0.90–0.60) yeasts and moulds are of significance in spoilage. However, most micro-organisms are inhibited in low moisture foods ($a_w < 0.60$). In Table 1 the minimal a_w required for the growth of a number of genera of bacteria, yeasts

Table 1

Minimal a_w for multiplication of micro-organisms associated with meat and meat products

a_w	Bacteria	Yeasts	Moulds
0.98	*Clostridium* (1), *Pseudomonas**	– –	–
0.97	*Clostridium* (2)	–	–
0.96	*Flavobacterium, Klebsiella, Lactobacillus*, Proteus,* Pseudomonas*, Shigella*	–	–
0.95	*Alcaligenes, Bacillus, Citrobacter, Clostridium* (3), *Enterobacter, Escherichia, Proteus, Pseudomonas, Salmonella, Serratia, Vibrio*	–	–
0.94	*Lactobacillus, Microbacterium, Pediococcus, Streptococcus*, Vibrio**	–	–
0.93	*Lactobacillus*, Streptococcus*	–	*Rhizopus, Mucor*
0.92	–	*Rhodotorula, Pichia*	–
0.91	*Corynebacterium, Straphylococcus* (4), *Streptococcus**	–	–
0.90	*Lactobacillus*, Micrococcus, Pediococcus, Vibrio**	*Hansenula Saccharomyces*	–
0.88	–	*Candida, Debaryomyces, Hanseniaspora, Torulopsis*	*Cladosporium*
0.87	–	*Debaryomyces**	–
0.86	*Staphylococcus* (5)	–	*Paecilomyces*
0.80	–	*Saccharomyces**	*Aspergillus, Penicillium*
0.75	Halophilic bacteria	–	*Aspergillus**
0.70	–	–	*Eurotium*
0.62	–	*Saccharomyces**	*Eurotium**

* Some strains; (1), *Clostridium botulinum* type C; (2), *Cl. botulinum* type E and some strains of *Cl. perfringens*; (3), *Cl. botulinum* type A and B and *Cl. perfringens*; (4), anaerobic; (5), aerobic.

and moulds which are commonly recovered from foods, especially meat and meat products, are listed. This Table has been compiled from the data reported by the following authors: Snow (1949), Burcik (1950), Bullock & Tallentire (1952), Christian & Scott (1953), Scott (1953, 1957), Williams & Purnell (1953), Christian (1955), Beers (1957), Wodznski & Frazier (1960), Christian & Waltho (1962, 1964), Lanigan (1963), Riemann (1963), Blanche Koelensmid & van Rhee (1964), Gough & Alford (1965), Hobbs (1965), Kim (1965), Matz (1965), Brownlie (1966), Ohye & Christian (1966), Ohye et al. (196), Segner et al. (1966, 1971), Baird-Parker & Freame (1967), McLean et al. (1968), Pitt & Christian (1968), Pivnick & Thatcher (1968), Kang et al. (1969), Mossel (1969), Bem & Leistner (1970a), Strong et al. (1970), Troller (1971, 1972), Rödel et al. (1973), Tomčov et al. (1974), Pitt (1975).

The a_w requirements of some micro-organisms reported by different authors are not always in agreement; therefore, the data in Table 1 represent a compromise and have to be regarded as incomplete. When the authors reported only the NaCl tolerance of the organisms studied, the corresponding a_w was calculated. Most of the data in this Table were obtained by investigating the a_w tolerance of the studied organisms under otherwise optimal growth conditions using an artificial substrate. Presumably the a_w tolerance of the organisms will decrease if other factors, such as temperature, redox-potential and pH, are not optimal as if often the case if e.g. meats are the substrate. In addition, meat products often contain nitrite and sometimes competitive organisms, these might cause a further increase in the a_w sensitivity of the listed micro-organisms.

From Table 1 it is possible to conclude that substrates with a_w of <0.95 inhibit multiplication of most Gram negative bacteria as well as spore forming bacteria of the genera *Bacillus* and *Clostridium,* and also apparently inhibit germination of bacterial spores. Some Gram positive bacteria which are desirable for the fermentation of meats, i.e. representatives of the genera *Lactobacillus, Pediococcus* or *Micrococcus,* tolerate a much lower a_w than 0.95. Certainly, well-adapted strains of bacteria, for instance in curing brines, are able to grow at an a_w at which most representatives of their genera would be inhibited. Yeasts and moulds of the genera *Debaryomyces* and *Penicillium,* which are important for the fermentation of certain sausages, multiply and are metabolically active at remarkably low a_w levels too.

Of the food-poisoning bacteria, representatives of the genus *Shigella* are inhibited by an $a_w < 0.96$ (Tomčov et al., 1974) and other Gram negative rods by <0.95; the latter was demonstrated for *Salmonella* (Christian & Scott, 1953; Christian, 1955; Hansen & Riemann, 1962; Blanche Koelensmid & van Rhee, 1964; Tomčov et al., 1974), enteropathogenic *Escherichia coli* (Tomčov et al., 1974) and *Vibrio parahaemolyticus* (Beuchat, 1974). However, some strains of *V. parahaemolyticus* require an $a_w < 0.94$ for inhibition (Rödel et al., 1973). The growth and toxin production of *Clostridium botulinum* type C is inhibited below an a_w of 0.98 (Segner et al., 1971), of Cl. botulinum type E, <0.97 (Segner et al., 1966; Ohye & Christian, 1966; Pivnick & Thatcher, 1968), and of Cl. botulinum types A and B (Greenberg et al., 1959; Anderton, 1963; Ohye & Christian, 1966) and Cl. perfringens (Kang et al., 1969), <0.95. Undoubtedly, of the food-poisoning bacteria the lowest a_w is tolerated by *Staphylococcus aureus*. Under anaerobic conditions this organism is reported to be inhibited at $<a_w$ 0.91 (Scott, 1953), but aerobically only at <0.86 (Scott, 1953; Christian & Waltho, 1962, 1964; Matz, 1965).

Some strains of *Staph. aureus* produce enterotoxins; their classification (A, B, C_1, C_2, D, E, F) is based on reactions of these toxins with specific antibodies. Types A and D cause most outbreaks of food-poisoning. Genigeorgis & Sadler (1966), McLean et al. (1968), Genigeorgis et al. (1969), Hojvat & Jackson (1969), Markus & Silverman (1970), Genigeorgis et al. (1971), and Troller (1971, 1972) related both staphylococcal growth and enterotoxin production to sodium chloride concentration or a_w. It was found that the production of enterotoxin B (Genigeorgis et al., 1966, 1969; McLean et al., 1968; Troller, 1971) and enterotoxin C (Genigeorgis et al., 1971) by *Staph. aureus* ceases at an a_w of 0.94. However, enterotoxin A might also be produced with an $a_w < 0.90$ (Troller, 1972). Preliminary data of Troller (1972) indicate that the effect of a_w on enterotoxin C production is similar to that on enterotoxin A, and the author concluded that reliance cannot be placed on the reduction of a_w alone as a means of preventing the formation of enterotoxins A and C by *Staph. aureus*.

Mycotoxins (aflatoxins, citreoviridin, citrinin, cyclopiazonic acid, ochratoxins, patulin, rugulosin, and sterigmatocystin) produced by moulds of the genera *Aspergillus* and *Penicillium* occur in meats (Bullerman et al. 1969a, 1969b; Burmeister & Leistner, 1970; Leistner & Tauchmann, 1970; Tauchmann & Leistner, 1971; Alperden et al., 1973a, 1973b; Escher et al., 1973; Halls & Ayres, 1973; Wu et al., 1974; Manabe et al., 1975). Information on the a_w requirements for mycotoxin production is scarce. In peanuts aflatoxins are produced within 14 days at 25° at an a_w of 0.86 (Sanders et al., 1968); as limiting RH (relative humidity) for aflatoxin production in peanuts at 20° Diener & Davis (1970) reported 83% (a_w 0.83). In a study by Bacon et al. (1973) with poultry feed inoculated with *Aspergillus ochraceus* it was observed that penicillic acid began to accumulate at an a_w of 0.80 at 22° and 30°, whereas ochratoxin A began to accumulate at an a_w of 0.85 at 30°; maximum production of penicillic acid occurred at 22° and a_w of 0.90 and of ochratoxin A at 30° and an a_w of 0.95.

It was suggested that this contribution should include some data on meats, especially on continental sausages. Therefore, some examples of the importance of the a_w for certain meat processes and the involved micro-organisms will be given. However, since there would be no understanding of the water activity of meat and meat products without appropriate methods to measure a_w, a comprehensive review of the available methods is presented first.

3. Measurement of a_w of Meats

An inexpensive procedure to determine the a_w of foods is the graphical interpolation method described by Landrock & Proctor (1951), since no instruments, except a balance, are required. The a_w measurement of meats obtained by this method proved to be quite accurate, provided that the a_w of the product was <0.98; however, the procedure is a little cumbersome and time-consuming (Bem & Leistner, 1970b).

Other methods for measuring the a_w of meats are based on the determination of the equilibrium RH of the sample, i.e. the sample is put into a closed space, and after an equilibrium between the a_w of the sample and the RH of the enclosed air is established, the RH is measured with different devices. The equilibrium RH divided by 100 is assumed to be the a_w of the sample. A simple method to determine the equilibrium RH of meats was described by Kvaale & Dalhoff (1963) who employed strips of filter paper impregnated with certain salts. The state—dry or wet—of these test papers is used as an indication of the a_w of a sample left together with the strips in a closed space for 20 h at 20°.

Hair hygrometers have been used for a long time to measure the RH of the atmosphere. Most of these instruments are reasonably priced, but not very accurate. However, the accuracy is improved considerably when human hair is replaced by selected, artificially aged, plastic threads which have a sensitive

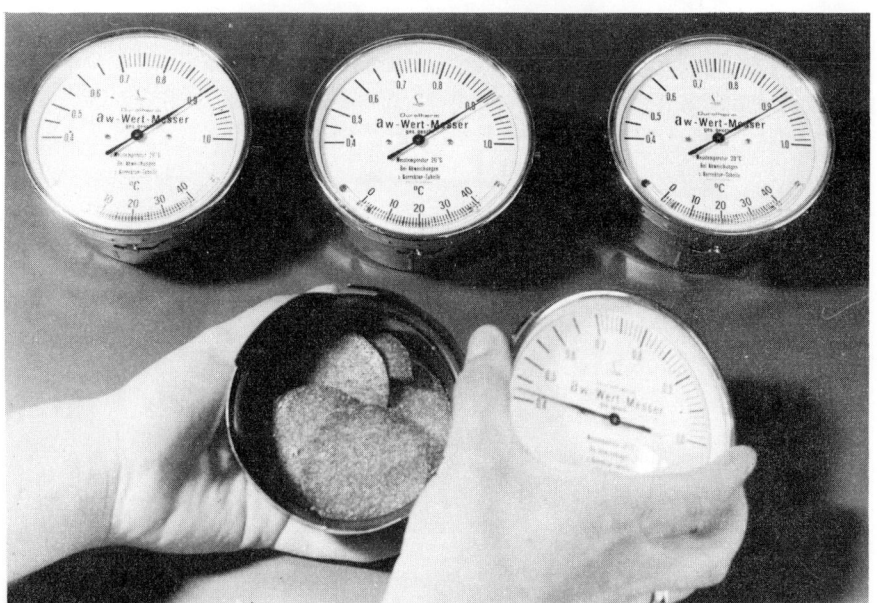

Plate 1. a_w-Wert-Messer LUFFT. One box is open; the lid of each box contains the humidity measuring element and a capillary thermometer.

response in the humidity range of interest. An instrument in which a polyamide thread with a distinctive precision in the range of 85–100% RH is incorporated, was modified by our laboratories to measure the a_w of meat and meat products (Rödel & Leistner, 1971; Rödel et al., 1975a). This a_w-Wert-Messer is manufactured by Firma LUFFT, Stuttgart, West Germany, and is widely used in Germany for quality control of meats (Plate 1).

Some instruments for measuring the a_w of foods are based on the conductivity of lithium chloride, which changes with the RH of the atmosphere. In our laboratories a sensor manufactured by the SINA AG, Zürich, Switzerland, is in use (Karan-Djurdjić & Leistner, 1970; Vrchlabský & Leistner, 1970a). This SINA sensor consists of a small chamber for the sample closed with a lid which contains the measuring element. Since the conductivity of the stabilized lithium chloride in the measuring element changes according to the RH of the enclosed

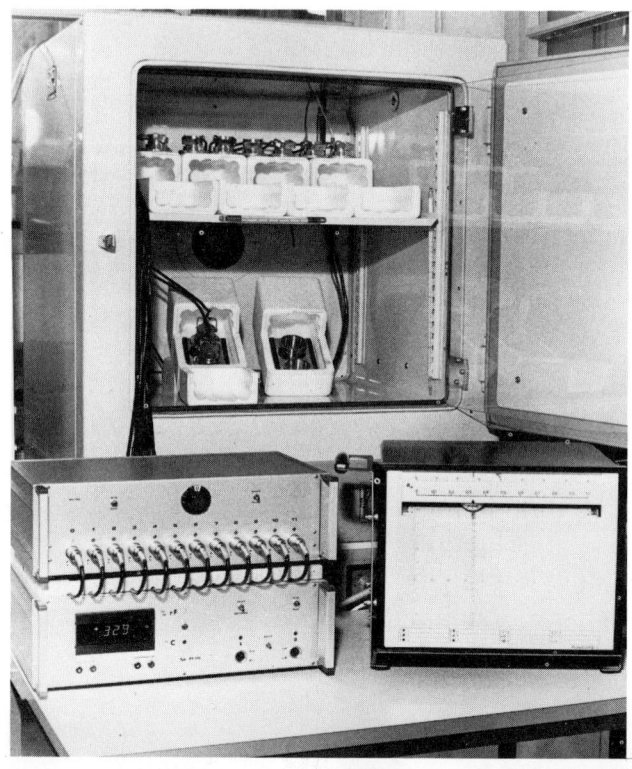

Plate 2. SINA sensors combined with ISO instrument. In the background are 12 SINA sensors (kept in styrofoam boxes) in a constant temperature incubator set at 25°. In the foreground is the ISO instrument (left) which automatically calls on the sensors and has a digital read-out, and the recorder (right).

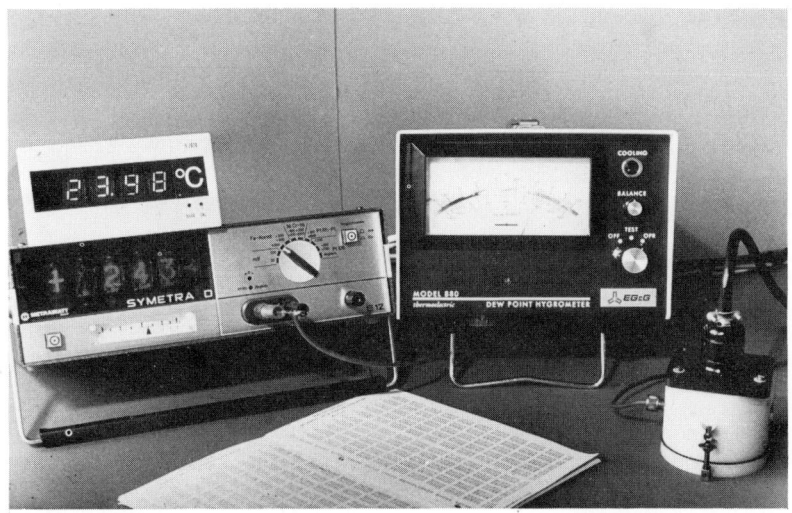

Plate 3. Modified EG&G instrument. In front of the instrument is the sample chamber used in our laboratories, and on the left the digital temperature read-out.

air, the a_w of the sample may be determined with this element. In a newly developed modification (Rödel et al., 1974) 12 samples can be measured simultaneously and the obtained data are recorded (Plate 2). Other instruments for the a_w determination of foods are based on the principle of dew-point hygrometers. Our laboratories (Rödel & Leistner, 1972; Rödel et al., 1974) have modified a dew-point hygrometer of EG&G, Cambridge Systems, Mass., U.S.A. (Plate 3).

The recommendations of our laboratories for determination of the a_w of meat and meat products are that the multiple SINA sensor or the modified EG&G instrument should be used for research, and the a_w-Wert-Messer LUFFT should be used for routine examinations.

4. Significance of a_w for Meats

Table 2 lists the minimal and maximal as well as the modal a_w of fresh meat and representative meat products. Fresh meat has in the lean portion a water activity of 0.99. The surface of a carcass or a muscle might have, if it is dried up, a lower a_w than 0.99; this improves the shelf life. Meat products have a lower a_w than fresh meat. The factors which are of most importance in changing a_w during the processing of meat products are (i) the withdrawal or addition of water, (ii) the addition of sodium chloride or other salts and (iii) the content or addition of fat. The fat only influences the a_w of meats indirectly since it contains little water and therefore favours the concentration of water soluble substances, such as added salts, in the water of the lean portion of the product. As Table 2 indicates, the meat products with a generally high a_w, i.e. 0.97, are Bologna type sausages, including frankfurters, because these items contain a considerable amount of water. However, some of these products have a remarkably low a_w, and thus an exceptional shelf life.

Table 2

The a_w range of fresh meat and representative continental meat products

Product	Minimal	Maximal	Modal
Fresh meat	0.98	0.99	0.99
Bologna sausage	0.93*	0.98	0.97
Liver sausage	0.95	0.97	0.96
Blood sausage	0.86†	0.97	0.96
Raw ham	0.80††	0.96	0.92
Dried beef‡	0.80	0.94	0.90
Fermented sausage	0.65§	0.96§§	0.91

* Italian Mortadella; † Speckwurst; †† Country cured ham; ‡ Bündner Fleisch; § Hard sausage; §§ Frische Mettwurst.

This is, for example, true of genuine Italian Mortadella. The a_w of liver sausages, due to the relatively large amount of fat added, is in general, 0.96. The same a_w is typical for blood sausages, but some products of this type, i.e. German Speckwurst, are sometimes dried to an a_w of 0.86, and thus have a prolonged shelf life. In order to secure the expected stability and safety, the a_w of meat products, which are not heated during the processing and are consumed raw, has to be much lower than in the heated products mentioned previously. However, here again there is much variation according to the type of product. The a_w of raw ham is in the range of 0.80–0.96; a low a_w is found, for example, in some country cured hams, especially on the surface of the products, while a raw ham which is not ripened or dried for months but only for a few weeks has a much higher a_w. In general, dried beef, i.e. Bündner Fleisch, typical of Switzerland, has an a_w of $c.$ 0.90, but again the actual a_w of the products depends on the drying process used. Fermented sausages too, are dried to quite different a_w levels. In products processed and consumed within a week the a_w is just below 0.96, but products with a ripening time of several months have an a_w which is much lower, for example, the a_w of Hungarian salami is $c.$ 0.83.

Traditional processes, such as salting, curing, drying and freezing, are empirical methods to adjust the a_w of meats. However, if these processes are understood, the a_w of the products might be improved, and the traditional processes rationalized.

From the investigations of Moran (1936), Storey & Stainsby (1970) and

Table 3
Water activities (a_w) of meat at various subfreezing temperatures

Temp. (°)	a_w	Temp. (°)	a_w	Temp. (°)	a_w
−1	0.990	−11	0.899	−21	0.815
−2	0.981	−12	0.889	−22	0.807
−3	0.971	−13	0.881	−23	0.799
−4	0.962	−14	0.873	−24	0.792
−5	0.953	−15	0.864	−25	0.784
−6	0.943	−16	0.856	−26	0.776
−7	0.934	−17	0.847	−27	0.769
−8	0.925	−18	0.839	−28	0.761
−9	0.916	−19	0.831	−29	0.754
−10	0.907	−20	0.823	−30	0.746

* Calculated data from Moran (1936), Storey & Stainsby (1970), Fennema & Berny (1974).

Fennema & Berny (1974) the a_w of meat at subfreezing temperatures is known (Table 3). Scott (1957) suggested that the growth of some micro-organisms in frozen foods is not limited by temperature but by the a_w. In Table 4 the minimal a_w for growth of bacteria, yeasts and moulds (see Table 1) is correlated with the known minimal growth temperatures (Schmidt-Lorenz, 1970) of these organisms. Even though the agreement between the minimal a_w and the a_w corresponding to the minimal growth temperature of these organisms is not perfect, a relationship is evident. Therefore, indeed for some micro-organisms not the temperature but the a_w of frozen foods might inhibit their multiplication.

Beside the pH the a_w is of utmost importance for the processing of fermented sausages (Plate 4). This product is preserved by a withdrawal of water, i.e. a decrease of the a_w, and a lowering of the pH, caused by lactic acid bacteria and sometimes fostered by the addition of glucono-δ-lactone. The fermentation takes places in a temperature range of 25–15°, and the processing requires, depending on the type of product, one week to several months. Fermented sausages are smoked or unsmoked products, some are mould-fermented, most

Table 4
Correlation between minimal a_w and minimal temperature for growth of micro-organisms

	Minimal a_w	Minimal temperature
Bacteria	0.90	−10° (i.e. a_w 0.91)
Yeasts	0.88	−12° (i.e. a_w 0.89)
Moulds	0.70	−18° (i.e. a_w 0.84)

are consumed raw. An investigation of the water activity of fermented sausages (Leistner et al., 1971), and the a_w changes during the ripening process (Rödel, 1973) revealed that the process control of these products could be much improved. For instance due to the introduction of modern, efficient equipment for stuffing the sausage meat into casings, the products enter the ripening chamber with a temperature close to 0°. Since it was customary to start the process in the ripening chamber with a temperature of c. 25° and a RH of c. 95%, immediate condensation of water on the cold sausages resulted in an uptake of moisture, i.e. an increase of the a_w of the sausage meat. Rödel (1973) has recommended keeping the RH in the ripening chamber low until the temperature of the sausages is equal to that of the environment. This process takes about 6 h and is followed by a drying process in which the RH of the ripening room is adjusted to four units below the a_w of the sausage meat (for example if the a_w of the sausage reaches 0.96 the RH in the ripening chamber should be 92%). This has rationalized the ripening process of fermented sausages and improved the products considerably. Manufacturers of fermented sausages are now aware of the importance of the water activity, and use the a_w-Wert-Messer (Rödel et al., 1975a) to control the a_w of their products. The official food inspection in Germany also uses the a_w of fermented sausages as a criterion for judging the stability of these products (Leistner & Rödel, 1974).

Plate 4. German meat products consumed raw: fermented sausages (1, Plockwurst; 2, schimmelpilzgereifte Salami; 3, geräucherte Salami; 4, Cervelatwurst; 5, Mettwurst) and hams (from pigs) (6, Katenschinken; 7, Schinkenspeck; 8, Lachsschinken; 9, Rollschinken).

In the manufacture of raw ham (Plate 4) as well it is essential to effect a decrease of the a_w to a low level. In the production of these hams a dry cure and/or cover brine is used for the withdrawal of water and the penetration of curing salts. The curing process, at a temperature of 8–10°, takes, depending on the product, two weeks to several months and is often succeeded by smoking and drying processes which lead to a further decrease of the a_w of the hams, which are consumed raw. The decrease of the a_w is necessary for the stability of the product and could be important for safety also, e.g. by inhibiting clostridia (including *Cl. botulinum*) in undeboned ham or for inactivating *Trichinella spiralis*. An investigation in our laboratories indicated that trichinae are no longer invasive when the a_w of fermented sausages is decreased to <0.93 (Lötzsch & Rödel, 1974; Lötzsch et al., 1975). This is true for fermented sausages with a pH range of 5.1–5.7; since the pH of raw ham is higher than that of fermented sausages, the a_w required to inactivate trichinae in raw ham might be lower.

Heated emulsion products of the Bologna type (Plate 5) have a rather high a_w and therefore a limited shelf life. Of these products > 100 varieties are distinguishable in Europe, ranging from Lyoner, Göttinger, Bierschinken, Gelbwurst to frankfurters and Luncheon meat. Most of these items are easily

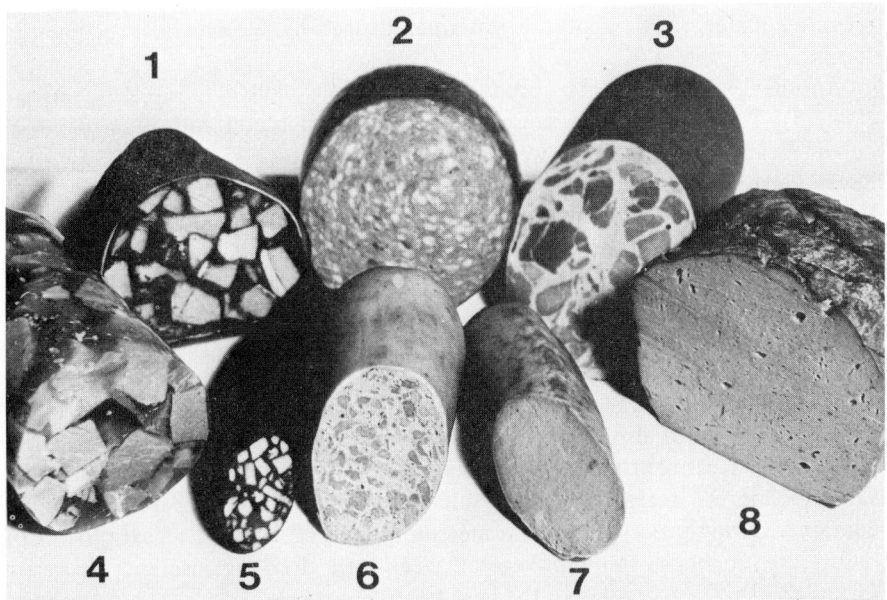

Plate 5. German meat products heat processed: blood sausage (1, Thüringer Rotwurst; 5, Speckwurst), Bologna type sausage (2, Göttinger; 3, Bierschinken; 8, Fleischkäse), liver sausage (6, grobe Leberwurst; 7, feine Leberwurst), and jelly sausage (4, Sülzwurst).

perishable and have to be kept under refrigeration. However, there are a few exceptions, such as genuine Italian Mortadella, which is produced with precooked raw material, relatively high amounts of salt and fat, and little addition of water; this results in an a_w in the range of 0.93–0.96. Due to this relatively low a_w the Italian Mortadella could be and often is stored without refrigeration for several days. It is feasible to decrease the a_w of Bologna type sausages and other meat products not only by the addition of salt and fat, but also by other additives. In Table 5 the depression of the a_w of meat caused by 1% of the additives is listed. This 1% value can be used to calculate the resulting depression of a_w when these additives are used in the usual amounts in meat manufacturing. If the usual dosage of the legally permitted substances used for meats is taken into account, then the most effective depression of the a_w is to be expected using sodium chloride and fat. The addition of 3% of sodium chloride causes about the same depression of the a_w of meat as 30% of fat. However, most effective would be lithium chloride (which is not a legally permitted additive), since it depresses the a_w of meat even more than sodium chloride. By the addition of 1% of lithium chloride the a_w of vacuum-packaged Bologna was depressed from 0.986–0.967, and this was sufficient to inhibit the multiplication of enteropathogenic *E. coli*, even if no nitrite was added to the product (Leistner *et al.*, 1973). A decrease of the a_w of meat products would allow a decrease in the amount of nitrite added to cured meats, which is recommendable in view of the potential danger of forming nitrosamines in foods.

Another easily perishable product, due to the high a_w, is blood sausage (Plate 5). These are cooked sausages which contain blood and often pieces of cured meat and tongue as well as fat. Some blood sausages, i.e. German Speckwurst, contain not only many fat pieces but are also dried to a low a_w, i.e. 0.86–0.93, and thus have an exceptional shelf life. Similar to blood sausages are liver sausages (Plate 5), which, due to a generally high fat content, have an a_w in the range of 0.95–0.97. If the a_w of liver sausage is decreased below 0.95—which is accomplished by adding 2.5% of sodium chloride and 44% of fat—then some micro-organisms e.g. clostridia, are not able to germinate or multiply in the product. This was confirmed experimentally by Leistner & Karan-Djurdjić (1970) with canned liver sausage (Table 6). The data in Table 6 indicate that not only the growth but also the survival of clostridia was influenced by the a_w.

The heat treatment received by a canned meat product is expressed as an F value. F_s is the integrated lethal value of heat received by all points in a container during processing; it is a measure of the capacity of a heat process to reduce the number of spores or vegetative cells of a given organism in a container (Stumbo, 1965). According to the heat treatment received, four types of canned meat products are distinguishable, i.e. semi-preserved, three-quarter-preserved, fully-preserved, and tropical-preserved canned meats (Leistner *et al.*, 1970). A

Table 5
Depression of the a_w of meat caused by additives

1%-value*	Additive	Depression of a_w caused by using additives at (%)						
		0.1	0.3	2.0	3.0	5.0	10	30	50
0.0100	Lithium chloride†	0.0010	0.0030	—	—	—	—	—	—
0.0062	Sodium chloride	0.0006	0.0019	0.0124	0.0186	—	—	—	—
0.0061	Polyphosphate	0.0006	0.0018	—	—	—	—	—	—
0.0050	1,2-Propylene glycol†	0.0005	0.0015	—	—	—	—	—	—
0.0047	Sodium citrate × 5.5 H_2O	0.0005	0.0014	0.0100	—	—	—	—	—
0.0041	Ascorbic acid	0.0004	—	—	—	—	—	—	—
0.0040	Glucono-δ-lactone	0.0004	0.0012	—	—	—	—	—	—
0.0037	Sodium acetate, cryst.	0.0004	—	—	—	—	—	—	—
0.0033	Sodium hydrogentarttrate	0.0003	—	—	—	—	—	—	—
0.0030	Glycerol†	0.0003	0.0009	0.0060	0.0090	0.0150	0.0300	0.0900	—
0.0026	Potassium sorbate†	0.0003	0.0008	—	—	—	—	—	—
0.0024	Glucose	0.0002	0.0006	—	—	—	—	—	—
0.0022	Lactose	0.0002	0.0006	0.0044	0.0066	—	—	—	—
0.0019	Sucrose	0.0002	0.0006	—	—	—	—	—	—
0.0013	Milk protein	0.0001	0.0004	0.0026	0.0039	—	—	—	—
0.00062	Fat	0.0001	0.0002	0.0012	0.0019	0.0031	0.0062	0.0186	0.0310

* Depression of a_w caused by 1% of additive.
† Not legally permitted in West Germany.

Table 6
Stability of canned liver sausage in relation to water activity

Liver sausage a_W	*Clostridium sporogenes* PA 3679 organisms/g		
	After processing	14 days storage*	30 days storage*
0.970	35	>1,000,000	Spoiled
0.967	35	>1,000,000	Spoiled
0.962	110	>1,000,000	5,400,000
0.961	17	32	0
0.959	24	10	0
0.957	92	2	0
0.954	54	1	0
0.947	170	2	0

* Storage temperature 37°, Leistner & Karan-Djurdić (1970).

depression of the a_w below 0.95 is of advantage for the stability of three-quarter-preserved canned meats (F_s value 0.65–0.80); it is not necessary for fully-preserved canned meats (F_s 5.0–6.0), which receive a heat treatment sufficient to inactivate mesophilic Bacillaceae. Also, for semi-preserved canned meats, such as pasteurized hams, which are heated only to an internal temperature of c 68°, a depression of the a_w is not advisable. This is due to the fact that enterococci, i.e. *Streptococcus faecium* and *Strep. faecalis*, frequently survive this heat treatment; therefore these organisms are the major cause of spoilage in pasteurized hams. It was demonstrated that the heat resistance of enterococci might increase with decreasing a_w. If the water activity was adjusted with sodium chloride the highest heat resistance of the enterococci was observed at a_w 0.95 (Vrchlabský & Leistner, 1970b). Also the heat resistance of lactobacilli increases with decreasing a_w and proved maximal in the presence of sodium chloride at an a_w of 0.985–0.975; this may have implications for the preservation of meat products too (Vrchlabský & Leistner, 1971). However, in spite of a possible increase of the heat resistance of some

Table 7
Storage categories of meat products based on the a_w and the pH of the product, with corresponding storage temperatures

Category	Criteria	Temperature of storage
Storable	$a_w \leq 0.95$ *and* pH ≤ 5.2 *or* $a_w \leq 0.91$ *or* pH ≤ 5.0	no refrigeration required
Perishable	$a_w \leq 0.95$ *or* pH ≤ 5.2	$\leq +10°$
Easily perishable	$a_w > 0.95$ *and* pH > 5.2	$\leq + 5°$

spoilage organisms in meats with a reduced a_w, the stability of meat products would be improved substantially if the multiplication of micro-organisms which cause spoilage or food-poisoning could be inhibited by a depressed water activity.

Meat manufacturers are well aware of the fact that some meat products are more perishable than others, and therefore need better refrigeration during storage. The meats which are less perishable are in general characterized by a low a_w. However, the multiplication of micro-organisms on or in meats does not depend solely on the a_w, but on other factors too, such as temperature, pH, E_h, nitrite, and the competitive flora (Leistner, 1974). Even in most instances the shelf life of a meat product is related to the combined effect ('Hürdeneffekt') of several factors which inhibit microbial growth. Nevertheless, it is feasible to predict the shelf life of meat products if only the a_w, pH, and temperature, which are easy to measure in routine work, are taken into consideration. In our laboratories the a_w of representative German meat products as well as their usual shelf life was studied, and related to the a_w, pH and temperature requirements reported in the literature of food-poisoning and spoilage organisms associated with meats. From this study derived a concept for grouping meat products into three categories based on the a_w and pH of the product (Rödel, 1975; Rödel et al., 1975b). Every category demands on appropriate storage temperature. This concept covers bacteria which cause spoilage as well as food-poisoning, but not yeasts and moulds, which usually grow slower than bacteria, and may, where necessary, be inhibited by fungistatic substances, such as potassium sorbate (Leistner et al., 1975). Table 7 indicates, that according to the concept of Rödel (1975) 'easily perishable' meat products have an $a_w < 0.95$ and a pH > 5.2, and must be stored at or below $+5°$. The perishable meat products have either an $a_w < 0.95$ or a pH < 5.2, and must be stored at or below $+10°$. The 'storable' meat products have an $a_w < 0.95$ and a pH < 5.2 or an $a_w > 0.91$ or a pH < 5.0; these products need no refrigeration, and their shelf life is often not limited by bacterial but by chemical or physical spoilage, especially rancidity and discoloration. Concepts of this type, which are based on measurable factors important for food-poisoning and spoilage bacteria, and not on empirical experience alone, could probably be helpful for meat manufacturers as well as food inspection services, and would improve the appropriate storage of meats.

5. References

ALPERDEN, I., MINTZLAFF, H.-J., TAUCHMANN, F. & LEISTNER, L. (1973a). Untersuchungen über die Bildung des Mykotoxins Patulin in Rohwurst. *Die Fleischwirtschaft* **53**, 566-568.

ALPERDEN, I., MINTZLAFF, H.-J., TAUCHMANN, F. & LEISTNER, L. (1973b). Bildung von Sterigmatocystin in mikrobiologischen Nährmedien und in Rohwurst durch *Aspergillus versicolor*. *Die Fleischwirtschaft* **53**, 707-710.

ANDERTON, J. I. (1963). Pathogenic organisms in relation to pasteurized cured meats. *Scientific and Technical Surveys* No. 40. The British Food Manufacturing Industries Research Association, Leatherhead, Surrey, England, pp. 1-158.

BACON, C. W., SWEENEY, J. G., ROBBINS, J.D. & BURDICK, D. (1973). Production of penicillic acid and ochratoxin A on poultry feed by *Aspergillus ochraceus* : temperature and moisture requirements. *Applied Microbiology* 26, 155-160.

BAIRD-PARKER, A. C. & FREAME, B. (1967). Combined effect of water activity, pH and temperature on the growth of *Clostridium botulinum* from spore and vegetative cell inocula. *Journal of Applied Bacteriology* 30, 420-429.

BEERS, R. J. (1957). Effect of moisture activity on germination. In *Spores,* ed. Halvorson, H. O. Washington, U.S.A.: American Institute of Biological Science.

BEM, Z. & LEISTNER, L. (1970a). Die Wasseraktivitätstoleranz der bei Pökelfleischwaren vorkommenden Hefen. *Die Fleischwirtschaft* 50, 492-493.

BEM, Z. & LEISTNER, L. (1970b). Bestimmung der Wasseraktivität von Fleisch und Fleischwaren mit der Methode von Landrock und Proctor. *Die Fleischwirtschaft* 50, 1412-1414.

BEUCHAT, L. R. (1974). Combined effects of water activity, solute, and temperature on the growth of *Vibrio parahaemolyticus. Applied Microbiology* 27, 1075-1080.

BLANCHE KOELENSMID, W. A. A. & VAN RHEE, R. (1964). Salmonella in meat products. *Annales de l'Institut Pasteur de Lille* 15, 85-97.

BROWNLIE, L. E. (1966). Effect of some environmental factors on psychrophilic microbacteria. *Journal of Applied Bacteriology* 29, 447-454.

BULLERMAN, L. B., HARTMAN, P. A. & AYRES, J. C. (1969a). Aflatoxin production in meats. I. Stored meats. *Applied Microbiology* 18, 714-717.

BULLERMAN, L. B., HARTMAN, P. A. & AYRES, J. C. (1969b). Aflatoxin production in meats. II. Aged dry salamis and aged country cured hams. *Applied Microbiology* 18, 718-722.

BULLOCK, K. & TALLENTIRE, A. (1952). Bacterial survival in systems of low moisture content. Part IV. The effects of increasing moisture content on heat resistance, viability and growth of spores of *B. subtilis. Journal of Pharmacy and Pharmacology* 4, 917-931.

BURCIK, E. (1950). Über die Beziehungen zwischen Hydratur and Wachstum bei Bakterien und Hefen. *Archiv für Mikrobiologie* 15, 203-235.

BURMEISTER, H. R. & LEISTNER, L. (1970). Aflatoxinbildung in Fleischwaren. *Die Fleischwirtschaft* 50, 685.

CHRISTIAN, J. H. B. (1965). The influence of nutrition on the water relations of *Salmonella oranienburg. Australian Journal of Biological Sciences* 8, 75-82.

CHRISTIAN, J. H. B. & SCOTT, W. J. (1953). Water relations of salmonellae at 30°C. *Australian Journal of Biological Sciences* 6, 565-573.

CHRISTIAN, J. H. B. & WALTHO, J. A. (1962). The water relations of staphylococci and micrococci. *Journal of Applied Bacteriology* 25, 369-377.

CHRISTIAN, J. H. B. & WALTHO, J. A. (1964). The composition of *Staphylococcus aureus* in relation to the water activity of the growth medium. *Journal of General Microbiology* 35, 205-213.

DIENER, U. L. & DAVIS, N. D. (1970). Limiting temperature and relative humidity for aflatoxin production by *Aspergillus flavus* in stored peanuts. *Journal of the American Oil Chemists' Society* 47, 347-351.

ESCHER, F. E., KOEHLER, P. E. & AYRES, J. C. (1973). Production of ochratoxins A and B on country cured ham. *Applied Microbiology* 26, 27-30.

FENNEMA, O. & BERNY, L. A. (1974). Equilibrium vapor pressure and water activity of food at subfreezing temperatures. *Proceedings of the 4th International Congress of Food Science and Technology, Madrid, 1974.* Work Documents Topic 2, 12-14.

GENIGEORGIS, C. & SADLER, W. W. (1966). Effect of sodium chloride and pH on enterotoxin B production. *Journal of Bacteriology* 92, 1383-1387.

GENIGEORGIS, C., RIEMANN, H. & SADLER, W. W. (1969). Production of enterotoxin B in cured meats. *Journal of Food Science* 334, 62-68.

GENIGEORGIS, C., FODA, M. S., MANTIS, A. & SADLER, W. W. (1971). Effect of

sodium chloride and pH on enterotoxin C production. *Applied Microbiology* **21**, 862-866.
GOUGH, B. J. & ALFORD, J. A. (1965). Effect of curing agents on the growth and survival of food-poisoning strains of *Clostridium perfringens. Journal of Food Science* **30**, 1025-1028.
GREENBERG, R. A., SILLIKER, J. H. & FATTA, L. D. (1959). The influence of sodium chloride on toxin production and organoleptic breakdown in perishable cured meat inoculated with *Clostridium botulinum. Food Technology* **13**, 509-511.
HALLS, N. A. & AYRES, J. C. (1973). Potential production of sterigmatocystin on country-cured ham. *Applied Microbiology* **26**, 636-637.
HANSEN, N. H. & RIEMANN, H. (1962). Mikrobiologische Beschaffenheit von vorverpacktem Fleisch und vorverpackten Fleischwaren. *Die Fleischwirtschaft* **14**, 861-868.
HOBBS, B. C. (1965). *Clostridium welchii* as a food poisoning organism. *Journal of Applied Bacteriology* **28**, 74-82.
HOJVAT, S. A. & JACKSON, H. (1969). Effects of sodium chloride and temperature on the growth and production of enterotoxin B by *Staphylococcus aureus. Journal of the Canadian Institute of Food Science and Technology* **2**, 56-59.
KANG, C. K., WOODBURN, M., PAGENKOPF, A. & CHENEY, R. (1969). Growth, sporulation, and germination of *Clostridium perfringens* in media of controlled water activity. *Applied Microbiology* **18**, 798-805.
KARAN-DJURDJIC, S. & LEISTNER, L. (1970). Messung der Wasseraktivität von Fleisch und Fleischwaren mit dem SINA-Gerät. *Die Fleischwirtschaft* **50**, 1104-1106.
KIM, C. H. (1965). Substrate factors for growth and sporulation of *Clostridium perfringens* in selected foods and in simple systems. *Dissertation Abstracts* **26**, 1288.
KVAALE, O. & DALHOFF, E. (1963). Determination of the equilibrium relative humidity of foods. *Food Technology* **17**, 659-661.
LANIGAN, G. W. (1963). Silage bacteriology. I. Water activity and temperature relationships of silage strains of *Lactobacillus plantarum, Lactobacillus brevis,* and *Pediococcus cerevisiae. Australian Journal of Biological Sciences* **16**, 606-615.
LANDROCK, A. H. & PROCTOR, B. E. (1951). A new graphical interpolation method for obtaining humidity equilibria data, with special reference to its role in food packaging studies. *Food Technology* **5**, 332-337.
LEISTNER, L. (1974). Mikrobiologie der Verpackung von Fleisch und Fleischwaren. *Die Fleischwirtschaft* **54**, 1036-1039.
LEISTNER, L. & KARAN-DJURDJIĆ, S. (1970). Beeinflussung der Stabilität von Fleischkonserven durch Steuerung der Wasseraktivität. *Die Fleischwirtschaft* **50**, 1547-1549.
LEISTNER, L. & RÖDEL, W. (1974). Die Wasseraktivität (a_w-Wert) als Kriterium für die Beurteilung von Rohwurst. *Die Fleischwirtschaft* **54**, 1039-1040.
LEISTNER, L. & TAUCHMANN, F. (1970). Aflatoxinbildung in Rohwurst durch verschiedene Aspergillus flavus-Stämme und einen Aspergillus parasiticus-Stamm. *Die Fleischwirtschaft* **50**, 965-966.
LEISTNER, L., WIRTH, F. & TAKÁCS, J. (1970). Einteilung der Fleischkonserven nach der Hitzebehandlung. *Die Fleischwirtschaft* **50**, 216-217.
LEISTNER, L., HERZOG, H. & WIRTH, F. (1971). Untersuchung über die Wasseraktivität (a_w-Wert) von Rohwurst. *Die Fleischwirtschaft* **51**, 213-216.
LEISTNER, L., HECHELMANN, H., BEM, Z. & ALBERTZ, R. (1973). Untersuchungen zur Reduktion des Nitrizusatzes zu Fleischerzeugnissen. *Die Fleischwirtschaft* **53**, 1751-1754.
LEISTNER, L., MAING, I. Y. & BERGMANN, E. (1975). Verhinderung von unerwünschtem Schimmelpilzwachstum auf Rohwurst durch Kaliumsorbat. *Die Fleischwirtschaft* **55**, 559-561.
LÖTZSCH, R. & RÖDEL, W. (1974). Untersuchungen über die Lebensfähigkeit von Trichinella spiralis in Rohwürsten in Abhängigkeit von der Wasseraktivität. *Die Fleischwirtschaft* **54**, 1203-1208.
LÖTZSCH, R., PONERT, H., MEYER, W. & RAUH, D. (1975). Einfluβ der Wasseraktivität

(a_w-Wert) von Rohwursterzeugnissen auf das Überleben von Trichinen. *Jahresbericht der Bundesanstalt für Fleischforschung, Kulmbach.* 1975 (in press).

MANABE, M., HALLS, N.A., TAUCHMANN, F. & LEISTNER, L. (1975). Bildung von Citreoviridin, Cyclopiazonsäure und Rugulosin in Fleischerzeugnissen. *Jahresbericht der Bundesanstalt für Fleischforschung, Kulmbach,* 1975 (in press).

MARKUS, Z. H. & SILVERMAN, G. J. (1970). Factors affecting the secretion of staphylococcal enterotoxin A. *Applied Microbiology* **20**, 492-496.

MATZ, S. A. (1965). *Water in Foods.* Westport, Connecticut: The AVI Publishing Company, Inc. pp. 249-261.

McLEAN, R. A., LILLY, H. D. & ALFORD, J. A. (1968). Effects of meat-curing salts and temperature on production of staphylococcal enterotoxin B. *Journal of Bacteriology* **95**, 1207-1211.

MORAN, T. (1936). The state of water in tissues. *Report of the Food Investigation Board for the year 1935* London: H. M. Stationery Office, pp. 20-24.

MOSSEL, D. A. A. (1969). Nahrungsmittel als Umwelt für Mikroorganismen, die Lebensmittel gesundheitsschädlich machen. *Alimenta* **8**, 8-16.

OHYE, D. F. CHRISTIAN, J. H. B. (1966). Combined effects of temperature, pH and water activity on growth and toxin production by *Cl. botulinum* types A, B and E. *Proceedings of the 5th International Symposium of Food Microbiology, Moscow, 1966.* 217-223.

OHYE, D. F., CHRISTIAN, J. H. B. & SCOTT, W. J. (1966). Influence of temperature on the water relations of growth of *Cl. botulinum* type E. *Proceedings of the 5th International Symposium of Food Microbiology, Moscow, 1966.* 136-143.

PITT, J. I. (1975). Xerophilic fungi and the spoilage of foods of plant origin. In *Water Relations of Foods,* ed. Duckworth, R. B. London & New York: Academic Press.

PITT, J. I. & CHRISTIAN, J. H. B. (1968). Water relations of xerophilic fungi isolated from prunes. *Applied Microbiology* **16**, 1853-1858.

PIVNICK, H. & THATCHER, F. S. (1968). Microbial problems in food safety with particular reference to *Clostridium botulinum.* In *The Safety of Foods,* eds Ayres, J. C., Blood, F. R., Chichester, C. O., Graham, H. D., Miccutcheon, R. S., Powers, J. J., Schweigert, B. S., Stevens, A. D. & Zweig, G. Westport Connecticut: The AVI Publishing Company, Inc.

RIEMANN, H. (1963). Safe heat processing of canned cured meats with regard to bacterial spores. *Food Technology* **17**, 39-42, 45-46, 49.

RÖDEL, W. (1973). Messung der Wasseraktivität unter Praxisbedingungen. *Die Fleischwirtschaft* **53**, 27-31.

RÖDEL, W. (1975). Einteilung von Fleischerzeugnissen in leicht verderbliche, verderbliche und lagerfähige Produkte auf Grund des pH-Wertes und a_w-Wertes. Thesis. Freie Universität Berlin.

RÖDEL, W. & LEISTNER, L. (1971). Ein einfacher a_w-Wert-Messer für die Praxis. *Die Fleischwirtschaft* **51**, 1800-1802.

RÖDEL, W. & LEISTNER, L. (1972). Messung der Wasseraktivität (a_w-Wert) von Fleisch und Fleischwaren mit einem Taupunkt-Hygrometer. *Die Fleischwirtschaft* **52**, 1461-1462.

RÖDEL, W., HERZOG, H. & LEISTNER, L. (1973). Wasseraktivitäts-Toleranz von lebensmittelhygienisch wichtigen Keimarten der Gattung Vibrio. *Die Fleischwirtschaft* **53**, 1301-1303.

RÖDEL, W., PONERT, H. & LEISTNER, L. (1974). Verbesserung von Geräten zur Messung der Wasseraktivität (a_w-Wert) von Fleisch und Fleischwaren. *Jahresbericht der Bundesanstalt für Fleischforschung, Kulmbach.* 1974, C 41-C 43.

RÖDEL, W., PONERT, H. & LEISTNER, L. (1975a). Verbesserter a_w-Wert-Messer zur Bestimmung der Wasseraktivität (a_wWert) von Fleisch und Fleischwaren. *Die Fleischwirtschaft* **55**, 557-558.

RÖDEL, W., PONERT, H. & LEISTNER, L. (1975b). Einstufung von Fleischerzeugnissen in leicht verderbliche, verderbliche und lagerfähige Produkte. *Proceedings of the 21st European Meeting of Meat Research Workers, Bern, Switzerland.* 77-78.

SANDERS, T. H., DAVIS, N. D. & DIENER, U. L. (1968). Effect of carbon dioxide, temperature, and relative humidity on production of aflatoxin in peanuts. *Journal of the American Oil Chemists' Society* **45**, 683-685.
SCHMIDT-LORENZ, W. (1970), Psychrophile Mikroorganismen und tiefgefrorene Lebensmittel. *Alimenta* **9**, 32-45.
SCOTT, W. J. (1953). Water relations of *Staphylococcus aureus* at 30°C. *Australian Journal of Biological Sciences* **6**, 549-564.
SCOTT, W. J. (1957). Water relations of food spoilage microorganisms. In *Advances in Food Research* Vol. 7, eds Mrak, E. M. & Stewart, G. F. New York & London: Academic Press.
SEGNER, W. P., SCHMIDT, C. F. & BOLTZ, J. K. (1966). Effect of sodium chloride and pH on the outgrowth of spores of type E *Clostridium botulinum* at optimal and suboptimal temperatures. *Applied Microbiology* **14**, 49-54.
SEGNER, W. P., SCHMIDT, C. F. & BOLTZ, J. K. (1971). Minimal growth temperature, sodium chloride tolerance, pH sensitivity, and toxin production of marine and terrestrial strains of *Clostridium botulinum* type C. *Applied Microbiology* **22**, 1025-1029.
SNOW, D. (1949). The germination of mould spores at controlled humidities. *Annals of Applied Biology* **36**, 1-13.
STOREY, R. M. & STAINSBY, G. (1970). The equilibrium water vapour pressure of frozen cod. *Journal Food Technology* **5**, 157-163.
STRONG, D. H., FOSTER, E. F. & DUNCAN, C. L. (1970). Influence of water activity on the growth of *Clostridium perfringens*. *Applied Microbiology* **19**, 980-987.
STUMBO, C. R. (1965). *Thermobacteriology in Food Processing*. New York & London: Academic Press.
TAUCHMANN, F. & LEISTNER, L. (1971). Aflatoxinbildung in Rohwurst und Reis in Abhängigkeit von Zeit und Temperatur. *Die Fleischwirtschaft* **51**, 77-79.
TOMČOV, D., BEM, Z. & LEISTNER, L. (1974). Minimalne a_w vrednosti odabranih bakterijskih vrsta. *RIM* **6**, 3-9.
TROLLER, J. A. (1971). Effect of water activity on enterotoxin B production and growth of *Staphylococcus aureus*. *Applied Microbiology* **21**, 435-439.
TROLLER, J. A. (1972). Effect of water activity on enterotoxin A production and growth of *Staphylococcus aureus*. *Applied Microbiology* **24**, 440-443.
VRCHLABSKÝ, J. & LEISTER, L. (1970a). Erprobung eines SINA-Gerätes mit eingebautem Schreiber zur Messung der Wasseraktivität von Fleisch und Fleischwaren. *Die Fleischwirtschaft* **50**, 1106-1107.
VRCHLABSKÝ, J. & LEISTNER, L. (1970b). Hitzeresistenz der Enterokokken bei unterschiedlichen a_w-Werten. *Die Fleischwirtschaft* **50**, 1237-1238.
VRCHLABSKÝ, J. & LEISTNER, L. (1971). Hitzeresistenz von Laktobazillen bei unterschiedlichen a_w-Werten. *Die Fleischwirtschaft* **51**, 1368-1370.
WILLIAMS, O. B. & PURNELL, H. G. (1953). Spore germination, growth, and spore formation by *Clostridium botulinum* in relation to the water content of the substrate. *Food Research* **18**, 35-39.
WODZINSKI, R. J. & FRAZIER, W. C. (1960). Moisture requiremens of bacteria. I. Influence of temperature and pH on requirements of *Pseudomonas fluorescens*. *Journal of Bacteriology* **79**, 572-578.
WU, M. T., AYRES, J. C. & KOEHLER, P. E. (1974). Toxigenic aspergilli and penicillia isolated from aged, cured meats. *Applied Microbiology* **28**, 1094-1096.

The Inactivation of Vegetative Bacterial Cells by Ionizing Radiation

R. DAVIES

National College of Food Technology, University of Reading, St. George's Avenue, Weybridge, England

CONTENTS

1. Introduction 239
2. Radiation resistance of wild-type bacterial cells 240
   - (a) Survival nomenclature 240
   - (b) Species differences 241
   - (c) The influence of environmental factors 244
3. Development of radioresistant mutants 247
4. Practical implications 249
5. References 252

1. Introduction

IONIZING RADIATION occurs naturally as geological radioactivity, it is generated artificially in the form of X-rays and in particle accelerators and increasing benefit is derived from its technological applications as a power source and as a microbicidal agent. The fact that it is also mutagenic has caused concern for some time that its indiscriminant use may lead to a selective ecological enrichment of resistant microbial mutants. Not only would these require more severe irradiation doses for inactivation but fears have been expressed that pathogens may be mutated beyond routine recognition or, more alarmingly, to new levels of virulence. It is always opportune, therefore, to monitor and appraise the rapidly advancing concepts and developments of radiobiology in the applied context. Reference will be limited in this discussion, however, to factors affecting the resistance and general response of vegetative bacterial cells to ionizing radiations of the γ-, X-ray and high energy electron types.

The basic nature of ionizing radiation, in relation to its interaction with living cellular systems, has been reviewed elsewhere (Silverman & Sinskey, 1968; Pollard, 1969; Goldblith, 1971). To summarize briefly, however: fast charged particles, usually electrons, are generated which collide with atoms in the irradiated material and dislodge further electrons along random tracks thus causing primary and secondary ionization events. An average event, though occupying a mere 10^{-7} sec or less, results in the deposition of 60 eV of energy which is a quantity up to 20 times greater than the average binding energy of a covalent chemical bond. A bacterial cell can be affected or inactivated (i.e. lose its capacity for indefinite reproduction in optimal conditions) if these disruptive events occur 'directly' within a vital molecular target or if they 'indirectly' create transient but highly reactive chemical radicals within diffusion distance (2-10 nm) of such a target. The vital target, in prokaryotic micro-organisms, is widely recognized as deoxyribonucleic acid (DNA) or some closely associated functional component of its replicative process (Ginoza, 1967; Altman et al., 1970). Indirect effects, which frequently account for 53-82% of the overall radiobiological effect, largely result from the interaction of radiation with water molecules though organic radicals can also participate (Blok & Loman, 1973).

The response of a bacterial cell and hence its resistance to ionizing radiation, depends therefore: (i) on the nature and amount of direct damage produced within its vital target; (ii) on the number, nature and longevity of radiation-induced, reactive chemical species and the indirect damage that they may cause; (iii) on the inherent ability of the cell to tolerate or accurately repair either of the above and (iv) on the influence of intra- and extracellular environment on any of the above. Clearly, therefore, in the applied context where we are particularly concerned with environmental factors, any attempt to categorize and compare the radioresistances of important micro-organisms is only meaningful when all related conditions are precisely defined and understood.

2. Radiation Resistance of Wild-type Bacterial Cells

Many categories of interrelated factors influence the radiation survival characteristics of vegetative bacterial cells, including species differences and environmental conditions before, during and after irradiation. These have been reviewed extensively by Bridges & Horne (1959); Thornley (1963a, b); Silverman & Sinskey (1968) and Goldblith (1971). Consideration here, therefore, is deliberately restrictive and selective.

(a) Survival nomenclature

Radiation survival is conveniently represented graphically with the logarithm of the number of surviving organisms (or a related function) plotted against dose to give a 'survival curve'. Typical radiation survival curves, as shown in Fig. 1, are essentially of three types as described by Silverman & Sinskey (1968): type A exponential survival curves, quite common for the more radiation sensitive organisms, are simply characterized by their D_{10} values — dose increments (rad) to inactivate 90% of the viable cells present at any time period during irradiation.

Type C curves described as concave with resistant 'tails', reflect either heterogeneity in resistance within a population or a shift towards environmentally protective conditions during prolonged irradiation. These curves are frequently encountered, for example, in the case of microbial survival within irradiated food systems and are difficult to quantitate mathematically. Survival here is often best represented by quoting an 'inactivation dose', for example: LD_{90}—the dose (rad) to inactivate 90% of the initial viable cell population.

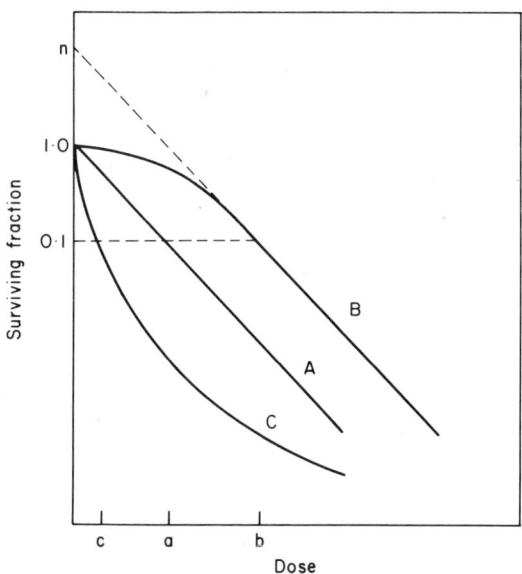

Fig. 1. Types of survival curve for irradiated bacteria. The LD_{90} values of curves A, B & C are represented by doses a, b & c, respectively. (Modified from Silverman & Sinskey, 1968.)

Type B curves, however, characterized by an initial 'shoulder' region, are probably the most common and are described by the general equation: surviving fraction = $N/N_0 = 1-(1-e^{-KN})^n$; where N organisms survive out of an initial population N_o after exposure to a dose of X(rad) at a death rate represented by a constant K(rad^{-1}) with n interpreted merely as an 'extrapolation number'. The D_{10} and n values are sufficient to characterize a curve but n values alone do not necessarily define the extent of a shoulder region. More elaborate mathematical analyses of survival data are presented by Fowler (1964) and Miller (1970).

(b) *Species differences*

The relative radiation sensitivities of various representative, wild-type, vegetative bacterial cells are given in Tables 1 and 2. Considering only reported survival data which have been obtained under essentially similar conditions (mostly aerated, aqueous, buffered systems at temperatures between 3 and 25°), it can be seen that the range of resistance is extremely large. For example, the D_{10} and LD_{90} values for *Micrococcus radiodurans*, are 52 and 200 times greater, respectively, than those for *Serratia marcescens*.

In general, the most sensitive species are those Gram negative rods (Table 1) in the genera *Aeromonas*, *Bacteroides*, *Proteus*, *Pseudomonas*, *Serratia* and

Table 1
Radiation resistance of some Gram negative bacteria

Organism	Irradiation conditions	D_{10} value (Krad)	'n'	Reference
Escherichia coli Pol A1	PB air	2.2	8	Tyrrell (1974)
Serratia marcescens	PB air	3.7	1	Watts *et al.* (1975)
Pseudomonas sp.	PB air	4.5	1	Lewis *et al.* (1971)
E. coli B_{S-1}	MM air	5.0	1	Horan *et al.* (1972)
Vibrio parahaemolyticus	Seawater air	5.1	1	Matches & Liston (1971)
Proteus vulgaris	PB air	6.7	1	Lewis *et al.* (1971)
Aeromonas hydrophila	PB air	7.0	1	Lewis *et al.* (1971)
Bacteroides putridus	Anaerobic diluent	7.8	1	Yamazaki *et al.* (1974)
E. coli B/r	MM air	11.0	1	Horan *et al.* (1972)
Salmonella typhimurium	PB air	12.0	1	Davies & Sinskey (1973)
E. coli K12	PB air	13.2	1	Tyrrell (1974)
Salm senftenberg	PB sealed tubes	19.3	1	Underdal & Rossebo (1972)
Acinetobacter sp.	PB air	34.0	1	Ito *et al.* (1973)
Alcaligenes sp. (*Acinetobacter*)	PB air	31.0	40	Matsuyama *et al.* (1964)

PB, phosphate buffer; MM, minimal medium.

Vibrio which exhibit D_{10} values < 10 Krad; whereas wild-type *Escherichia*, *Salmonella* and *Shigella* are only marginally more resistant. It is interesting to note that within these groups of Gram negative organisms, the survival curves are generally exponential ($n = 1$) with the notable exception of the most sensitive strain tested, namely *Escherichia coli* Pol A1. This mutant, lacking Kornberg DNA polymerase I and hence deficient in rapid (type II) repair of irradiated DNA, as described by Town *et al.* (1973), nevertheless exhibits a miniscule shoulder in its survival curve with an estimated $n = 8$ (Tyrrell, 1974). This serves as a reminder that the extrapolation number quoted alone can be a deceptive indicator of the extent and significance of a shoulder region within a survival curve.

The more resistant organisms, among the Gram negative genera considered, are members of the psychrotrophic *Acinetobacter-Moraxella* group. These organisms exhibit D_{10} values in excess of 30 Krad in aerated buffer systems and Thornley (1963a) and Matsuyama (1964) have demonstrated substantial shoulder characteristics. It is recognized and understandable, therefore, that this group can dominate the spoilage microflora of irradiated flesh foods such as

Table 2
Radiation resistance of some Gram positive bacteria

Organism	Irradiation conditions	D_{10} value (Krad)	'n'	LD_{90} (Krad)	Reference
Eubacterium lentum	Diluent	6.0	1	6	Yamazaki *et al.* (1974)
Corynebacterium sp.	Dileunt	7.0	1	7	
Lactobacillus plantarum	PB air	8.0	300	28	Dupuy & Tremeau (1961)
Bacillus pumilus	PB oxygen	9.6	5.2	16	Ewing (1973)
B. megaterium	PB oxygen	14.5	2.5	25	Ewing (1973)
Staphylococcus aureus	PB (not specified)	20.0	200	86	Kudryasheva *et al.* (1973)
Microbacterium thermosphactum	PB air	20.0	1	20	Oka *et al.* (1973)
B. pumilus	Saline/sealed tubes	21.0	10	55	Parisi & Antoiné (1974)
Streptococcus faecium	PB sealed tubes	30.0	1	30	Anellis *et al.* (1973)
Micrococcus sodonensis	PB air	30.0	400	80	Watts *et al.* (1975)
M. luteus	PB air	80.0	1.2	87	Lewis *et al.* (1971)
L. brevis	PB air	120.0	1.3	140	Dupuy & Tremeau (1961)
M. radiodurans	PB oxygen	210.0	600	800	Moseley (1967)
Strep. faecium	SB dried	–	–	1500	Emborg (1972)
M. radiophilus	TGYM air	280.0	10,000	1600	Lewis (1971)

PB, phosphate buffer; SB, serum broth; TGYM, tryptone-glucose-yeast extract media.

poultry (Thornley, 1966), fish (Laycock & Regier, 1970), meat (Maxcy et al., 1972) and Wiener sausage (Ito et al., 1973).

On the other hand, the wild-type Gram positive, vegetative bacterial cells (Table 2) represent a more diverse spectrum with respect to variations in radiation resistance. Some are relatively sensitive, including organisms within the genera *Eubacterium, Corynebacterium, Peptostreptococcus, Microbacterium, Staphylococcus, Streptococcus* and *Bacillus*. The latter two being interesting in that both *Strep. faecium* in the dry state (Christensen, 1964) and the spores of *B. pumilus* (Parisi & Antoine, 1974) are highly radioresistant and are therefore used as standard indicators for the radiation sterilization of medical products, yet the vegetative cells in aerated aqueous systems are quite sensitive.

The group also includes various Gram positive cocci which are exceptionally resistant to ionizing radiation. The best characterized of these is *M. radiodurans,* isolated originally from irradiated meat by Anderson et al. (1956), which typically requires LD_{90} doses of c. 800 Krad (Moseley, 1967) and which consequently is frequently used in model studies of microbial radiation-resistance mechanisms. Its resistance has been variously attributed to: intracellular radioprotective compounds (Bruce, 1964); cell wall structural features Thornley et al. (1965); highly efficient repair of radiation-induced single and double-strand breaks in DNA (Dreidger & Grayston, 1971; Kitayama & Matsuyama, 1973) and to DNA-membrane association (Dreidger, 1970; Burrell et al., 1971; Dardalhon-Samsonoff & Rebeyrotte, 1975).

Similar Gram positive, tetrad forming, red pigmented cocci, displaying even higher levels of resistance to radiation than *M. radiodurans*, have been isolated from irradiated fish (Lewis, 1971) and from irradiated sutures (Osterberg, 1974). The fish isolates, designated *M. radiophilus,* have exceptionally high D_{10} and LD_{90} values of 280 and 1600 Krad, respectively, (Table 2) which are indepedent of the presence or absence of carotenoid pigments (Lewis et al., 1974) but which are markedly decreased in the presence of iodine radiosensitizers (Lewis & Kumta, 1975).

(c) *The influence of environmental factors*

It is widely accepted that the presence of oxygen increases the lethal effect of ionizing radiation on microbial cells. In conditions of complete anaerobiosis, for example, the resistance levels quoted previously for vegetative bacteria (Tables 1 & 2) may be expected to increase by factors ranging from 2.5-4.7 (Thornley, 1963*b*). This effect is typically demonstrated for *E. coli* B/r (Fig. 2) which, when irradiated in mixtures of oxygen or nitrogen (Epp et al., 1968), displays a dose modification factor of 3.1. The experimental technique featured in these studies involves the use of high intensity electron radiation applied as short pulses (30 nsec) which cause oxygen depletion within the irradiated cells. Similar

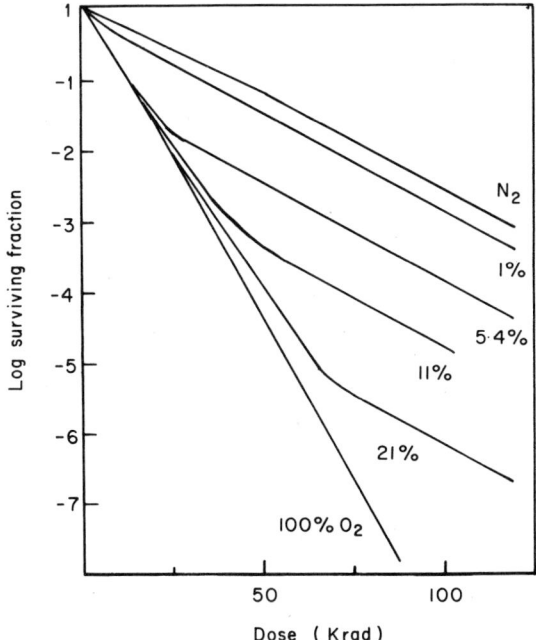

Fig. 2. Survival of *E. coli* B/r irradiated in the presence of various concentrations of oxygen (in pure N_2): high intensity electron radiation dose delivered in single pulses of 30 nsec. (From Epp *et al.*, 1968.)

experiments with *Serratia marcescens* indicate that oxygen has a true, but transient (2 msec), postirradiation sensitizing effect (Michael *et al.*, 1973; Weiss *et al.*, 1975).

The oxygen effect clearly has considerable practical significance particularly when attempting to compare the relative radiosensitivities of micro-organisms. For example, data reported for bacterial cell suspensions irradiated in sealed tubes can frequently be plotted as type C survival curves. These tailing effects, which probably represent a shift to anaerobic conditions, were included in regression analyses of D_{10} values for irradiated *Salmonella senftenberg* (Underdal & Rossebø, 1972) and were attributed to bi-model death-kinetic mechanisms in the case of *Strep. faecium* (Anellis *et al.*, 1973). In the latter case, it is assumed here (Table 2) that residual traces of air probably remained within the irradiated menstrua and thus affected the observed, initial inactivation rates (as represented in Fig. 3).

The temperature maintained during irradiation can undoubtedly, beyond certain limits, also influence the radiosensitivity of bacterial cells (Bridges & Horne, 1959; Goldblith, 1971). Elevated temperature treatments, generally in

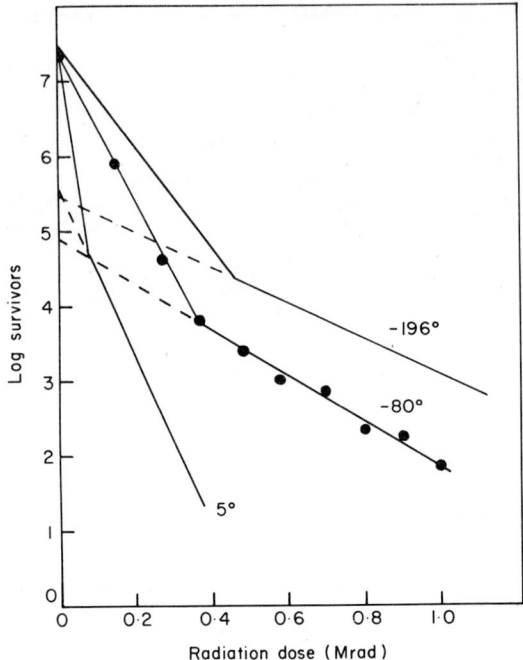

Fig. 3. Effect of freezing on the resistance of *Strep. faecium* α21 to γ-rays. Cells were suspended in phosphate buffer, vacuum sealed in tubes and irradiated at various temperatures. (From Anellis *et al.*, 1973.)

the sublethal range above 45°, synergistically enhance the bactericidal effects of ionizing radiation particularly when applied simultaneously. Licciardello (1964) observed marked increases in the death rates of salmonellae at irradiation temperatures of 45-55° and Tullis *et al.* (1973) demonstrated that spores of the radiosterilization indicator organism *B. pumilus* can be inactivated by 60% reduced doses of irradiation when exposed at 60°.

Freezing, on the other hand, protects cells during irradiation by abolishing at least 50% of the indirect effects, as was shown for *Escherichia coli* B in anoxic conditions by Sanner & Pihl (1969). Analysis of data such as that reported by Matsuyama *et al.* (1964) for *Pseudomonas* and *Acinetobacter*, reveals that freezing can increase radioresistance (D_{10} values) by factors of up to 6.7 whereas a combination of freezing and anoxia increases resistance by a factor of 8.5. A similar pattern with factors of 3.1 and 10.5 respectively, is demonstrated for *Strep. faecium* irradiated in aqueous buffer at 5° or at −80° (Anellis *et al.*, 1973) as shown in Fig. 3. Extrapolation of the resistant portion of these survival curves fails to reveal a common initial population heterogeneity and it seems reasonable to assume that anoxic conditions develop during prolonged irradiation as discussed earlier.

It is also generally conceded that bacteria are more resistant to ionizing radiation when irradiated in dry conditions. As with freezing, the physical removal of water also diminishes the contribution of indirect effects to cell inactivation (Goldblith, 1971). The protective effects of drying are clearly demonstrated in the case of *Strep. faecium* where cells of the reference-standard strain $A_2 1$ dried from serum broth and irradiated in glass cells (Christensen, 1964) are more than 40 times more resistant (LD_{90} values) than when irradiated in aqueous suspensions (Emborg, 1972).

3. Development of Radioresistant Mutants

The early resistant mutant *E. coli* B/r, isolated by Witkin (1947) after a single UV-irradiation treatment of *E. coli* B, has served as a model for radioresistance studies for some time but, as pointed out by Adler (1966), its level of resistance is no higher than that of *E. coli* K12. Horan et al. (1972), using chemical mutagenesis, have obtained *E. coli* PHHl, a DNA-degradation-resistant mutant which is slightly more radioresistant than *E. coli* B/r. Its parent *E. coli* B_{s-1}, is normally regarded as a radiosensitive mutant of *E. coli* B.

Several radioresistant organisms, currently accorded the status of wild-type and thus listed in Table 2, were also isolated originally among low numbers of survivors from large irradiation doses. *Micrococcus radiodurans* and *M. radiophilus*, for example, are in this category. Indeed radioresistant mutants (Table 3) include some developed deliberately in this way, from less resistant parents, namely: *Strep faecium* $\phi 12$ (Christensen & Kjems, 1965) and *M. radiodurans* $R_{II} 5$ (Emborg & Eriksen, 1971) which together constitute the most radiation resistant, vegetative bacterial cells known at the present time. The $R_{II} 5$ mutant, for instance, requires 9 Mrad in dry conditions to inactivate 90% of the initial population.

Most attempts to develop and select radioresistant mutants, however, have generally used the principle of 'growth-irradiation cycles' introduced by Gaden & Henley (1953). Organisms are exposed repeatedly to sublethal doses of irradiation with intervening periods of outgrowth to express mutations which may eventually enrich the population with resistant progeny. Erdman et al. (1961) applied this procedure to several food-borne pathogenic bacteria using a constant repeated dose and reported slight increases in resistance to modest plateau levels with no serious concomitant changes in recognition characteristics. This has been the experience of other workers, using cycles of single doses, as shown in Table 3.

Erdman subsequently (cited in Pontefract & Thatcher, 1965) used 12 increasing cyclic-irradiation doses over a 120 day period to obtain a stepped

Table 3
Radiation resistant strains

Method	Organism	Increased resistance cf parent	Reference
Direct isolation	*E. coli* PHH1	$LD_{90} \times 3.6$	Horan *et al.* (1972)
	Strep. faecium ∅ 12r	$LD_{90} \times 2$	Christensen & Kjems (1965)
	M. radiodurans $R_{II}5$	$LD_{90} \times 4$	Emborg & Eriksen (1971)
Single-dose cycles	*E. coli* I t7	$LD_{90} \times 2$	Wright & Hill (1968)
	E. coli K12/3OR	$LD_{90} \times 2.3$	Mouton & Tremeau (1970)
	Salm. typhimurium	$D_{10} \times 3$	Corry & Roberts (1970)
	Salm. newport	$D_{10} \times 3$	Licciardello *et al.* (1969)
Ascending-dose cycles	*Salm. typhimurium* S_{II}	$D_{10} \times 2$	Epps & Idziak (1970)
	B. pumilus V-23	$D_{10} \times 4.5$	Parisi & Antoiné (1974)
	E. coli 12γ	$LD_{99,999} \times 14$	Pontefract & Thatcher (1965)
	Salm. typhimurium D21R6	$LD_{90} \times 22$	Davies & Sinskey (1973)

series of highly resistant derivatives of *E. coli* I designated 1γ to 12γ. These have since been characterized and found to be multinucleate, highly auxotrophic and able to resist postirradiation DNA degradation (Idziak & Thatcher, 1964; Pontefract & Thatcher, 1965; Robern & Thatcher, 1968; Stavrić, 1969a, b).

A similar stepped series of γ-radiation resistant derivaties were developed from *Salm. typhimurium*, as reported by Davies & Sinskey (1973). An ascending dose pattern was applied during a series of 84 × 24h cycles, (summarized in Table 4) which increased the irradiation survival characteristics (Fig. 4) in terms of D_{10} and LD_{90} values by factors of up to 12 and 22 times, respectively. The most radioresistant mutant, designated *Salm. typhimurium* D21R6, has been

Table 4
Effect of increasing cyclic doses of irradiation on the radioresistance of Salmonella typhimurium

No. of cycles*	Cyclic dose (Krad)	Culture designation	Radioresistance LD_{90} (Krad)	
			in TSY broth (growth medium)	in buffer
0		DB21	30	12
1-14	100	D21R1	112	75
15-28	225	D21R2	230	95
29-42	350	D21R3	250	115
43-56	420	D21R4	345	140
57-70	520	D21R5	450	210
71-84	780	D21R6	550	275

* A cycle consists of: exposure of 1 ml of stationary phase TSY (tryptone-soy-yeast extract) broth culture to irradiation dose to inactivate 99.9% followed by 24 h incubation at 37°C in fresh TSY broth.

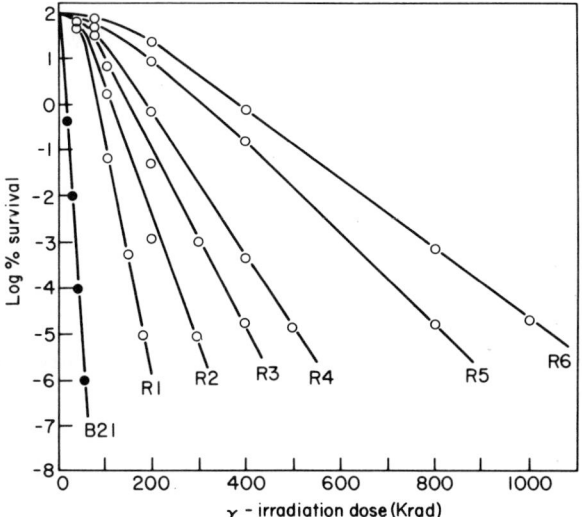

Fig. 4. γ-irradiation survival curves for parent and radioresistant cultures of *Salm. typhimurium*. Cells previously grown to stationary phase in TSY broth were irradiated in phosphate buffer at 0°. Curve B21 represents the survival of the parent strain DB21; curves R1 and R6 represent survival data for the resistant mutants D21R1 to D21R6 derived from DB21. (From Davies & Sinskey, 1973.)

studied further and compared with its original parent (Davies & Sinskey, 1973; Davies *et al.*, 1973). The mutant is genetically stable and displays most of the identification characteristics of *Salm. typhimurium* (Table 5). Its high level of radioresistance has been attributed to several additive mutations which have led to an exceptional ability to repair strand-breaks in irradiated DNA. At least two of these mutations, mapping close to *Uvr* and *Rec* gene clusters on the genetic map, have been transferred genetically into wild-type recipient strains of *Salm. typhimurium* producing radioresistant recombinants (Ibe *et al.*, 1972).

The radiation survival curves of the radioresistant mutants discussed above, together with those of other organisms listed in Tables 1, 2 and 3, are grouped for comparative purposes in Fig. 5.

4. Practical Implications

The interrelated influences of environment on the radiation resistance of micro-organisms are well recognized by applied radiobiologists but are difficult to reproduce realistically in the laboratory. In determining actual radiation process requirements, therefore, it is considered (Goldblith, 1971) that there is probably no substitute for 'inoculated pack' studies under commercial conditions, using meaningful numbers of the most relevant, radioresistant

Table 5
Summary of general characters of parent and radioresistant cultures of
Salmonella typhimurium

Characters that remained unchanged	Characters that changed with increased γ-radiation resistance
Growth in TSY media	Increased resistance to UV irradiation
Gram stain	Cell size increased (× 2)
Motility	Less detectable H_2S released in differential media
Antibiotic spectra	Requirement for amino acids
Serological reactions: antigens 4, 5, 12 : i–1, 1,2	Restricted range of carbon sources utilized
Phage type	Increased mutator activity
Qualitative reactions on selective media	Increased reactivation of gamma irradiated phage
Virulence to mice	Elevated DNA polymerase I activity
DNA content per cell	Elevated DNA ligase activity
Heat sensitivity	More efficient repair of DNA single-strand breaks

Davies & Sinskey (1973).

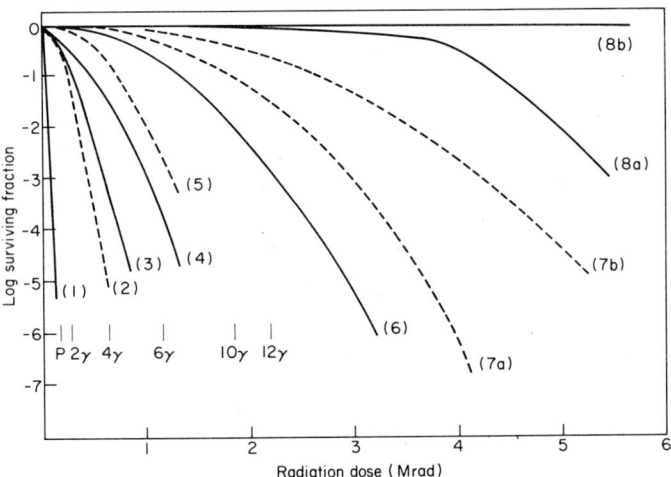

Fig. 5. Comparative radiation survival curves of various radioresistant bacteria. (1) *E. coli* B/r (Horan *et al.,* 1972); (2) *B. pumilus* V23 (Parisi & Antoine, 1974); (3) *Strep. faecium* $A_2 1$ (Emborg, 1972); (4) *Salm. typhimurium* D21R6 (Davies & Sinskey, 1973); (5) *M. radiodurans* (Moseley, 1967); (6) *M. radiophilus* (Lewis, 1971); (7a) *Strep. faecium* $\phi 12$ (wet), (7b) *Strep. faecium* $\phi 12$/R (dry) (Christensen & Kjems, 1965); (8a) *M. radiodurans* $R_{II}5$ (wet), (8b) *M. radiodurans* $R_{II}5$ (dry) (Emborg & Eriksen, 1971). The short vertical lines series P to 12 represent data points quoted for the survival of the parent and resistant cultures derived from *E. coli* I (Pontefract & Thatcher, 1965).

contaminants. In this context, the radioresistant mutants described in this paper begin to assume awesome potential significance. In the case of radicidation processes designed to eliminate salmonellae from dried or frozen feeds, for example, if D21R6 was a major contaminant then even the most conservative estimates of the most probable effective dose (MPED: Mossel et al., 1967) would be technically and economically unrealistic (Table 6).

Table 6
Estimated processes to inactivate radiation-resistant Salmonella typhimurium *D21R6 in food and feed products*

Product	Suggested radicidation process (Mrad)	Estimated MPED (Mrad)
Fish meal	0.6*	5.4
Frozen meat	0.6†	5.4
Poultry feed	1.0‡	9.0
Herring meal	1.3§	11.0

MPED, most probable effective dose; assuming the presence of D21R6 with D_{10} values increased by a factor $\geqslant 9$ (Davies & Sinskey, 1973).
* Mossel et al. (1968).
† Ley et al. (1970).
‡ Epps & Idziak (1972).
§ Underdal & Rossebø (1972).

In normal processing circumstances, however, it is unlikely that the elaborate conditions of the growth-irradiation cycles would be reproduced. Attempts to simulate recycling situations with food products have demonstrated, in one case a reduction in radioresistance for salmonellae in frozen horsemeat recycled six times at 0.65 Mrad (Ley et al., 1970) and in another, only minor increases in resistance with a loss of virulence for various salmonella serotypes recycled 36 times at increasing doses in poultry meat (Epps & Idziak, 1970). The opportunity for the outgrowth of survivors between irradiation doses is particularly limited and can be eliminated relatively easily be the application of conventional control procedures. Cross-contamination can likewise be avoided by such measures as product packaging, controlled ventilation, restricted personnel movement and by careful design and hygienic operation of process flow lines.

The alternative, 'control' possibility of destroying rather than avoiding radiation resistant survivors has, as yet, been relatively unexplored. It is known that *M. radiodurans* is quite heat sensitive (Duggan et al., 1963) and that its radioresistance can be reduced to that of *E. coli* if irradiated cells are subsequently held at a restrictive temperature of 42° for 6 h (Kitayama &

Matsuyama, 1973). Thermal survival data, however, are not currently available for most other radioresistant strains but it seems reasonable to assume that heat and irradiation applied consecutively or simultaneously, or indeed other stress combination-processes, should be effective against them.

Finally, though it can be argued that hyper-radioresistant mutants are radiobiological curiosities and unlikely to occur in practical circumstances, the fact that they exist warns that it would be prudent to guard against them in any long-term microbicidal application of ionizing radiation. In particular the demonstration that radioresistance can be transferred genetically into previously sensitive recipients has potentially ominous environmental implications.

5. References

ADLER, H. I. (1966). The genetic control of radiation sensitivity in microorganisms. *Advances in Radiation Biology* 2, 167-191.
ANDERSON, A. W., NORDAN, H. C., CAIN, R. F., PARRISH, G. & DUGGAN, D. (1956). Studies on radioresistant micrococcus. *Food Technology* 10, 575-577.
ANELLIS, A., BERKOWITZ, D. & KEMPER, D. (1973). Comparative resistance of nonsporogenic bacteria to low-temperature gamma irradiation. *Applied Microbiology* 25, 517-523.
ALTMAN, K. I., GERBER, G. B. & OKADA, S. (1970). *Radiation Biochemistry* London & New York: Academic Press.
BLOK, J. & LOMAN, H. (1973). The effects of γ-irradiation in DNA. *Current Topics in Radiation Research* Q9, 165-245.
BRIDGES, B. A. & HORNE, T. (1959). The influence of environmental factors on the microbicidal effect of ionizing radiations. *Journal of Applied Bacteriology* 22, 96-115.
BRUCE, A. K. (1964). Extraction of the radio-resistant factor of *Micrococcus radiodurans*. *Radiation Research* 22, 155-164.
BURRELL, A. D., FELDSCHREIBER, P. & DEAN, C. J. (1971). DNA-membrane association and the repair of double breaks in X-irradiated *Micrococcus radiodurans*. *Biochimica et biophysica acta: Previews* 247, 38-53.
CHRISTENSEN, E. A. (1964). Radiation resistance of enterococci dried in air. *Acta Pathologica et Microbiologica Scandinavica* 61, 483-492.
CHRISTENSEN, E. A. & KJEMS, E. (1965). The radiation resistance of substrains from *Streptococcus faecium* selected after irradiation of two different strains. *Acta pathologica et microbiologica scandinavica* 63, 281-290.
CORRY, J. E. L. & ROBERTS, T. A. (1970). A note on the development of resistance to heat and gamma radiation in salmonella. *Journal of Applied Bacteriology* 33, 733-737.
DARDALHON-SAMSONOFF, M. & REBEYROTTE, N. (1975). Rôle de l'attachement du DNA à la membrane dans la réparation des radiolésions chez *Micrococcus radiodurans*. *International Journal of Radiation Biology* 27, 157-169.
DAVIES, R. SINSKEY, A. J. (1973). Radiation-resistant mutants of *Salmonella typhimurium* LT2: development and characterisation. *Journal of Bacteriology* 113, 133-144.
DAVIES, R., SINSKEY, A. J. & BOTSTEIN, D. (1973). Deoxyribonucleic acid repair in a highly radiation-resistant strain of *Salmonella typhimurium*. *Journal of Bacteriology* 114, 357-366.
DREIDGER, A. A. (1970). Are there multiple attachments between bacterial DNA and the cell membrane? *Canadian Journal of Microbiology* 16, 881-882.

DREIDGER, A. A. & GRAYSTON, M. J. (1971). Demonstration of two types of DNA repair in X-irradiated *Micrococcus radiodurans. Canadian Journal of Microbiology* **17**, 495-499.
DUGGAN, D. E., ANDERSON, A. W. & ELLIKER, P. E. (1963). Inactivation rate studies on a radiation resistant spoilage microorganism. III Thermal inactivation rates in beef. *Journal of Food Science* **28**, 130-134.
DUPUY, P. & TREMEAU, O. (1961). Resistance aux radiations ionisontes de quelques souches de *Lactobacilllus. International Journal of Applied Radiation Isotopes* **11**, 145-151.
EMBORG, C. (1972). The influence of preparation technique, humidity and irradiation conditions on the radiation inactivation of *S. faecium*, strain A_2 1. *Acta pathologica microbiologica scandinavica* **80**, 367-372.
EMBORG, C. & ERIKSEN, W. H. (1971). Radiation induced radiation resistance in *M. radiodurans* strain R_1. In *Proceedings of the 8th Annual Meeting of the European Society of Radiation Biology*, Basko Polje, Yugoslavia, pp. 39-40.
EPP, E. R., WEISS, H. & SANTOMASSO, A. (1968). The oxygen effect in bacterial cells irradiated with high-intensity pulsed electrons. *Radiation Research* **34**, 320-325.
EPPS, N. A. & IDZIAK, E. S. (1970). Radiation treatment of foods II. Public Health significance of irradiation-recycled *Salmonella. Applied Microbiology* **19**, 338-344.
EPPS, N. A. & IDZIAK, E. S. (1972). Poultry feed radiation. *1.* Microbiological aspects of poultry feed irradiation. *Poultry Science* **51**, 277-282.
ERDMAN, I. E., THATCHER, F. S. & MacQUEEN, K. F. (1961). Studies on the irradiation of microorganisms in relation to food preservation. II. Irradiation resistant mutants. *Canadian Journal of Microbiology* **7**, 207-215.
EWING, D. (1973). Anoxic radiosensitization of two *Bacillus* species by p-nitroacetophenone. *International Journal of Radiation Biology* **24**, 505-515.
FOWLER, J. F. (1964). Differences in survival curve shapes for formal multi-target and multi-hit models. *Physics in Medicine and Biology* **9**, 177-188.
GADEN, E. L. & HENLEY, E. J. (1953). Induced resistance to gamma irradiation in *Escherichia coli. Journal of Bacteriology* **65**, 727-732.
GINOZA, W. (1967). The effects of ionizing radiation on nucleic acids of bacteriophages and bacterial cells. *Annual Review of Microbiology* **21**, 325-368.
GOLDBLITH, S. A. (1971). The inhibition and destruction of the microbial cell by radiations. In *Inhibition and Destruction of the Microbial Cell* ed. Hugo, W. B. London & New York: Academic Press.
HORAN, P. K., HIRD, K. & POLLARD, E. C. (1972). A strain of *Escherichia coli* with minimum postirradiation degradation properties. *Radiation Research* **52**, 291-300.
IBE, S., DAVIES, R. & SINSKEY, A. J. (1972). Genetic transfer of gamma-radiation resistance in *Salmonella typhimurium. Bacteriological Proceedings* **72**, 30.
IDZIAK, E. S. & THATCHER, F. S. (1964). Some physiological aspects of mutants of *Escherichia coli* resistant to gamma irradiation. *Canadian Journal of Microbiology* **10**, 683-697.
ITO, H., SATO, T. & IIZUKA, H. (1973). Study of *Acinetobacter* as contaminants of wiener sausages and their radio-sensitivities. *Food Irradiation, Japan* **8**, 51-57.
KITAYAMA, S. & MATSUYAMA, A. (1973). Enhancement of radiation lethality of *Micrococcus radiodurans* by incubation at restrictive temperature. *Food Irradiation, Japan* **8**, 48-50.
KUDRYASHEVA, A. A., VOLOKHINA, M. I. & EMEL'YANOV, I. S. (1973). Effect of the ^{60}Co gamma-irradiation dose on *Staphylococcus aurens. Voprosy pitaniya, Moskva* **32**, 72-76.
LAYCOCK, R. A. & REGIER, L. W. (1970). Pseudomonads and Achromobacters in the spoilage of irradiated haddock of different preirradiation quality. *Applied Microbiology* **20**, 333-341.
LEWIS, N. F. (1971). Studies on radio-resistant coccus isolated from Bombay Duck. *Journal of General Microbiology* **66**, 29-35.
LEWIS, N. F. & KUMTA, U. S. (1975). Radiosensitization of *Micrococcus radiophilus. Radiation Research* **62**, 159-163.

LEWIS, N. F., ALUR, M. D. & KUMTA, U. S. (1971). Radiation sensitivity of fish micro-flora. *Indian Journal of Experimental Biology* **9**, 45-57.

LEWIS, N. F., ALUR, M. D. & KUMTA, U. S. (1974). Role of carotenoid pigments in radio-resistant micrococci. *Canadian Journal of Microbiology* **20**, 455-459.

LEY, F. J., KENNEDY, T. S. & KAWASHIMA, K. (1970). The use of gamma radiation for the elimination of *Salmonella* from frozen meats. *Journal of Hygiene, Cambridge* **68**, 293-311.

LICCIARDELLO, J. J. (1964). Effect of temperature on radiosensitivity of *Salmonella typhimurium*. *Journal of Food Science* **29**, 469-474.

LICCIARDELLO, J. J., NICKERSON, J. T. R., GOLDBLITH, S. A., SHANNON, C. A. & BISHOP, W. W. (1969). Development of radiation resistance in *Salmonella* cultures. *Applied Microbiology* **18**, 24-30.

MATCHES, J. R. & LISTON, J. (1971). Radiation destruction of *Vibrio parahaemolyticus*. *Journal of Food Science* **36**, 339-340.

MATSUYAMA, A., THORNLEY, M. J. & INGRAM, M. (1964). The effect of freezing on the radiation sensitivity of vegetative bacteria. *Journal of Applied Bacteriology* **27**, 110-124.

MAXCY, R. B., TIWARI, N. P. & ANAGNOSTIS, C. C. (1972). Study of the control of some public-health pathogens in meat. *Isotopes and Radiation Technology* **9**, 292-294.

MICHAEL, B. D., ADAMS, G. E., HEWITT, H. B., JONES, W. B. G. & WATTS, N. E. (1973). A posteffect of oxygen in irradiated bacteria : A sub-millisecond fast mixing study. *Radiation Research* **54**, 239-251.

MILLER, D. R. (1970). Theoretical survival curves for radiation damage in bacteria. *Journal of Theoretical Biology* **26**, 383-398.

MOSELEY, B. E. B. (1967). The isolation and some properties of radiation sensitive mutants of *Micrococcus radiodurans*. *Journal of General Microbiology* **49**, 293-300.

MOSSEL, D. A. A., van SCHOTHORST, M. & KAMPELMACHER, E. H. (1967). Comparative study of decontamination of mixed feeds by radiation and by pelletisation. *Journal of the Science of Food and Agriculture* **18**, 362-367.

MOSSEL, D. A. A., van SCHOTHORST, M. & KAMPELMACHER, E. H. (1968). Prospects for the Salmonella radicidation of some foods and feeds with particular reference to the estimation of the dose required. In *Elimination of Harmful Organisms from Food and Feed by Irradiation*. Vienna: International Atomic Energy Agency.

MOUTON, R. F. & TREMEAU, O. (1970). Evolution microbiologique sous rayonnement : phenotype d'un mutant radioresistant d'*Escherichia coli* K12 induit et selectionne par expositions successives en rayonnement gamma du ^{60}Co. *International Journal of Radiation Biology* **17**, 237-248.

OKA, M., GOTOH, A. & OZAWA, S. (1973). Studies on *Microbacterium thermosphactum* and on its radiosensitivity. *Food Irradiation, Japan* **8**, 38-47.

OSTERBERG, B. (1974). Radiation sensitivity of the microbial flora present in suture material prior to irradiation. *Acta pharmaceutica Suecica* **11**, 53-58.

PARISI, A. & ANTOINÉ, A. D. (1974). Increased radiation resistance of vegetative *Bacillus pumilus*. *Applied Microbiology* **28**, 41-46.

POLLARD, E. C. (1969). The biological action of ionizing radiation. *American Scientist* **57**, 206-236.

PONTEFRACT, R. D. & THATCHER, F. S. (1965). A cytological study of normal and radiation resistant *Escherichia coli*. *Canadian Journal of Microbiology* **11**, 271-278.

ROBERN, H. & THATCHER, F. S. (1968). Nutritional requirements of mutants of *Escherichia coli* resistant to gamma-irradiation. *Canadian Journal of Microbiology* **14**, 711-715.

SANNER, T. & PIHL, A. (1969). Significance and mechanism of the indirect effect in bacterial cells. The relative protective effect of added compounds in *Escherichia coli* B irradiated in liquid and frozen suspension. *Radiation Research* **37**, 216-227.

SILVERMAN, J. G. & SINSKEY, A. J. (1968). The destruction of microorganisms by

ionizing radiation. In *Disinfection, Sterilization and Preservation,* eds Lawrence, C. A. & Block, S. S. Philadelphia: Lea & Febiger.
STAVRIC, S., DICKIE, N. & THATCHER, F. S. (1969a). Effects of γ-irradiation on *Escherichia coli* wild type and its radiation resistant mutants. I. Post-irradiation synthesis of DNA. *International Journal of Radiation Biology* **14**, 403-410.
STAVRIC, S., DICKIE, N. & THATCHER, F. S. (1969b). Effects of γ-irradiation on *Escherichia coli* wild type and its radiation resistant mutants. II. Post-irradiation degradation of DNA. *International Journal of Radiation Biology* **14**, 411-416.
THORNLEY, M. J. (1963a). Radiation resistance among bacteria. *Journal of Applied Bacteriology* **26**, 334-345.
THORNLEY, M. J. (1963b). In *Radiation Control of Salmonellae in Food and Feed Products.* Technical Reports Series 22. Vienna: International Atomic Energy Agency, pp. 81-106.
THORNLEY, M. J. (1966). Irradiation of poultry and egg products. In *Food Irradiation.* Vienna: International Atomic Energy Agency.
THORNLEY, M. J., HORNE, H. W. & GLAUBART, A. M. (1965). The fine structure of *Micrococcus radiodurans. Archiv für Mikrobiologie* **51**, 267-273.
TOWN, C. D., SMITH, K. C. & KAPLAN, H. S. (1973). The repair of DNA single-strand breaks in *E. coli* K12 X-irradiated in the presence or absence of oxygen; the influence of repair on cell survival. *Radiation Research* **55**, 334-345.
TULIS, J. J., FOGARTY, M. G. & SLIGER, J. L. (1973). Thermoradiation as a sterilization method. *Developments in Industrial Microbiology* **14**, 49-56.
TYRRELL, R. M. (1974). The interaction of near UV and X-radiations on wild-type and repair deficient strains of *Escherichia coli* K12: Physical and Biological measurements. *International Journal of Radiation Biology* **25**, 373-390.
UNDERDAL, B. & ROSSEBO, L. (1972). Inactivation of strains of *Salmonella senftenberg* by gamma irradiation. *Journal of Applied Bacteriology* **35**, 371-377.
WATTS, M. E., WILLIAMS, D. W. & ADAMS, G. E. (1975). Studies of the mechanisms of radiosensitization of bacterial and mammalian cells by diamide. *International Journal of Radiation Biology* **27**, 259-270.
WEISS, H., LING, C. C., EPP, E. R., SANTOMASSO, A. & HESLIN, J. M. (1975). Irradiation of *Serratia marcescens* by single and double pulses of high intensity electrons. *Radiation Research* **61**, 355-365.
WITKIN, E. M. (1947). Genetics of resistance to radiation in *Escherichia coli. Genetics* **32**, 221-248.
WRIGHT, S. J. L. & HILL, F. C. (1968). The development of radiation resistance cultures of *E. coli* by a process of 'growth-irradiation' cycles. *Journal of General Microbiology* **51**, 97-106.
YAMAZAKI, K., GOTOH, A. & OKA, M. (1974). Radiosensitivity of the anaerobic bacteria causing spoilage of foods. *Food Irradiation, Japan* **9**, 35-42.

Some Aspects of the Effects of Hydrostatic Pressure on Micro-organisms

G. J. DRING

*Unilever Research Laboratories, Colworth House,
Sharnbrook, Bedford, England*

CONTENTS

1. Introduction 257
2. The effects of hydrostatic pressure on the gross morphology, cell integrity and motility of bacteria 258
   (a) Cell morphology 258
   (b) Cell integrity 259
   (c) Cell motility 260
3. The interaction of hydrostatic pressure with other environmental parameters . 261
   (a) Temperature and the growth, reproduction and viability of bacteria . . 262
   (b) The effects of pH 268
   (c) Composition of the growth medium 270
4. The application of hydrostatic pressure inactivation of microorganisms in food products 272
5. References 275

1. Introduction

THE TOPIC is of long historical standing. Awareness of the existence of organisms living at the bottom of deep oceans, and therefore existing at high hydrostatic pressure, came from the now celebrated ecological ocean dredging operations of the *Talisman* in 1882-1883. Organisms were gathered from ocean depths of up to 6000 m where the hydrostatic pressure is in excess of 600 atmospheres. Whilst the opportunities presented to study the organisms were rapidly seized upon by such notable physiologists as Regnard and Certes, absence at that time of a satisfactory kinetic treatment for the complex effects of hydrostatic pressure even on relatively simple chemical reactions made interpretation of effects with complex biological material impossible. Thus, not unexpectedly, only during fairly recent times do we see older notions relating inactivation of microbial activities by hydrostatic pressure to simple protein denaturation being replaced by interpretation approaching the molecular level of cell organization.

An extensive literature now exists of pressure studies made on biological material including many employing micro-organisms as model systems. The subject has been recently reviewed at length elsewhere, the interested reader being referred to Zimmerman (1970) and ZoBell & Kim (1972).

For this review essentially three aspects of hydrostatic pressure studies on micro-organisms are included. First, attention is given to the gross effects which hydrostatic pressure can exert on the structure and morphology of micro-organisms. Secondly, important interactions of other environmental parameters such as temperature, pH, medium composition and presence of inhibitors in relation to the growth and death of micro-organisms under hydrostatic pressures are discussed. The concluding section of the review looks at the question of the application of hydrostatic pressure treatment as a method for the inactivation of micro-organisms in food material.

The mechanism of inactivation of microbial growth by hydrostatic pressure, the question as to why some deep ocean bacteria exhibit barotolerance in comparison with their terrestrial counterparts and the subject of the effects of high pressure oxygen and other gases on micro-organisms are considered to be outside the scope of this review.

2. The Effects of Hydrostatic Pressure on the Gross Morphology, Cell Integrity and Motility of Bacteria

(a) *Cell morphology*

(i) *Pressure-induced filamentation*

Probably one of the most striking, gross effects exhibited by bacteria actually cultivated at increased pressure is the development in some instances of an abnormal and often quite bizarre morphology (ZoBell & Cobet, 1964), ZoBell & Oppenheimer (1950) observed that whilst cells of *Serratia marcescens* grew at ordinary pressures in seawater broth as short rods c. 1.0 µm in length, at a pressure of 600 atm there was a 200-fold increase in cell length and septation was not evident. However, when restored to normal pressures at 25° reversion to the short habit was completed within a few hours.

Berger (1959) showed that with both *Pseudomonas perfectomarinus* and *Escherichia coli* cross-wall formation and cell division were inhibited by pressures in the order of 50-150 atm, elongation to rods up to 100 µm in length taking place. At high pressure (200-400 atm) the cells were generally shorter but of increased diameter (10 µm). These larger cells underwent lysis, presumably by mechanical rupture. It was further shown that during growth at 300 atm a compound, designated M and thought to be a uridine nucleotide precursor essential for cell wall synthesis, the incorporation of which into the wall was inhibited by pressure, was released into the medium. Perhaps surprisingly, addition of diaminopimelic acid increased the release of M at 300 atm and led to decrease in cell size, but was without effect at 1 atm. ZoBell & Cobet (1964) studied pressure-induced filamentation in three strains of *E. coli* and examined the cells for nucleic acid and protein content. It was shown at all pressures that the cells had the same protein and nucleic acid content but whereas RNA/cell increased with cell length the amount of DNA was virtually unaltered for pressure-induced filaments compared with normal cells grown at 1 atm. It was concluded that pressure inhibited DNA replication thus resulting in repression of cell division and filament formation. Addition of pantoyl lactone, an inhibitor of filament formation induced in various *Erwina* species by D-amino acids, penicillin and UV light (Grula & Grula, 1962), did not inhibit pressure-induced filamentation in the strains of *E. coli* tested.

Further evidence implicating pressure-mediated cessation of DNA synthesis in the phenomena of filamentation resulted from studies by Pollard & Weller (1966) and Yayanos (1967). Using incorporation of ^{14}C-labelled thymine in *E. coli*, suppression of DNA synthesis by pressure was demonstrated whilst RNA synthesis was less markedly affected.

Boatman (1967) demonstrated pressure-induced filamentation and accompanying absence of septation for *E. coli, Bacillus mycoides* and species of *Aerobacter* and *Vibrio* grown at pressures of 270-400 atm but some other bacteria including a barophobic species of *Pseudomonas* did not form filaments when cultivated at 400 atm. Non-filament formation during growth at 400-600 atm for *Vibrio haloplanktis* and *Achromobacter stationis* was described by ZoBell and Oppenheimer (1950).

Studies therefore, seem to indicate that the tendency to a filamentous habit under pressure is shared by some but not all bacteria. However, could it mean that failure to demonstrate it simply reflects that the optimum pressure treatment was not achieved experimentally?

(ii) *Non-filamentous pleomorphism*

As an alternative to filamentation when cultivated at pressures near to their maximum level of tolerance some organisms form grossly pleomorphic cells. Oppenheimer & ZoBell (1952) described pleomorphism accompanying cell-volume increase in *Micrococcus aquiviveus* at high pressure whilst *V. phytoplanktis* was found to grow as elongated, granular pleomorphic rods. *Bacillus borborokoites* developed a multibulbous habit. It is commonly observed that cells showing pleomorphism under pressure usually also develop a thickened and in some instances, convoluted cell wall. Such effects were shown for several bacteria including *E. coli*, vibrios and corynebacteria (Boatman, 1967).

(b) *Cell integrity*

(i) *Rapid compression-decompression effects on bacterial cells*

As pointed out (q.v.) the period of duration of exposure and the amount of pressure applied influences the extent to which cells are affected by hydrostatic pressure. ZoBell (1964) studied the effects of pressure shock by a rapid compression-decompression treatment of cells of *E. coli*. During a period of 60 min, cells were subjected to 10 successive pressurizations to 1000 atm, within one to two minutes, in nutrient medium at 25°, followed by immediate decompression. Following such treatment, there was no reduction in the numbers of viable cells. Similarly ZoBell (1970) reported that many marine bacteria brought to the surface within a relatively short period of time from depths of at least 7000 m and *in situ* pressures of the order of 700-1000 atm were able to grow when recompressed to these pressures.

In contrast, studies by Seki & Robinson (1969) showed that adverse effects had taken place on some bacteria which were unable to grow when recovered from depths of 400 m where the pressure would have been only 40 atm.

(ii) *Cell disruption by violent decompression*

Although some bacteria are able to withstand compression-decompression cycling without injury, under certain conditions sudden release of pressure can rupture bacteria, yeasts and other microbes. Such studies have usually been made by compressing cells to *c.* 60 atm under atmospheres of argon, nitrogen, nitrous oxide or carbon dioxide and then rapidly releasing the pressure, (Fraser, 1951). Foster *et al.* (1962) ruptured cells of *Brucella abortus, Staphylococcus aureus* and *Serr. marcescens* under nitrogen by rapid decompression from 120 atm. It is considered that under pressure, gas is forced into solution and hence gains entry into cells. When the pressure is released, gas, escaping from solution as bubbles, bursts the cell walls and disrupts the cytoplasm.

(c) *Cell motility*

Although Regnard (1891), during his studies of the effects of hydrostatic pressure in the range 650-700 atm on bacteria in cheese, milk and urine, first observed that motility was affected, even today the basis for the loss of motility remains obscure. ZoBell (1970) reported that the flagellate bacterium *E. coli*, and various species of *Vibrio* and *Pseudomonas,* when pressurized to 400 atm and then examined by phase contrast and electron microscopy, did not have flagella although they were formed during growth at 100 atm. Whether the lack of flagella results from their being lost during compression or decompression or possibly because they do not form at all at high pressures is not known.

Marquis (1973) examined the effect of hydrostatic pressure on the motility of varous bacteria growing in soft agar. It was shown (Table 1) that motility decreased with increased pressure and whilst the bacteria all grew at 306 atm they were immotile. Motility in the marine isolate *Serr. marcescens* was affected to the same degree as that of the so-called terrestrial species examined. Marquis (1973) concluded that the major cause of immotility under pressure lies in the failure of the bacteria to form flagella at increased pressures. However since cells can grow at pressures totally inhibitory to motility they must be able to synthesize ATP, and therefore immotility could indicate a restriction of the supply of ATP to any flagella already present by a conservation mechanism.

Table 1
Bacterial motility under hydrostatic pressure

Organism	Motility at pressure of (atm)				
	1	102	204	306	408
Escherichia coli	+++	+++	++	–	–
Serratia marcescens	+++	+++	++	–	–
Proteus vulgaris	+++	+++	+	–	–
Salmonella typhi	+++	++	+	–	–

+, ++, +++, increasing degree of motility; –, non-motile.
Bacteria were inoculated into tubes of soft agar with needles to depth 1.4 cm. 1.0 ml, 2% agar overlayered and allowed to set. Space above agar filled with sterile water; tubes sealed with rubber caps prior to pressurization. Tubes incubated at room temperature for 52 h. Growth was apparent in all tubes. From Marquis (1973).

3. The Interaction of Hydrostatic Pressure with Other Environmental Parameters

The manner and extent to which the growth and viability of micro-organisms are affected by hydrostatic pressure depends upon the organism and its stage of growth (i.e. lag phase cells, log phase cells, stationary phase cells), the extent and duration of compression, the growth temperature and the chemical composition of the medium. Generally speaking it is found that bacteria will grow at higher than normal temperatures if the pressure is increased (ZoBell & Cobet, 1962; Morita, 1972). Thus if the temperature is very low and the pressure high, it follows that growth will probably not occur, a situation which led Jannasch *et al.* (1971) to ask whether bacteria on deep ocean beds actually grow or if they exist in a state of suspended animation.

Early studies ignored the role of environmental factors such as the temperature, which are now known to play such a fundamental part in determining the degree of barotolerance. Also these studies were concerned almost exclusively with establishing the pressure-time combinations that totally inactivate bacteria. Even today there are few really definitive studies (ZoBell &

Johnson, 1949; ZoBell & Cobet, 1962, 1964) where such important aspects as the interactions of temperature and pressure, particularly on the various stages of a bacterial growth cycle, have been carefully controlled and studied in any detail.

(a) *Temperature and the growth, reproduction and viability of bacteria*

(i) *Growth at constant temperature*

Data from comprehensive studies (ZoBell & Johnson, 1949) is shown in Table 2. For each of the five groups of organisms examined the organisms are tabulated in order of decreasing resistance to pressure in terms of whether they could grow during 48 h at 30° at the various pressures and also if, following decompression, they retained viability. None of the organisms was able to grow at 600 atm and generally speaking growth was markedly affected at 400 atm. Very few of the organisms were capable of growth at 500 atm, only *B. mesentericus, E. coli* and *Streptococcus lactis* grew. Neither *Serr. marcescens* nor *Schizosaccharomyces octosporus* grew at 300 atm. The yeasts appeared to be more sensitive to pressure than the bacteria and whilst none of the bacteria which failed to grow at 400 atm were actually killed, most of the yeasts were.

(ii) *Growth at various temperatures*

The fact that bacteria will grow at high pressure when temperature is increased was established for several species of bacteria by ZoBell & Johnson (1949). Table 3 illustrates data obtained. Cultures were incubated at various hydrostatic pressures at temperatures of 20, 30 or 40° for periods of four, two and one day, respectively. Whilst, in comparison with controls at 1 atm, most of the cultures either grew poorly or not at all at 300 atm at 20° growth was generally considerably enhanced at 30°. There was little improvement in the degree of growth at 40° but *Clostridium septicum* did respond thus. This relationship was again apparent at 400 atm being particularly evident for *Cl. bifermentans* and *Cl. septicum*, neither of which grew at 400 atm at 20 or 30°, but which did so at 40°. At 500 atm few of the cultures grew at all at any of the temperatures studied, however it was noticeable at this pressure that *B. subtilis, B. mesentericus, E. coli, Ps. fluorescens* and *Strep. lactis* all of which failed to grow at 20° did so at 30 and 40°. With the exception of *Ps. fluorescens*, the other four organisms of this group all grew at 40° at 600 atm although none grew at either 20 or 30°. Zobell & Cobet (1962) extended their initial studies to include a selection of marine bacteria and found that the same relationship held.

Johnson & Lewin (1946) reported similar effects for the growth of *E. coli*, at

Table 2
Growth and inactivation of bacteria and yeasts at various hydrostatic pressures

Organism	Extent of growth (as % turbidity) following incubation for 48 h at 30° at (atm)				
	1	300	400	500	600
Torula cremoris	100	100	50	D*	D
Saccharomyces cerevisiae	100	100	D	D	D
Hansenula anomala	100	50	D	D	D
Sacch. ellipsoides	100	25	D	D	D
Sporobolomyces salmonicolor	100	25	D	D	D
Schizosaccharomyces octosporus	100	0	D	D	D
Escherichia coli	100	100	75	50	0
Pseudomonas fluorescens	100	100	75	0	0
Alkaligenes viscosus	100	75	50	D	D
Proteus vulgaris	100	50	0	0	D
Serratia marcescens	100	0	0	0	0
Streptococcus lactis	100	100	100	50	0
Staphylococcus albus	100	75	75	0	0
Mycobacterium phlei	100	75	50	D	D
Sarcina lutea	100	75	25	0	D
Myco. smegmatis	100	50	25	0	D
Micrococcus lysodeikticus	100	50	0	D	D
Staph. aureus	100	50	0	D	D
Clostridium histolyticum	100	100	75	0	D
Cl. chauvei	100	100	50	0	0
Cl. putreficum	100	100	50	0	0
Cl. sporogenes	100	75	50	0	0
Cl. welchii	100	100	25	0	0
Cl. septicum	100	50	25	0	0
Bacillus mesentericus	100	100	75	50	D
B. cereus	100	100	0	0	0
B. mycoides	100	100	25	0	D
B. megaterium	100	100	25	0	D
B. brevis	100	75	25	0	D
B. alvei	100	75	25	0	D
B. subtilis	100	50	25	0	0
B. circulans	100	100	0	0	0

* indicates the the organism was non-viable following decompression.
ZoBell & Johnson (1949).

Table 3
Growth of bacteria following incubation at 20° (4 days), 30° (2 days) and 40° (1 day) at various hydrostatic pressures

Extent of growth (as % turbidity) at various hydrostatic pressures and temperatures

Organism	300 atm			400 atm			500 atm			600 atm		
	20°	30°	40°	20°	30°	40°	20°	30°	40°	20°	30°	40°
Escherichia coli	50	100	100	0	75	100	0	50	100	0	0	100
Pseudomonas fluorescens	50	75	100	0	50	100	0	0	75	0	0	0
Alcaligenes viscosus	50	75	100	50	50	50	0	0	0	0	0	0
Streptococcus lactis	75	100	100	25	100	100	0	50	100	0	0	100
Staphylococcus albus	50	50	100	0	25	50	0	0	0	0	0	0
Staph. aureus	0	75	100	0	50	75	0	0	0	0	0	0
Sarcina lutea	50	50	100	0	25	50	0	0	0	0	0	0
Mycobacterium phlei	0	75	100	0	50	25	0	0	0	0	0	0
Myco. smegmatis	0	50	50	0	25	25	0	0	0	0	0	0
Clostridium bifermentans	50	100	100	0	0	75	0	0	0	0	0	0
Cl. chauvei	0	100	100	0	50	75	0	0	0	0	0	0
Cl. histolyticum	0	100	100	0	0	75	0	0	0	0	0	0
Cl. putreficum	0	100	100	0	50	50	0	0	0	0	0	0
Cl. septicum	0	25	50	0	0	25	0	0	0	0	0	0
Cl. sporogenes	0	100	100	0	50	75	0	0	0	0	0	0
Cl. welchii	0	100	100	0	25	50	0	0	0	0	0	0
Bacillus mesentericus	0	100	100	0	75	100	0	50	100	0	0	100
B. megaterium	0	100	75	0	25	50	0	0	0	0	0	0
B. brevis	0	75	50	0	25	25	0	0	0	0	0	0
B. subtilis	0	75	100	0	50	100	0	0	75	0	0	50

ZoBell & Johnson (1949).

above optimum temperatures, during the early logarithmic phase, when the pressure was increased.

(iii) *Cell viability*

Studies by ZoBell & Johnson (1949) showed that the actual viability of organisms within a population at an elevated hydrostatic pressure was affected rather more than might be concluded from the relative degree of turbidity achieved. It was found (see Table 4) that both *Alkaligenes viscosus* and *Proteus vulgaris* achieved, respectively, turbidities of 75% and 100% of unpressurized controls at 300 atm although when recovery counts were made both were found to contain < 100 viable cells/ml. How are we to explain this? It is most unlikely that viability-loss accompanied decompression, since elsewhere (ZoBell, 1964) rapid compression-decompression 'cycling' from 1000 atm did not affect viability of *E. coli*. Further, ZoBell (1970) has commented that stationary phase cells are more barotolerant than cells of the early logarithmic phases of growth; thus it is difficult to visualize how these cells could withstand elevated pressure and grow well only to be inactivated during the stationary phase. Alternatively, accumulation of waste metabolic products together with perhaps development of an adverse environmental pH could, at the increased pressure, favour inhibition and inactivation of the cells.

(iv) *The bacterial growth cycle*

It should be borne in mind, when considering pressure effects on bacteria, that growth, in terms of an increase in the overall biomass of a culture can take place in the absence of cell division. Thus as discussed elsewhere filament formation occurs in many bacteria at the threshold of their pressure tolerance whilst cell division is arrested (ZoBell & Cobet, 1964).

Studies by ZoBell & Cobet (1962) are probably the most definitive of examinations made of the effects of pressure and temperature on the growth cycle of a single bacterial species, *E. coli*. Here we shall consider how hydrostatic pressure and temperature affect the various phases of the growth cycle.

Lag phase of growth. ZoBell & Cobet (1962) showed, for *E. coli*, that as pressure was increased to 400 atm a doubling or trebling in the duration of the lag period occurred. Above 400 atm a significant increase in the duration of the lag period took place. The culture (initial count 2×10^4 cells/ml) was completely inactivated following pressurization at 525 atm for seven days.

Cell division. When cells of *E. coli* were incubated at 30°, in nutrient medium, under increasing hydrostatic pressures of 100-400 atm, the rate of cell division (Table 5) was slowed down (ZoBell & Cobet, 1962). Several days' incubation at 500 atm produced little growth and no division occurred at all at 525 atm. Fig. 1 (a) and (b) shows comparisons of the growth rates for *E. coli* at

Table 4
Growth of bacteria following 48 h incubation at 30° at various hydrostatic pressures

Organism	Initial count (ml)	1 atm (a)	1 atm (b)	300 atm (a)	300 atm (b)	400 atm (a)	400 atm (b)	500 atm (a)	500 atm (b)	600 atm (a)	600 atm (b)
Bacillus cereus	1.63×10^3	4.0×10^7	100	6.5×10^5	100	7.0×10^2	0	10^2	0	0	0
B. circulans	2.0×10^4	1.1×10^7	100	8.8×10^5	100	2.65×10^5	0	2.0×10^2	0	2.0×10^1	0
B. mesentericus	1.9×10^3	2.1×10^6	100	2.3×10^5	100	6.0×10^4	75	1.0×10^1	50	0	0(D)*
B. mycoides	2.25×10^3	2.0×10^6	100	1.6×10^5	100	1.14×10^5	25	7.0×10^2	0	0	0(D)
B. subtilis	1.6×10^3	1.4×10^7	100	6.0×10^5	50	4.2×10^4	25	3.0×10^1	0	$< 10^1$	0
Staphylococcus aureus	8.0×10^2	1.56×10^8	100	1.74×10^8	50	9.0×10^3	0	0	0	0	0
Staph. albus	4.4×10^2	7.7×10^7	100	8.0×10^7	75	5.7×10^4	75	7.0×10^2	0(D)	3.5×10^2	0
Streptococcus lactis	4.0×10^3	2.73×10^8	100	1.49×10^8	100	7.0×10^7	100	1.6×10^{15}	50	1.8×10^4	0
Sarcina lutea	4.7×10^3	3.6×10^7	100	6.7×10^4	75	2.2×10^4	25	4.0×10^2	0	1.2×10^1	0
Proteus vulgaris	2.6×10^3	1.42×10^8	100	$< 10^2$	50	$< 10^2$	0	$< 10^1$	0	0	0(D)
Alkaligenes viscosus	7.0×10^2	1.6×10^8	100	$< 10^2$	75	$< 10^2$	50	$< 10^1$	0(D)	0	0(D)
Serratia marcescens	3.0×10^2	6.4×10^7	100	$< 10^2$	0	$< 10^2$	0	$< 10^1$	0	0	0
Pseudomonas fluorescens	1.0×10^4	9.5×10^7	100	2.1×10^7	100	5.0×10^6	75	8.1×10^2	0	8.0×10^1	0

*(D) indicates that organism was non-viable following decompression.
ZoBell & Johnson (1949).

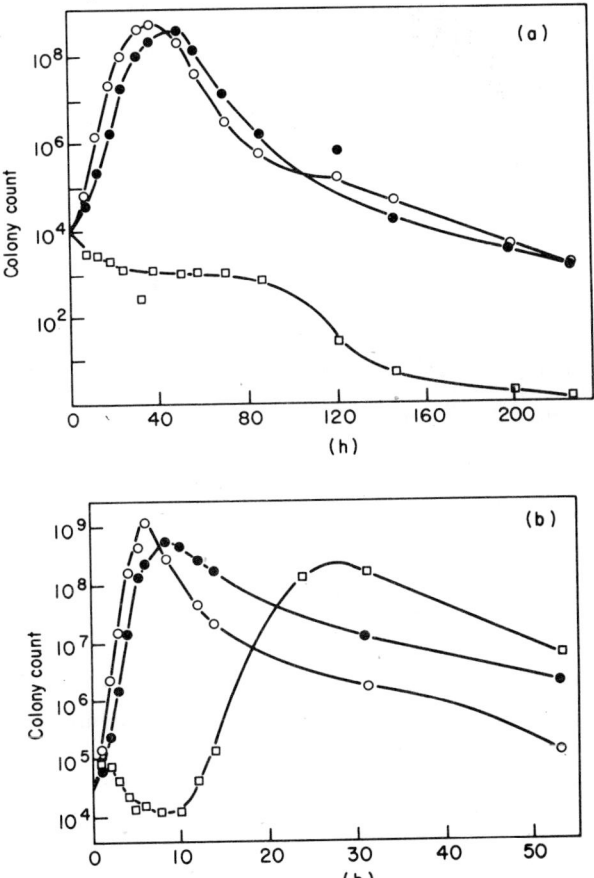

Fig. 1. Efect of hydrostatic pressure on growth of *Escherichia coli* at 20° (a) and 40° (b) ○, 1 atm; ●, 200 atm; ▫, 400 atm. (Data from ZoBell & Cobet, 1962.)

20 and 40° at various hydrostatic pressures. It is evident that at pressures up to 200 atm and, indeed at 1 atm too, the rate of cell division was more rapid at 40° than at 20° with the maximum level of growth at stationary phase being reached in 5-10 h at 40° and in 35-45 h at 20°. At 400 atm and 40° a small increase in the number of cells present was followed by a prolonged lag phase of some 10 h, during which time the number of viable cells decreased by some 90-99%. This lag phase was then followed by active cell division, the number of cells finally produced after 25-30 h being c. 50% of the total at 1 atm. At 20° growth was completely inhibited at 400 atm.

Stationary phase cells. The effect of pressure on the death rate for cells of *E. coli* taken from a stationary phase culture was studied by ZoBell & Cobet

Table 5
Effect of hydrostatic pressure on cell division of
Escherichia coli

Hydrostatic pressure (atm)	Log no. of cells*/ml after	
	5 h	10 h
1	3×10^7	1×10^9
100	8×10^6	7×10^8
200	5×10^6	4×10^8
300	1×10^6	8×10^7
400	2×10^5	3×10^6

* Initial count was 3×10^5 cells/ml.
Temperature 30°.
ZoBell & Cobet (1962).

(1962). Table 6 shows the rates of death for such cells held at 1 and 1000 atm at 30°. Within a period of 36 h the cells at 1000 atm were all killed.

(b) *The effects of pH*

Generally speaking, the effects of pH on microbial growth or biochemical reactions are complex and will depend upon the species or system and the temperature. Marquis (1973) has examined the effect of hydrostatic pressure at different pH values, at constant temperature, on the growth of a range of microbes including the following bacteria, *E. coli* B, *Serr. marcescens*, *Strep. faecalis*, *Staph. aureus* and *B. thalassokoites* and the yeast, *Sacch. cerevisiae*. These studies showed (Table 7) that the pH range normally tolerated by the

Table 6
Effect of hydrostatic pressure on death rate of
Escherichia coli* *at 30°*

Hours	Log no. survivors at (atm)	
	1	1000
0	9×10^8	9×10^8
6	4×10^8	2×10^7
12	8×10^7	3×10^5
18	8×10^6	6×10^3
24	4×10^6	1×10^2
30	2×10^6	4
36	1×10^6	No survivors

* Cells taken from stationary phase of growth.
ZoBell & Cobet (1962).

bacteria growing at 1 atm became restricted as the hydrostatic pressure was increased.

Studies with *E. coli* (Table 7) cultured in medium (trypticase-soy broth, 24°) of increasing acidity revealed that at 1 atm, growth was inhibited at pH 4.9. When the hydrostatic pressure was increased to 272 atm, growth was inhibited at

Table 7
The effect of hydrostatic pressure on the pH range of growth

Organism	Pressure (atm)	Turbidity (OD_{700})	Growth inhibited at pH	
Streptococcus faecalis	1	0.53	4.7	9.5
	272	0.33	4.7	9.5
	340	0.25	4.9	9.1
	408	0.08	5.0	8.4
Staphylococcus aureus, H	1	0.71	4.0	10.0
	340	0.29	5.0	10.0
Escherichia coli	1	0.82	4.9	10.0
	272	0.55	5.8	9.0
	340	0.16	6.0	8.7
Serratia marcescens	1	0.58	5.0	10.0
	408	0.08	6.0	9.0
Bacillus thalassokoites	1	0.63	4.0	10.0
	272	0.30	5.0	10.0
	408	0.22	5.0	9.0
	544	0.12	5.0	9.0
Saccharomyces cerevisiae	1	0.59	3.0	9.0
	136	0.45	3.0	9.0
	204	0.38	3.0	9.0

Marquis (1973).

pH 5.8, whilst at 340 atm growth ceased at pH 6.0. For growth in alkaline medium the effects of pressure were more marked. Thus it was shown that although growth was inhibited at pH 10.0 at 1 atm, as the pressure was increased growth ceased at pH 9.0 (272 atm) and pH 8.7 (340 atm). In essence, similar responses were demonstrated for the other bacteria examined. For the yeast *Sacch. cerevisiae* however, pressure did not markedly restrict the pH range of growth. All of the test organisms grew less well as the pressure was increased. Marquis (1973) has suggested that the generality of the potentiation of growth inhibitory effects of acids and bases by pressure may be of ecological importance particularly so for those organisms existing on surfaces of charged particles or in sediments in deep oceans where the pH may not be neutral.

In their study of the effect of hydrostatic pressure on luminescence of the organism *Photobacterium phosphoreum*, Johnson and his co-workers (Johnson et al., 1945) found that pH had a marked effect on the degree of luminescence produced at various hydrostatic pressures. Normally, at pH 7.0 and 1 atm, the degree of luminescence produced was 100 (arbitrary units). At 400 atm, pH 7.0, the degree of luminescence was 95 units whilst at pH 4.63 it was only 15 units. However increasing the pH to 8.04 resulted in an increase in the degree of luminescence to 120 units.

(c) *Composition of the growth medium*

(i) *Nutrients*

Although little comparative data is available it is generally found that the threshold pressure tolerance of bacteria is lower in non-nutrient salt solutions than in the presence of essential amino acids and vitamins, as for example, in a rich organic medium (ZoBell, 1970).

(ii) *Osmotic pressure*

As with nutrients there is but little definitive data here available. ZoBell (1970) reported that at high osmotic pressure bacteria were more sensitive to hydrostatic pressure both in mineral salts solution and nutrient media and suggested that cell wall permeability was affected.

(iii) *Ions*

The interaction of the ionic environment and hydrostatic pressure is not well understood. Although the ionization of weak electrolytes is increased under pressure the extent is probably not sufficient to cause large changes in the osmotic pressure. Palmer & Albright (1969) showed that the highest hydrostatic pressure at which *V. marinus* would grow was greatly affected by the sodium chloride concentration.

The pH or hydrogen ion concentration is affected by pressure, for example, seawater at 0° and at 1 atm has a pH value of 8.10 whilst at 1100 atm it becomes lowered to 7.87 (Park, 1966; Distèche & Distèche, 1967). Therefore, small but significant biological effects resulting from pH change might take place under increasing hydrostatic pressure.

(iv) *Narcotics*

The influences of certain narcotics under increased hydrostatic pressure have been studied using the luminescent bacterium *P. phosphoreum*. Johnson et al. (1942) found that although the addition of urethan (0.78 M) or chloroform (0.05 M) at 18° reduced the degree of luminescence by 50% the inhibition could be totally reversed by pressurization to 270-475 atm. Similarly the inhibitory effects of ethanol, ethyl ether and procaine were reversed by pressure whereas those caused by sulphanilamide, *p*-amino benzoic acid, chloral hydrate or barbital were not.

Johnson & Eyring (1948) found the latter group of compounds to be more toxic at 400 atm than at 1 atm. Johnson et al. (1945) found that ethanol (0.6 M) reduced luminescence by 50% at 1 atm but only 10% at 400 atm. When the ethanol concentration was increased to 1.5 M the inhibitory effects were considerably less at 476 atm than at 1 atm. Ethanol was a more effective inhibitor of luminescence at 29° than at 17.5°.

Johnson & Lewin (1946) studied the rate of death of *E. coli* cells incubated at 1 atm in the presence of 9×10^{-3} M quinine, when during a 3 h period 40% reduction in viability took place. At pressures up to 200 atm the death rate was reduced whilst at 300-400 atm it was accelerated.

(v) *Antibiotics*

Antibiotics of known mode of action have proved invaluable probes in many biochemical and physiological studies. Marquis (1973) has reported some preliminary studies where antibiotics have been used as aids in identifying specific sites of hydrostatic pressure inhibition. Table 8 shows results obtained

Table 8
The effect of hydrostatic pressure on the activity of various antibiotics

Test organism	Antibiotic	MIC (µg/ml) at (atm)	
		1	340
Escherichia coli B	Penicillin G	2	0.2
	Ampicillin	10	10
	Bacitracin	> 20	> 20
	Rifampin	20	10
	Streptomycin	10	0.2
Streptococcus faecalis (9790)	Penicillin G	2	0.2
	Ampicillin	1	1
	Bacitracin	10	10
	Rifampin	2	2
	Streptomycin	> 20	> 20
Staphylococcus aureus H	Penicillin G	0.02	0.02

Experimental conditions. Trypticase-soy broth for *E. coli*; Tryptone-glucose-marmite broth for *Strep. faecalis* and *Staph. aureus*; cultured in sealed plastic vials at 30°.
Marquis (1973).

for a variety of antibiotics on several different bacteria. Pressure of 340 atm did increase the inhibitory effects of some of the antibiotics but had no effect on the others. Pressure enhanced the effect of penicillin G, which interferes with the peptidoglycan cross-linking reaction for *E. coli* and *Strep. faecalis,* but did not do so for *Staph. aureus.* The effects of ampicillin and bacitracin, both of which also affect wall synthesis, were not enhanced under pressure, Rifampin which binds specifically to the β-subunit of RNA polymerase thus inhibiting

transcription, was not more potent under pressure than at 1 atm. However, for
E. coli streptomycin, an antibiotic which affects both protein synthesis and cell
membranes, was potentiated at 340 atm.

(vi) *Deuterium oxide*

Deuterium oxide (D_2O) has been found to overcome adverse effects of high hydrostatic pressure on animal cells. The protection effect is thought to involve stabilization of microtubular structures (Marsland *et al.*, 1971).

Recently Marquis (1973) has reported data for studies examining the effect of D_2O on bacteria at increased hydrostatic pressure. It was shown (Table 9) that for *E. coli* grown in trypticase-soy broth the presence of D_2O alone and

Table 9

Effects of pressure and D_2O on growth of E. coli *B*

Pressure (atm)	D_2O content (%)	Exponential growth rate constant (h^{-1})†	Max OD
1	0	0.224	0.543
340	0	0.097	0.522
1	50	0.182	0.348
340	50	0.093	0.402
1	100	0.154	0.188
340	100	0.159	0.330

Cultures were grown in trypticase-soy broth containing 0.1% KNO_3, 30°.
† Constants were estimated from slopes of plots of $\log_{10} OD_{700}$ versus time in hours.
Marquis (1973).

pressure alone slowed the growth rate whilst deuterium oxide by itself also reduced the final number of cells. However at 340 atm the growth rate was greater in medium containing deuterium oxide although again the total number of cells produced was reduced. Where the medium was totally made up with deuterium oxide, pressure reversed some of its adverse effects since the total number of cells achieved was greater at 340 atm than at 1 atm. Therefore, although interactions between deuterium oxide and pressure were demonstrated they were not very dramatic. Marquis (1973) concluded that micro-organisms differ from animal cells in their response to pressure.

4. The Application of Hydrostatic Pressure Treatment to the Inactivation of Micro-organisms in Foods

The first studies concerned with the pressure treatment of foods were made by Hite (1899) and were concerned with improving the keeping quality of milk. It was found that milk, previously subjected to pressures of 4500 atm or more for about an hour kept well for 24 h whilst that treated at c. 13,500 atm remained 'sweet' for about four days. Although it was intended to study pressure effects of milk preinoculated with cultures of *Salmonella typhi* the work was prematurely abandoned when disintegration of a test vessel under pressure led to an operator subsequently contracting typhoid. Eventually work in this direction ceased due to a combination of practical problems involved in constructing suitable apparatus for containing the high pressures involved and because pressure could not inactivate enzymes in the milk which led to slow organoleptic changes during storage. Some studies were also made relating to pressure treatment of meat. A pressure of 6000 atm applied for one hour at 126° F was shown to have beneficial effects on the storage of meat. Small pieces of meat so treated were reported as acceptable when examined after a three month storage period.

Hite et al. (1914) reported studies of the effects of high pressure treatment on a variety of fruits and vegetables. Table 10 summarizes the results obtained. Although it can be seen that some of the treatments were fairly successful it is now quite clear that best effects were for those fruits whose natural acidity inhibited the development of bacterial spore formers. These workers also studied several bacteria in pure culture, including *Serr. marcescens, Strep. lactis, Ps. fluorescens* and *Aerobacter aerogenes* and demonstrated that cultures were sterilized at 20-25° in 5 min at 5780-6800 atm., 10 min at 3400-4080 atm or 60 min at 2040-3060 atm. Furthermore the bacteria died more rapidly at either 5° or at 40-50° than at 20-25°. As a result of the studies Hite et al. (1914) thought that a pressure of 2040 atm at 50° for 2-3 h could not be used safely to kill spoilage organisms in canned fruit. Dow (1940), also studying milk, found that pressures of 5000-10,000 atm inhibited lactic fermentation and arrested the accompanying pH decrease. Timson & Short (1965) found that the number of viable bacteria in milk was reduced c. 10-fold in 30 min by a pressure of 2000 atm at 35° and from 10^6/ml to 10^2/ml in a similar period at 5000 atm.

Whilst the main theme of this review relates to the effects of pressure on the vegetative cells of bacteria and other microbes, any discussion of the practical application of hydrostatic pressure for preserving foods must also consider the problem of bacterial spores. How are spores affected by hydrostatic pressure? As pointed out, the failure by Hite et al. (1914) to arrest spoilage of vegetables using hydrostatic pressure was most likely due to the survival of spores. Larson et al. (1918) showed that spores had great resistance to fairly high hydrostatic pressures, not all spores of *B. subtilis* were killed even following 14 h at 12,000 atm. Basset & Macheboeuf (1932) found that a few spores of *B. subtilis* withstood pressurization to 17,600 atm for a period of 45 min. Johnson & ZoBell (1949) showed that *B. subtilis* spores at an initial level of 8×10^4/ml were all killed when held at 93.6° at 1 atm for 1 h although at 600 atm more than 4 h compression was required at this temperature to kill all the spores. Are then, spores so resistant to inactivation by hydrostatic pressure that they would always survive any practical sort of pressure treatment which might otherwise be successful in killing vegetative cells in food? Recently, studies by Clouston & Wills (1969, 1970) and Sale et al. (1970) have demonstrated that spores can be

Table 10
Effect of hydrostatic pressure on some fruits and vegetables

Test Material	Pressure (atm)	Time (min)	Temp. (°)	Comments
Sugar solution with yeasts	4080-6800	30-105	Room temp.	Did not ferment — yeast killed
Grape juice with yeasts	1700-5100	30-960	? Room temp.	Did not ferment
Apple juice	4080-5440	30		Sterile after one month
Peaches, pears	4080	30		Acceptable after 5 years
Tomatoes	6800	60 min on 7 successive days	Room temp.	Sterilized
Tomatoes				
25 samples	1904-3060	30-1440	10-21.1°	Spoiled
1 sample	2176	1440	15.6°	
1 sample	2720	60	21.1°	
1 sample	2720	160	18.3°	All acceptable
1 sample	2720	180	18.3°	
1 sample	3060	120	18.3°	
Peas, beans, beets + other vegetables	2040-3040		12.8-15.6°	Always spoiled

Hite *et al.* (1914).

inactivated by an 'indirect' mechanism by pressures in the order of 2000-4000 atm. Although spores are, as the earlier studies indicated, very resistant to hydrostatic pressures in excess of 8000 atm, when held at lower pressures, in some instances in the order of a few hundred atmospheres, they are induced to germinate. Whilst the levels of pressure which cause spores to germinate may not then be great enough to inactivate the spores this can then be done by relatively moderate temperature treatments of some 55-65°.

In concluding this section on the practical implications for the use of hydrostatic pressure we might speculate as to the kinds of processing treatments employing high pressure which could be devised. There now exists much data relating to the pressure: time relationships required to inactivate vegetative cells.

Together with the knowledge that spores will not germinate at low pH and the recent findings that spores can be germinated and made heat sensitive by low pressures it would seem likely that a method, employing either a combination of moderate hydrostatic pressure and low pH or alternatively involving a pasteurization stage during or following compression, could be expected to give reasonably satisfactory levels of protection against spoilage. Indeed, such a treatment has recently been patented by Clouston (1971).

However, as with all aspects of hydrostatic pressure studies, and especially where large scale processing is envisaged, the main stumbling block lies undoubtedly in the design and construction of safe and economical pressurization vessels capable of handling reasonably large quantities of food material.

5. References

BASSET, J. & MACHEBOEUF, M. A. (1932). Etude sur les effets biologiques des ultrapressions: Résistance des bactéries, des diastases et des toxines aux pressions très élevées. *Compte rendu hebdomadaire des séances de l'Académie des sciences* **195**, 1431-1433.

BERGER, L. R. (1959). The effect of hydrostatic pressure on cell wall formation. *Bacteriological Proceedings* **59**, 129.

BOATMAN, E. S. (1967). The effects of hydrostatic pressure on the structure of marine bacteria. Ph.D. Dissertation, University of Washington, Seattle, Washington.

CLOUSTON, J. G. (1971). Sensitization of bacterial spores to the lethal effects of certain treatment. Canadian Patent No. 871413.

CLOUSTON, J. G. & WILLS, P. A. (1969). Initiation of germination and inactivation of *Bacillus pumilus* spores by hydrostatic pressure. *Journal of Bacteriology* **97**, 684-690.

CLOUSTON, J. G. & WILLS, P. A. (1970). Kinetics of initiation of germination of *Bacillus pumilus* spores by hydrostatic pressure. *Journal of Bacteriology* **103**, 140-143.

DISTÈCHE, A. & DISTÈCHE, S. (1967). The effect of pressure on the dissociation of carbonic acid from measurements with buffered glass electrode cells. *Journal of the Electrochemical Society* **114**, 330-340.

DOW, R. B. (1940). High pressure in food chemistry. *Food Manufacture,* August. pp. 207-210.

FOSTER, J. W., COWAN, R. M. & MAAG, T. A. (1962). Rupture of bacteria by explosive decompression. *Journal of Bacteriology* **83**, 330-334.

FRASER, D. (1951). Bursting bacteria by release of gas pressure. *Nature, London* **167**, 33-34.

GRULA, E. A. & GRULA, M. M. (1962). Cell division in a species of Erwinia. III. Reversal of inhibition of cell division caused by D-amino acids, penicillin and ultraviolet light. *Journal of Bacteriology* **83**, 981-988.

HITE, B. H. (1899). The effect of pressure in the preservation of milk. *Bulletin of the West Virginia University Agricultural Experiment Station* **58**, 15-35.

HITE, B. H., GIDDINGS, N. J. & WEAKLEY, C. E. (1914). The effect of pressure on certain microorganisms encountered in the preservation of fruits and vegetables. *Bulletin of the West Virginia University Agricultural Experiment Station* **146**, 1-67.

JANNASCH, H. W., EIRMHJELLEN, K., WIRSEN, C. O. & FARMANFARMAIAN, A. (1971). Microbial degradation of organic matter in the deep sea. *Science, New York* **171**, 672-675.

JOHNSON, F. H. & EYRING, H. (1948). The fundamental action of pressure, temperature and drugs on enzymes as revealed by bacterial luminescence. *Annals of the New York Academy of Sciences* **49**, 376-396.

JOHNSON, F. H. & LEWIN, I. (1946). The influence of pressure, temperature and quinine on the rates of growth and disinfection of *E. coli* in the logarithmic growth phase. *Journal of Cellular and Comparative Physiology* 28, 77-97.

JOHNSON, F. H. & ZOBELL, C. E. (1949). The retardation of thermal disinfection of *Bacillus subtilis* spores by hydrostatic pressure. *Journal of Bacteriology* 57, 353-358.

JOHNSON, F. H., EYRING, H. & WILLIAMS, R. W. (1942). The nature of enzyme inhibitions in bacterial luminescence: Sulphanilamide, urethane, temperature and pressure. *Journal of Cellular and Comparative Physiology* 20, 247-268.

JOHNSON, F. H., EYRING, H., STEBLAY, R., CHAPLIN, H., HUBER, C. & GHERADI, G. (1945). The nature and control of reactions in bioluminescence with special reference to the mechanism of reversible and irreversible inhibitions by hydrogen and hydroxyl ions, temperature, pressure, alcohol, urethane and sulphanilamide in bacteria. *Journal of General Physiology* 28, 463-537.

LARSON, W. P., HARTZELL, T. B. & DIEHL, H. S. (1918). The effect of high pressure on bacteria. *Journal of Infectious Diseases* 22, 271-279.

MARQUIS, R. E. (1973). The physiological bases for microbial barotolerance. Office of Naval Research, Contract N00014-67-A-0398-0013. Work Unit No. NR 136-924. Technical Report Number 1.

MARSLAND, D., TILNEY, L. G. & HIRSCHFIELD, M. (1971). Stabilizing effects of D_2O on the microtubular components and needle-like form of heliozoan axopods; A pressure-temperature analysis. *Journal of Cellular and Comparative Physiology* 77, 187-194.

MORITA, R. Y. (1972). Pressure. Bacteria, fungi and blue green algae. In *Marine Ecology*, Vol. 1, part 3, ed. Kine, O. London: Wiley-Interscience.

OPPENHEIMER, C. H. & ZOBELL, C. E. (1952). The growth and viability of sixty-three species of marine bacteria as influenced by hydrostatic pressure. *Journal of Marine Research* 11, 10-18.

PARK, K. (1966). Deep sea pH. *Science, New York* 154, 1540-1541.

PALMER, D. S. & ALBRIGHT, L. J. (1969). Salinity effects upon maximum hydrostatic pressure for growth of the obligate psychrophiles *Vibrio marinus*. *Limnology and Oceanography* 15, 343-347.

POLLARD, E. C. & WELLER, P. K. (1966). The effect of hydrostatic pressure on the synthetic processes in bacteria. *Biochimica et biophysica acta* 112, 573-580.

REGNARD, P. (1891). *Recherches experimentales sur les conditions physiques de la view dans les eaux* Paris: Maason.

SALE, A. J. H., GOULD, G. W. & HAMILTON, W. A. (1970). Inactivation of bacterial spores by hydrostatic pressure. *Journal of General Microbiology* 60, 323-334.

SEKI, H. & ROBINSON, D. G. (1969). Effect of decompression on activity of micro-organisms in sea water. *Internationale Revue der gesamter Hydrobiologie und Hydrographie* 54, 201-205.

TIMSON, W. J. & SHORT, A. J. (1965). Resistance of micro-organisms to hydrostatic pressure. *Biotechnology and Bioengineering* 7, 139-159.

YAYANOS, A. A. (1967). A study of the effects of hydrostatic pressure on macromolecular synthesis and thiamineless death in *Escherichia coli*, and the compression of some solutions of biological molecules. Ph.D. Dissertation, Pennsylvania State University, University Park, Pennsylvania.

ZIMMERMAN, A. M. (1970). High pressure effects on cellular processes. New York & London: Academic Press.

ZOBELL, C. E. (1964). Hydrostatic pressure as a factor affecting the activities of marine microbes. In *Recent Researches in the Fields of Hydrosphere, Atmosphere and Nuclear Geochemistry*, eds Miyake, Y. & Komaya, T. Tokyo: Maruzen Co.

ZOBELL, C. E. (1970). Pressure effects on morphology and life processes of bacteria. In *High Pressure Effects on Cellular Processes* ed. Zimmerman, A. M. New York & London: Academic Press.

ZOBELL, C. E. & COBET, A. B. (1962). Growth, reproduction and death rates of

Escherichia coli at increased hydrostatic pressures. *Journal of Bacteriology* **84**, 1228-1236.

ZOBELL, C. E. & COBET, A. B. (1964). Filament formation by *E. coli* at increased hydrostatic pressure. *Journal of Bacteriology* **87**, 710-719.

ZOBELL, C. E. & JOHNSON, F. H. (1949). The influence of hydrostatic pressure on the growth and viability of terrestrial and marine bacteria. *Journal of Bacteriology* **57**, 179-189.

ZOBELL, C. E. & KIM, J. (1972). Effects of deep-sea pressures on microbial enzyme systems. In *Symposia of the Society for Experimental Biology XXVI; The Effects of Pressure on Organisms,* eds Sleigh, M. A. & MacDonald, A. G. London: Cambridge University Press.

ZOBELL, C. E. & OPPENHEIMER, C. H. (1950). Some effects of hydrostatic pressure on the multiplication and morphology of marine bacteria. *Journal of Bacteriology* **60**, 771-781.

Inactivation of Yeast

I. W. DAWES

Department of Microbiology, School of Agriculture, University
of Edinburgh, Edinburgh, Scotland

CONTENTS

1. Introduction 279
2. Physical methods 280
   - (a) Temperature 280
   - (b) Lowered water activity (a_w) 282
   - (c) Limits of pH 284
   - (d) Radiation 285
3. Chemical and biological inhibitors 288
4. Selective inactivation 290
   - (a) Inactivation of growing yeast 290
   - (b) Selective inactivation of vegetative yeast in sporulated cultures 294
5. Disruption of yeasts 296
   - (a) Physical rupture of the cell wall 296
   - (b) Enzymic and chemical methods 297
6. Summary 297
7. References 298

1. Introduction

IN REVIEWING the literature it has become apparent there is less information concerned specifically with inactivation of yeasts than with their growth. Partly, this may reflect a lack of interest by the applied microbiologist who (with the exception of the medical microbiologist) is more likely to be concerned with promoting the metabolic activity of *Saccharomyces* sp. or consuming the products of this activity. It is also in part due to there being relatively few instances in which yeasts are more resistant to environmental extremes than the most resistant bacteria or other fungi. Specific inactivation of yeasts is therefore usually a lesser problem in sterilization or preservation industries, and it is perhaps not surprising that a number of books devoted to yeasts, including a multivolume treatise, pay scant attention to the subject.

Despite this apparent lack of interest, there are particular instances in which yeasts cause problems. A number of yeasts or dimorphic fungi with a yeast-like phase are pathogenic to man; this includes *Cryptococcus neoformans* and *Candida albicans* as the two most important pathogenic yeasts, and species in the genera *Torulopsis, Trichosporon, Rhodotorula* and *Pityrosporum* (Gentles & La Touche, 1969). Mycoses due to these and other fungi are now more frequently encountered as a consequence of antibacterial antibiotic therapy and the use of immunosuppressive drugs. The application of antimicrobials in the inhibition or destruction of moulds and yeasts is an area which has been extensively reviewed by D'Arcy (1971) and overlap with this article has been avoided.

In food preservation there are several types of food which are susceptible to spoilage by yeasts rather than other micro-organisms; these have been discussed in reviews by Ingram (1958) and Walker & Ayres (1970). These reviews indicate, in the main, the established products in which yeast spoilage often occurs by virtue of the nature of the product and its traditional pattern of processing and storage. When spoilage does occur it can usually be traced to faulty practice, a major example being lack of proper hygiene: for this no inhibitory or inactivating process is of any use. The following discussion is therefore aimed at describing the general methods used to inactivate yeasts, identifying where possible the organisms most resistant to each particular treatment. From this it should be possible to assess the extent of problems which may arise from proposed alterations in traditional methods of preserving or presenting foods and other microbiologically unstable products.

Another major reason for considering general methods of inactivating yeast is the increased interest now manifest in these organisms as systems for basic research into the physiology and genetics of eukaryotes. Such laboratory investigations frequently depend on the availability of methods for inhibiting particular yeast processes, or cell inactivation or even breakage.

Yeast populations are usually heterogeneous, consisting either of cells in all stages of a cell budding cycle, or cells in various phases of the life cycle including a sporulated phase for sporogenous species. In a number of situations it may be important to inactivate one form of yeast, without affecting another or to be aware of a differential resistance between one phase and another. Various aspects of *selective* inactivation of particular phases are discussed, including the killing of growing cells in mixtures of growing and non-growing cells and the killing of vegetative organisms in sporulated cultures containing mixtures of spores and vegetative cells.

2. Physical Methods

(a) *Temperature*

Temperature effects on yeast have been recently reviewed by Stokes (1971). With regard to high temperatures, yeasts are more sensitive than the most resistant vegetative bacteria or other fungi and are usually inactivated in all heated products. Growth for all except obligate psychrophiles ceases at temperatures between 30 and 47° (Lund, 1951; see Stokes, 1971 for an extensive list of cardinal temperatures), and yeast populations are usually killed in a few minutes at temperatures between 55 and 65°, (Lund, 1951).

Yeast spores, unlike bacterial endospores, are only slightly more resistant to heat than vegetative cells for those species tested, mainly of the genera *Saccharomyces* (Lund, 1951; Fowell, 1969) and *Hansenula* (Wickerham & Burton, 1954). While differences between ascospores and vegetative cells are not usually extensive, as can be seen from thermal death time data summarized in Table 1, in some cases they could be significant where very mild heat treatments

Table 1
*Heat inactivation of vegetative cells and ascospores of yeast**

Organism		Temperature (°)	Time (min)	Reference
Sacch. cerevisiae	cells	58	20	Lund (1958)
var. *ellipsoideus*	spores	58	20	Lund (1958)
Sacch. odessa	cells	52	20	Lund (1958)
	spores	64	20	Lund (1958)
Sacch. cerevisiae	cells	54	5	see Stokes (1971)
	spores	62	5	see Stokes (1971)
Hansenula anomala	cells	58	3	Wickerham & Burton (1954)
	spores	58	5	Wickerham & Burton (1954)

* Data for initial cell concentration often not given, nor whether single-spore cultures were used.

are designed to inactivate yeasts. Most data given in the literature are thermal death times obtained using sporulated cultures rather than purified single spore suspensions. Since there are four spores in normal asci of *Saccharomyces,* and *Hansenula* spores are prone to agglutination, initial population numbers are not given, so these data are less reliable than they could be. There is however a recent clear demonstration of a significant difference in the D values for purified spores and vegetative cells of a spoilage yeast isolated from carbonated beverages (Put & Sand, 1975).

In *Saccharomyces,* incubation at temperatures near the maximum growth temperature leads to induction of respiratory deficient petites (Sherman, 1959). Petites lack functional mitochondria and are unable to grow on media containing non-fermentable substrates. They can be a considerable nuisance in many tests for survival since they appear on plates several days or more after grande strains.

Petite induction is a problem not restricted solely to high temperature treatments, almost any method of inactivating or inhibiting yeast growth favours their formation, including treatment with mutagens, antibiotics and chemical inhibitors (Williamson, 1970). The multiplicity of possible target sites for heat-induced injury in vegetative bacteria has been discussed elsewhere in this volume (Tomlins & Ordal, p. 153) and many other reviews (see e.g. Brown & Melling, 1971) and the same considerations apply to yeasts (Farrell & Rose, 1967). Enzyme denaturation, cell membrane damage, ribosome breakdown and

DNA strand breakage have all been implicated. The possible involvement of the cell membrane as the site for thermal damage ultimately leading to death (rather than inactivation) is discussed in the subsequent section on selective inactivation of growing cells. From kinetic studies on the effect of superoptimal temperatures on petite strains of *Saccharomyces cerevisiae*, van Uden (1971) concluded that exponential death occurs concurrently with exponential growth: this has been extended to other species of yeast. Moreover, a detailed kinetic analysis of the effects of shifting a culture of *Sacch. cerevisiae* to a supermaximal temperature indicated that growth continued for an initial period before the onset of exponential death. This has been interpreted in terms of there being more than one thermosensitive site, all of which must be inactivated for cell death to occur (van Uden, 1971).

The minimum growth temperature for most yeast is close to $0°$, but there are obligately psychrophilic strains which can grow at temperatures down to $-7°$ (Stokes, 1971) with a report of growth of one yeast down to $-34°$ in a suitable (non-frozen) medium (McCormack, 1951). Most psychrophilic yeasts found belong to the genus *Candida* (Stokes, 1971) but some have been found for *Cryptococcus, Rhodotorula, Torulopsis, Debaromyces* and *Pichia* (Hagen, 1971).

At temperatures below the minimum for growth, and above the freezing point of the medium, yeasts usually do not lose viability. There are a number of theories proposed to account for the inhibition of growth below the minimum. In general these rely on either loss of membrane permeability at low temperatures, or to loss of either enzyme activity, or regulatory mechanism governing enzyme synthesis. These aspects are clearly covered in articles by Farrell & Rose (1965, 1967) and Hagen (1971).

Freezing and subsequent thawing can lead to extensive loss of viability in some species of yeast (Tanner & Williamson, 1928). The extent of kill is dependent on the time for which the yeast is held. The freezing temperature also affects survival. Rose (1970) reported maximum survival for freezing at $-30°$ for *Sacch. carlsbergensis* and *Schizosacch. pombe*, below that temperature the viability of the thawed culture decreased. Other factors which affect survival include the rate of cooling, the pregrowth conditions, the storage medium, and the rate of thaw. In general it appears that *Sacch. cerevisiae* is damaged less by slow cooling than fast, whereas thawing should be done rapidly to minimize damage (Mazur & Schmidt, 1968; see Hagen, 1971). Freeze–thaw cycles (and freeze-drying) lead to marked changes in membrane permeability (Conway & Duggan, 1958) and are commonly used to render yeast permeable to substrates for assays of enzyme activity.

(b) *Lowered water activity* (a_w)

The ability of certain yeasts, variously described as osmophilic or sugar-tolerant, to grow under dry conditions or in high solute concentrations is one of the more important properties defining one group of products likely to be susceptible to spoilage by yeasts. Yeasts and other fungi generally have lower minimal a_w, for growth than bacteria (Table 2), and osmophilic yeasts have been reported to have minimal a_w for growth below 0.765 (in 70% fructose), see Anand & Brown (1968) down to levels in the range 0.60–0.65 (Mossel & Sand, 1968) commonly found as minimal for xerophilic fungi (Kushner, 1971). The range of minimal a_w

Table 2
Limiting water activities for growth of various yeasts

Organism	Minimal a_w for growth	Reference
Sacch. cerevisiae	0.940 (NaCl)	Onishi (1963)
	0.919 (glucose)	Onishi (1963)
	0.895 (NaCl)	see Kushner (1971)
	0.917 (sucrose)	Anand & Brown (1968)
Sacch. fragilis	0.963 (NaCl)	see Kushner (1971)
	0.935 (sucrose)	Anand & Brown (1968)
Schizosacch. octosporus	0.984 (NaCl)	see Kushner (1971)
Hansenula auavolens	0.970 (NaCl)	see Kushner (1971)
Candida pseudotropicalis	0.931 (NaCl)	see Kushner (1971)
Sacch. rouxii	0.860 (NaCl)	Onishi (1963)
	0.857 (sucrose)	Onishi (1963)
	0.845 (glucose)	Onishi (1963)
	0.765 (fructose)*	Anand & Brown (1968)
Sacch. mellis	0.765 (fructose)*	Anand & Brown (1968)
Torulopsis halonitratophila	0.765 (fructose)*	Anand & Brown (1968)
Sacch. acidifaciens	0.765 (fructose)*	Anand & Brown (1968)

* Lowest a_w tested, in 70% fructose.

for growth, the distribution and classification of osmophiles have been reviewed by Scott (1957), Ingram (1957) and Onishi (1963). Most of the sugar-tolerant strains are now classified into either *Sacch. rouxii* or *Sacch. melis*; some strains of *Debaromyces* and *Torulopsis* can also grow at low a_w.

The limits of water activity tolerated by these yeasts are affected greatly by the nature of the extracellular solutes present (Onishi, 1963; Brown, 1974). Those solutes for which yeasts have adapted most successfully to tolerance are usually sugars or other related non-electrolytes, but salt tolerance is also encountered (Onishi, 1963). The sugar-tolerant yeasts have an optimum growth rate at an a_w of 0.96, whereas non-osmophilic species generally have optima around 0.99 (Anand & Brown, 1968).

It is now apparent that the ability of sugar-tolerant species to grow at low a_w is due to their capacity for synthesis of high intracellular concentrations (up to at least 5 molal) of polyols: that most commonly found in *Sacch. rouxii* is arabitol (Brown, 1974). In addition to this synthesis of a compensating solute, other adaptations are necessary. In particular Brown (1975) has shown that *Sacch. rouxii* differs from non-tolerant species in showing no catabolite repression of respiration when grown in the presence of high concentrations of glucose.

Lowering of a_w below the minimum for growth does not lead to inactivation, rather, in many cases, to protection against other types of physical injury. This is particularly the case for heat treatment. For osmophilic yeasts, Gibson (1973) has reported a dramatic increase in D value on decreasing a_w from 0.995 to 0.7. For example, at 55° the D value for *Sacch. rouxii* was <0.01 at a_w of 0.995 rising to 54.44 at a_w of 0.85. Rose (1972) has shown that storage of strains of *Sacch. cerevisiae* and *Sacch. carlsbergensis* in high concentrations of sucrose led to survivals in the range 40–80% after more than a year's storage at 1°.

Water activities preventing yeast growth can be achieved by desiccation and freezing as well as high solute concentrations, all of these methods inhibiting growth without necessarily causing cell death. Freezing has already been discussed. Two other techniques are also used in preservation of yeast strains: freeze-drying and desiccation in silica gel. With these techniques the main aim is usually not for maximum survival, but maintenance of specific physiological or genetical characteristics of the original culture. Freeze-drying has been reviewed by Fry (1966) and discussed by Hagen (1971). In general for yeast it appears that the procedure should involve slow-freezing, to a temperature around −30° (Rose, 1970), protective substances such as serum or sugars may be useful, dried cultures should be stored under vacuum or inert atmosphere, and reconstitution should be carried out slowly (Mitchell & Enright, 1957). For storage of mutants for genetic studies the author has found silica gel satisfactory and convenient.

(c) *Limits of pH*

The pH range for growth of a number of yeasts is given in Table 3. Most yeasts can grow well at pH values below 4, the minimum for most bacteria, and the pH optimum for *Sacch. cerevisiae* has been given as 4.5. Under acid conditions yeasts and other fungi usually outgrow bacteria in heterogenous populations. The upper pH limit for yeast growth, where known, is not greatly in excess of that found for other groups of organisms (Kushner, 1971). Battley & Bartlett (1966) found that yeasts were capable of metabolizing certain media to a final pH value in excess of that inhibiting growth.

Ability to grow at low pH is probably due to the existence of systems for maintaining the internal cell pH at a value nearer neutrality (Conway & Downey, 1950). The minimum pH for growth is very dependent on the nature of the acid present in the growth medium. In particular, fatty acids have much greater effect on growth than inorganic acids at the same pH value, and exert their greatest inhibitory effect on intact cells as the undissociated molecule (Prince, 1959). Whether this is due to undissociated fatty acids causing internal acidification of the cell due to their ability to penetrate the cell membrane, or to direct inhibitory effects on cell enzymes is not clear since there is evidence to support both points of view. Neal *et al.* (1965) have shown that the internal pH of *Sacch. cerevisiae* is much lower for cells suspended in acetate than phosphate or succinate at pH 3.0, supporting qualitative data of Maesen & Lako (1952) on

Table 3
pH range for growth of various yeasts

Organism	Minimum pH for growth	Maximum pH for growth
Candida pseudotropicalis*	2.3	8.8
Hansenula canadensis*	2.15	8.6
Sacch. cerevisiae*	2.35	8.6
Sacch. fragilis*	2.4	9.05
Sacch. microellipsoides*	2.2	8.8
Sacch. pastori*	2.1	8.8
Schizosacch. octosporus*	5.45	7.05
Candida krusei†	1.5	—
Hanseniaspora melligeri†	1.5	—
Rhodotorula mucilaginosa†	1.5	—
Sacch. exiguus†	1.5	—

* Data from Battley & Bartlett (1966)
† Data from Recca & Mrak (1952); pH 1.5, minimum tested.

internal acidification in fatty acid solutions. In addition Neal *et al.* (1965) have suggested that the data of Hoffman *et al.* (1938) on variation of fungistatic effect of various fatty acids with chemical structure indicates membrane permeability is an essential part of this effect. On the other hand direct inhibitory effects of sorbic acid (2,4 hexadienoic acid) on dehydrogenases (Melnick *et al.*, 1954) and on enolase (Azukas *et al.*, 1961) and sorbyl-CoA complex formation (Palleroni & de Pritz, 1960) have been noted *in vitro* indicating possible sites of energy metabolism specifically affected by fatty acids.

pH is also an important parameter governing survival of yeast which have been subjected to just sublethal stress. This has relevance to the design of survival experiments, since Nelson (1972) has shown for a range of organisms, including *Sacch. cerevisiae, C. albicans, C. utilis, Endomyces magnusii, Rhodotorula* sp. and *Klyveromyces lactis,* that low pH in the plating medium is inimical to survival. Low pH media have been used to select against bacteria when enumerating yeast in mixed populations, and Nelson suggests the use of media containing rose bengal and an antibacterial antibiotic at higher pH (Overcast & Weakley, 1959).

(d) *Radiation*

The literature on the radiobiology of yeast is very extensive, due largely to the suitability of these organisms for radiation studies. This is particularly so for those studies which are concerned with the damage induced by radiation in micro-organisms, and their response to it. The subject has been reviewed by James & Werner (1965) but since then considerable progress has been made, particularly in the study of repair of radiation damage. More general reviews on microbial destruction, mainly by ionizing radiation, are those of Goldblith (1967, 1971) and Silverman & Sinskey (1968).

Yeasts are generally more resistant to ionizing radiation than other fungi and most of the sensitive bacteria, but less so than bacterial spores or very resistant bacterial species such as *Micrococcus radiodurans*. Data for a number of organisms are given in Table 4, those for yeasts come from Bridges *et al.* (1956).

Table 4
Approximate lethal doses of ionizing radiation

Organism	Inactivating Dose (Mrad)
Yeasts*	
Sacch. ellipsoideus	0.65
Sacch. cerevisiae	0.83
Schizosacch. octosporus	0.47
Torula cremoris	0.47
Torula histolytica	0.93
Torulaspora rosei	0.47
Cryptococcus neoformans	0.83
Candida albicans	0.93
Candida krusei	1.07
Candida parakrusei	1.75
Candida mortifera	0.55
Vegetative bacteria†	
Escherichia coli	0.07
Pseudomonas aeruginosa	0.02
Bacillus mesentericus	0.30
Micrococcus radiodurans	2.0 – 4.0
Bacterial spores†	
Bacillus mesentericus	1.8
Bacillus subtilis	1.2 – 1.8
Clostridium botulinum A	1.9 – 3.7

* Data on yeasts from Bridges *et al.* (1956) for suspensions containing *c.* 10^7 organisms, originally given in Mrep.

† Data from various sources, including estimates from that of Goldblith (1971) and Koh *et al.* (1956).

It can be seen that the most resistant species belong to the genus *Candida,* including some species in this very diverse group which are potentially pathogenic to man (*C. krusei,* see Gentles & La Touche, 1969). This may impose restrictions on the use of radiation at pasteurizing doses designed to inactivate clostridial spores. As an example, the spoilage flora of packaged frankfurters subjected to 460 Krad was predominantly composed of yeasts and *Bacillus* sp., the yeasts were commonly *Debaromyces subglobosus* and *Torulopsis candida* (Drake *et al.*, 1958, 1959).

Ionizing and UV radiation induces a wide range of lesions in yeast, including genetic damage such as mutation and chromosome breakage, and metabolic and permeability changes (James & Werner, 1965); most of these lesions are potentially lethal. DNA is undoubtedly the major primary target since cells of different ploidy show different patterns of sensitivity to radiation with haploids most, and diploids least sensitive, and for ploidy higher than diploid an increase in sensitivity with increase in ploidy (Mortimer, 1958). This has been interpreted to confirm the hypothesis that lethal radiation damage to haploids is predominantly due to induction of recessive-lethal mutations, while in diploids and polyploids it is mainly dominant-lethal mutations or chromosome damage. Survival curves comparing haploids and diploids in *Sacch. cerevisiae* indicate that not only do diploids have a higher D value, but unlike haploids they also have a pronounced shoulder in their inactivation curves (Laskowski, 1960).

Mutants with altered sensitivity to radiation have been isolated in both *Sacch. cerevisiae* and *Schizosacch. pombe* (Mortimer & Hawthorne, 1969). Moustacchi (1965) has reported that a high proportion of the survivors obtained after X-ray and UV treatment were resistant to radiation; the resistance was recessive and due to mutation at a number of loci. A large number of mutations leading to sensitivity to ionizing and/or UV radiations have been found and allelism tests have shown that more than 25 loci are involved (Game & Cox, 1971; see also Plischke *et al.*, 1975). Tests of constructed double mutants and of the differential sensitivity of mutants to various types of radiation (e.g. UV versus X-ray) have led to the suggestion that there are many pathways for the repair of radiation-induced damage. In *Sacch. cerevisiae* there are at least three dealing with UV-induced damage (Game & Cox, 1974) and at least two for X-ray induced damage (Game & Mortimer, 1974). For several mutants the pathway affected has been characterised. As an example, a photoreactivationless mutant lacks the enzyme for directly reducing pyrimidine dimers induced by UV (Resnick, 1969). Other mutants are deficient in the excision repair system removing UV-induced dimers from DNA (Unrau *et al.*, 1971; Resnick & Setlow, 1972).

A large number of factors influence the extent of survival of yeast populations after radiation treatment. A number of these are discussed extensively by James & Werner (1965); those tending to confer higher survival include: (i) absence of O_2 during radiation; (ii) freezing, drying or use of solutes such as glycerol or sugars to reduce a_w; (iii) postirradiation holding under starvation conditions (liquid holding recovery, see Parry, 1969); (iv) respiratory capacity (Bruce & Parker, 1974) and (v) the presence of free radical trapping compounds in the suspending medium during irradiation (Goldblith, 1971). An additional factor of considerable interest to studies of the yeast cell cycle is the effect of cell age from formation by division. Budding cells show considerably higher resistance to ionizing radiation and UV (Beam et al., 1954). Moreover, such cell cycle effects are not restricted to lethality since sublethal treatments lead to a delay in cell division (Burns, 1956; Swann, 1962), and cell age-dependent responses with respect to mitotic recombination (Esposito, 1968).

3. Chemical and Biological Inhibitors

The literature on chemical compounds inactivating yeast is extensive. General disinfectants have been covered in detail by Sykes (1965) and Hugo (1971), and antibiotics and other chemical inhibitors discussed extensively from the medical viewpoint by Gentles & La Touche (1969) and D'Arcy (1971). In general there are only a few antifungal antibiotics used extensively in clinical practice. Of these the polyene amphotericin B is active against systemic infections, but can show serious side effects, nystatin is suitable for topical treatment and griseofulvin is active only against the ringworm fungi (Gentles & La Touche, 1969). One polyene, the N-acetyl derivative of candidin has been used in prevention of yeast and mould contamination of tissue culture media (D'Arcy, 1971).

Few compounds are permitted for use in foods. Those which have been used under circumstances favouring yeast or mould spoilage include organic acids such as benzoate and its derivatives, fatty acids including acetic acid, propionic acid and sorbic acid and inorganic acids including sulphur dioxide. All of these are known to be effective as the non-ionic species, present in higher proportion at lower pH (Jay, 1970). In carbonated products the main component inhibiting yeast growth is probably CO_2 rather than pH or sugar levels (Witter et al., 1958).

For laboratory studies of yeast physiology there is now available a fairly wide range of antimicrobial substances for which some knowledge of the site of action is available. These inhibitors are too numerous to discuss in detail here, but some are listed in Table 5 with their proposed site of action, and references which have been chosen to provide ready access to the specific literature. Reviews or extensive articles on protein synthesis inhibitors in yeast are given by Battaner & Vazquez (1971) and Jiménez et al. (1972), on inhibitors of mitochondrial function by Avner et al. (1973) and Wolf et al. (1973) and on the polyene macrolide antibiotics affecting membranes (Hamilton-Miller, 1973). Mutations to resistance have been found for many of the inhibitors listed in Table 5, particularly those affecting protein synthesis in either the cytoplasm or mitochondrion and for mitochondrial ATPase inhibitors. These have been useful research tools for investigating structure function relationships for the sites affected. For example, Sacch. cerevisiae mutants resistant to cryptopleurine and other phenanthrene alkaloids map at a site near the mating-type locus and cause an alteration in the 40S ribosomal subunit (Skogerson et al., 1973; Grant et al.,

Table 5
Compounds inhibiting or inactivating yeast, or in vitro yeast systems

Compound	Site of inhibition	Reference
Protein synthesis		
Cryptopleurine, and other phenanthrene alkaloids	Affects translocation by acting on 40S ribosomal subunit	Skogerson et al. (1973); Grant et al. (1974)
Edeine Polydextran sulphate Aurintricarboxylic acid	Act at codon-anticodon recognition; 40S subunit. inhibit mRNA attachment	Battaner & Vasquez (1971)
Trichodermin	60S subunit	Schindler et al. (1974)
Cycloheximide	60S subunit	Cooper et al. (1967)
Anisomycin Sparsomycin Blastocidin S	Inhibit peptide bond formation	Battaner & Vazquez (1971)
RNA synthesis		
Lomofungin	All RNA synthesis	Gottlieb & Nicholas (1969)
8-Hydroxyquinoline	All RNA synthesis	Fraser (pers. comm.)
Actinomycin D	All RNA synthesis	
α-Amanitin	*In vitro* mRNA synthesis	
DNA synthesis		
Hydroxyurea	Inhibits ribonucleotide reductase	Slater (1973)
Function of plasma membrane		
Nystatin, and other polyenes	Binds to membrane sterols	Lampen (1966)
Cell wall synthesis		
2-Deoxyglucose	Glucan synthesis	Johnson (1968)
Fluorodeoxyglucoses	Glucan synthesis	Biely et al. (1973)
Mitochondrial function		
Oligemycin, triethyltin, ossamycin, venturicidin, rutamycin	Inhibit mitochondrial ATPase activity	Walter et al. (1967)

1974). On the other hand resistance to cycloheximide can arise by mutation at any one of nine loci, at least one of which affects the 60S ribosomal subunit (Cooper et al., 1967).

Mutants resistant to nystatin have been isolated in *Candida* sp. (Hamilton-Miller, 1973) and in *Sacch. cerevisiae* one mutation to nystatin resistance was found to be dominant (Patel & Johnston, 1968). If the results with *Saccharomyces* can be extrapolated to the case of *Candida* these mutants may eventually be a potential problem in polyene antibiotic therapy.

One biological method of inactivating yeast of some recent interest is the killer character found in a number of strains of *Sacch. cerevisiae*. There are three types of yeast cell, killer cells, sensitive cells which are killed when mixed with killer cells, and neutral cells which do not have killer activity and are resistant to killer activity (Bevan & Makower, 1963). The killer factor is a protein released into the culture supernatant liquid by killer cells (Woods & Bevan, 1968), and its synthesis is determined by a cytoplasmically inherited factor and a nuclear gene controlling its maintenance (Somers & Bevan, 1969). The cytoplasmic factor is probably double-stranded RNA whose presence correlates with killer activity (Bevan *et al.*, 1973). The mode of action of the killer factor protein is not clear, although it appears to cause leakage of molecules such as ATP and a general inhibition of macromolecular synthesis (Bussey *et al.*, 1973). Log phase cells are most sensitive to killer protein, resting cells on entering the log phase are killed immediately (Woods & Bevan, 1968). If the killer protein acts on membranes as suggested by Bussey *et al.* (1973) it may find considerable application in processes requiring release of cell constituents. So far the author is unaware of reports on the distribution of sensitivity to killer factor in species other than *Sacch. cerevisiae* or of other killer systems in other yeasts.

4. Selective Inactivation

(a) *Inactivation of growing yeast*

The main reason for wishing to inactivate growing yeast selectively in populations containing growing and non-growing organisms is in selecting yeast mutants with either an auxotrophic requirement or sensitivity to an antibiotic. Moreover, these studies on the inactivation of growing cells have particular relevance to the analysis of mechanisms of cell death.

There are a number of methods available for selective killing of vegetative cells. Snow (1966) introduced the antibiotic nystatin for the enrichment of auxotrophic mutants in analogy with penicillin selection in bacteria. Other selection methods depend on the induction of 'unbalanced growth'. This phenomenon has been found in inositol-requiring mutants of *Neurospora* and biotin auxotrophs of *Aspergillus*. If conidia are starved for the vitamin requirement they lose viability on germinating in an otherwise complete growth medium. If a second mutation is present preventing macromolecular synthesis the germinating conidia do not grow, but do not lose viability either (Lester & Gross, 1959; Pontecorvo *et al.*, 1953). Inositol-requiring strains have been isolated in *Sacch. cerevisiae* (Culbertson & Henry, 1975) and used for selection of spontaneous mutants (Henry *et al.*, 1975). Similar effects of inositol starvation have been seen in requiring strains of *Kloerckea apiculata* and *Sacch. carlsbergensis* (Ridgway & Douglas, 1958) and for biotin or pantothenate starvation in *Sacch. cerevisiae* (Kuraishi *et al.*, 1971; Shimada *et al.*, 1972).

Death due to unbalanced growth may be the result of uncoupling membrane phospholipid synthesis from macromolecular synthesis (Henry, 1973; Culbertson & Henry, 1975) since mutants requiring fatty acids also die when deprived of these, but survive in the absence of fatty acids if protein synthesis is arrested either by adding cycloheximide or by withholding an auxotrophic requirement (Henry, 1973). Moreover, the vitamins biotin and pantothenate are required for fatty acid synthesis; in aerobically grown *Sacch. cerevisiae* biotin can be replaced by addition of fatty acids, fatty acid esters and aspartate (Suomalainen & Oura, 1971). Inositol is a constituent of a limited number of membrane phospholipids in yeast (Culbertson & Henry, 1975).

Another method used for mutant enrichment depends on a temperature-sensitive (*ts*) mutant of *Sacch. cerevisiae* which dies at the restrictive temperature unless protein synthesis is blocked. This 'kamikazi' mutant has been used by Littlewood (1972) to select yeast sensitive to an antibiotic not affecting intact wild-type cells, and mutants with similar suicide phenotype have been used by the author for efficient selection of auxotrophs in *Sacch. cerevisiae.* Figure 1 illustrates the effect on cell viability of holding a uracil auxotroph with this lesion at the restrictive temperature in the presence and absence of added uracil. Optimal conditions for auxotroph selection in liquid and solid media have been determined (Hardie & Dawes, in preparation).

So far the specific metabolic defects in *ts* suicide mutants have not been traced, although it seems likely from the known biochemical lesion in other suicide mutants discussed above that they have a defect in membrane function or biosynthesis. Very few mutants from a large collection of *ts Sacch. cerevisiae* strains lose viability at the restrictive temperature as measured by their inability to recover and grow on subsequent incubation at a permissive temperature (Hartwell, 1967); as an example, from complementation tests of 48 generally selected *ts* mutants only four suicide mutants were found and these all fell into the same complementation group (Dawes, unpublished). It seems clear that cell death due to unbalanced growth can only be induced by disturbing very few of the many essential functions in the cell.

It is tempting to speculate and extend the above considerations by suggesting that loss of cell viability under other circumstances than unbalanced growth is ultimately the result of macromolecular synthesis when membrane biosynthesis is impaired. This could even apply to radiation-induced lethality where the primary target for damage is largely DNA, and may account for the phemomenon of liquid holding recovery in which starvation after irradiation leads to an increased survival. The interpretation here is that irradiation induces

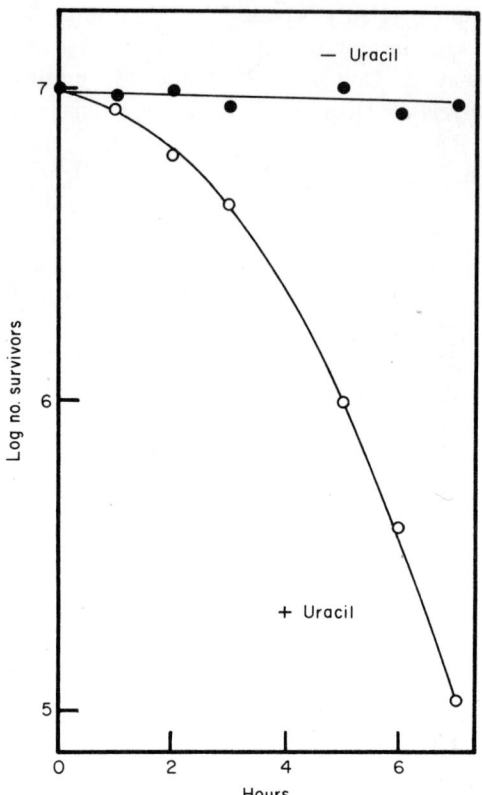

Fig. 1. Growth-dependent loss of viability of *ts* suicide mutant (α *ade* 6 *ura* 1 *ts* 78). The effect on viability of the strain indicated of holding at the restrictive temperature in the presence (o) and absence (●) of the growth requirement is indicated.

damage which, in the short term before repair is possible, prevents membrane synthesis while some macromolecular synthesis can occur. Starvation would prevent loss of viability by allowing repair to occur without extensive macromolecular synthesis taking place and generating a situation of imbalance.

Tritium suicide is another method which has been used in bacterial and fungal systems to discriminate against growing cells (Lubin, 1959; Donkersloot & Mateles, 1968; Littlewood & Davies, 1973). Since the mean path length for the β-particle from [^{3}H] decay is of a similar order to the diameter of microbial cells, incorporation of tritiated precursors into the macromolecules of a cell leads to induction of radiation damage mainly within that cell. The method depends for success on the availability of very high specific activity tritiated precursors of macromolecular synthesis. Currently it is possible to obtain tritiated amino acids, purines and pyrimidines with specific activities in the range

15–60 Ci/mmole; these have been found adequate for selecting auxotrophic and *ts* mutants defective in RNA or protein synthesis (Littlewood & Davies, 1973). The technique poses a problem in yeast inactivation since it it essential to find conditions of storage for periods up to three weeks which inhibit both growth and macromolecular degradation without affecting the viability of unlabelled yeast. A study of conditions affecting viability of *Sacch. cerevisiae* indicated that storage in 10% glycerol at −20° or at 0° in a minimal growth medium is satisfactory (Dawes & Henry, unpublished). Figure 2 illustrates the time course

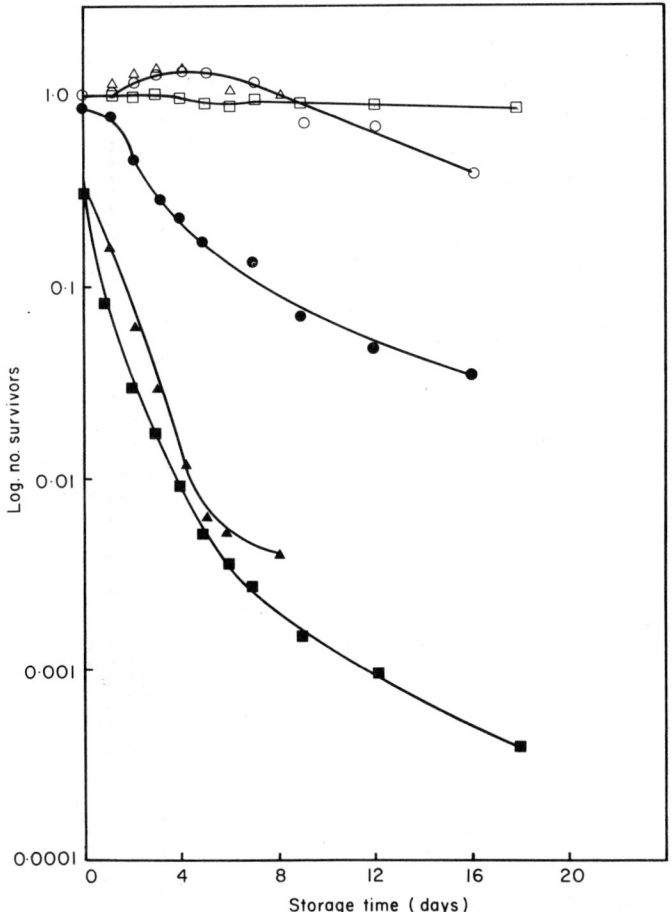

Fig. 2. Tritium suicide of ethylmethanesulphonate treated *Sacch. cerevisiae* cultures labelled with ³H-uracil: ●, uracil at 18 Ci/nmole, 1.5 μg/ml, storage at 4°; ○, control. ▲, uracil at 18 Ci/mmole, 2 μg/ml, storage at 4°, △, control. ■, uracil at 56 Ci/mmole, 1.25 μg/ml, storage at 0°; □, control.

of tritium suicide of EMS-mutagenized cultures of a uracil auxotroph of *Sacch. cerevisiae* which were labelled with [5,6-^{3}H] uracil at different specific activities while selecting for *ts* mutants (Dawes & Henry, unpublished).

While tritium suicide procedures are restricted to small scale laboratory experiments, other methods discussed above may have application in industry where cell disruption or leakage of cellular components is necessary. For example the use of *ts* suicide mutants, or strains with a requirement for biotin, pantothenate, inositol or saturated fatty acids may facilitate the rather difficult task of disrupting yeast cells. There are probably many other methods which could be used for killing growing cells specifically. For example, there are differences in the susceptibility of dividing and stationary phase yeast to the glucose analogues 2-deoxy-D-glucose and 2-deoxy-2-fluoro-D-glucose (Megnet, 1965; Johnson, 1968; Biely *et al.*, 1973) to the effect of killer factor protein (Woods & Bevan, 1968) and to lysis by snail gut enzyme (Eddy & Williamson, 1957, 1959; Holter & Ottolenghi, 1960).

(b) *Selective inactivation of vegetative yeast in sporulated cultures*

Yeast life cycles usually include a number of different morphogenetic states including spores in perfect diploid species. It is in one sense fortunate that haploid spores produced during meiosis and sporulation in yeasts are unlike bacterial spores in not showing extreme resistance to most chemical and physical inactivation treatments. Usually no particular attention is paid during sterilizing practice to survival of yeast spores, although it should be emphasized that recent data on a spoilage yeast isolated from carbonated beverages indicate that significant differences in survival between vegetative cells and ascospores of particular yeast strains do occur (Put & Sand, 1975; Lund, 1958). Under circumstances in which low heat treatments are given specifically to inactivate yeasts, the somewhat greater heat resistance of ascospores should be taken into account.

There are circumstances in which it is quite important to inactivate vegetative cells without killing all spores in a population. These are mainly concerned with genetic manipulation such as hybridization, or mapping by random spore analysis (Fowell, 1969; Mortimer & Hawthorne, 1969) in which available procedures for physically separating spores from vegetative cells (Emels & Gutz, 1958; Resnick *et al.*, 1967; Rousseau & Halvorson, 1969) are inapplicable. A number of methods have been used for this purpose. Wickerham & Burton (1954) used the slightly greater heat resistance of spores of *Hansenula* spp. to obtain haploids, and this procedure has been used for strains of *Saccharomyces* showing poor spore viability (Fowell, 1969). Heat treatment does not lead to complete inactivation of diploids, while useful for isolation of haploids it is not suitable for random spore analysis or selection of mutants altered in their ability to sporulate.

Other methods, using solvents, have been found to lead to almost complete kill of vegetative yeast with high spore survival. Zakharov & Inge-Vechtomov (1964) used 33% ethanol for this purpose with *Sacch. cerevisiae,* while biphasic diethyl ether–buffer mixtures have been used by Dawes & Hardie (1974). *Schizosaccharomyces pombe* spores are differentially resistant to 30% ethanol (Leupold, 1957). These methods have the advantages of being rapid, convenient and complete in their kill of diploids but in *Sacch. cerevisiae* they lead to the induction of petites. In our experience the ether method is less prone to induce petites and by treating the cultures at $0°$ for short times (5 min) the proportion of petites among survivors is low and spore survival is enhanced. Figure 3

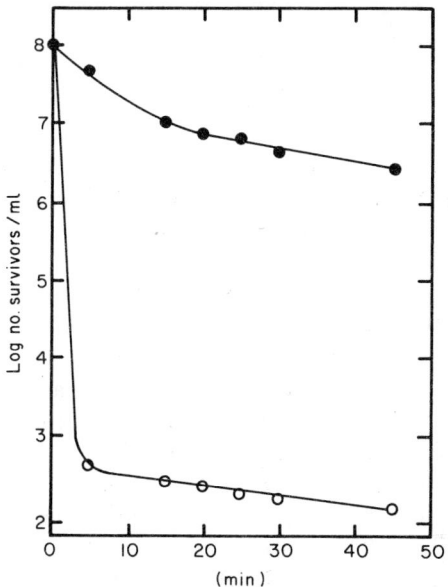

Fig. 3. Sensitivity of sporulated (●) and non-sporulated (○) cultures of *Sacch. cerevisiae* to treatment with biphasic diethyl ether–buffer mixtures.

illustrates the extent of differences in survival between spores and vegetative cells treated with ether–buffer mixtures for varying times. Ether treatment has the additional advantage that plate cultures can be exposed to ether vapour, and it is now possible to screen for many classes of developmental mutants including those altered in their ability to sporulate, germinate or outgrow (Dawes, 1975).

In routine random spore analysis the method usually used requires the introduction of a recessive mutation to drug resistance (usually to canavanine) in heterozygous state (Sherman & Roman, 1963). Canavanine is an analogue of arginine and resistant mutants are known to be deficient in an arginine permease (Grenson *et al.,* 1966). Heterozygous diploids and half of the spore progeny are unable to grow on canavanine-containing media.

5. Disruption of Yeasts

A special case in inactivation is the physical disruption of cells to release cytoplasmic contents. This can be a problem with some yeast species since the cell wall is often fairly resistant to some methods commonly used to break open micro-organisms. A number of techniques are available, each with advantages and disadvantages for particular purposes. Some are designed to rupture cells as gently as possible to avoid shear or denaturation of cell components such as DNA or enzymes, or damage to organelles. For others the stress during rupture is of lesser importance to considerations of weights to be processed or cost. General methods for breaking micro-organisms have been reviewed extensively (Wood, 1966; Hughes et al., 1971) and the following discussion is restricted to the applicability of various methods to yeast.

(a) *Physical rupture of the cell wall*

On a laboratory scale, cell extracts for biochemical analysis are frequently obtained by use of either a Hughes Press (Hughes, 1951) or the similar Eaton Press (Eaton, 1962). Both work on the principle of forcing a frozen cell paste through an orifice; modifications to the Hughes Press can handle up to 80 g wet weight of cells (Hughes et al., 1971). A similar effect has been obtained using modifications to a commercially available French Pressure Cell and freezing the mixture before passage (Bhargava & Halvorson, 1971). These freeze–shear methods are limited in terms of capacity and are unsuitable for isolating enzymes susceptible to freezing. Rupture is, however, almost complete and the method is widely used in the laboratory for preparation of cell extracts and cell walls, moreover, it can be adapted to the isolation of fragile organelles from yeast. By careful attention to the composition of the homogenizing medium and the freezing protocol it is possible to obtain intact nuclei from *Sacch. cerevisiae* (Bhargava & Halvorson, 1971), *Sacch. carlsbergensis* (Rozijn & Tonino, 1964) and *Schizosacch. pombe* (Duffus, 1969). The availability of intact nuclei has enabled isolation and characterization of high molecular weight DNA (Cramer et al., 1972) from yeast. Another method for DNA extraction using the Eaton Press depends for success on the use of buffers to maintain nuclear material in a condensed state less susceptible to shear (Williamson et al., 1971).

For many purposes, including isolation of enzymes and cell wall material, yeast can be homogenized by shaking or grinding in suspension with glass beads (Garver & Epstein, 1959; Küenzi & Fiechter, 1969; Réháćek et al., 1969). Probably the simplest method involves a modification of that of Novotny (1964) described by Beck & von Meyenburg (1968). It entails vibrating a cell suspension mixed with ballotini beads. With such methods there are disadvantages due to foaming and adhesion of components to glass beads, nonetheless it is possible to isolate relatively intact mitochondria (Winterberger & Winterberger, 1970). Membrane disruption to permeabilize cells for enzyme assay can also be achieved by freezing cycles of freeze-drying (Tauro et al., 1968).

(b) *Enzymic and chemical methods*

By far the gentlest method of disrupting yeast involves enzymic digestion of the cell wall under hypertonic conditions, followed by osmotic lysis or detergent disruption of the membrane. There are a number of sources of enzymes active against yeast cell walls, that most commonly used from the digestive fluid of the snail (*Helix pomatia*) is available commercially. The original method of Eddy & Williamson (1957, 1959) has been optimized through the work of several groups so that it is now fairly rapid and gives rise to fairly stable protoplasts. Thiols are required for optimal rates of enzyme activity, and the stability of protoplasts is enhanced in hypertonic sugars, polyols or $MgSO_4$ (Phaff, 1971). Effective lysis has been reported for a number of yeast genera including *Saccharomyces, Candida* and *Schizosaccharomyces*.

This approach has a number of limitations of which the major one is the small quantity of yeast that can be handled due to the large amounts of enzyme needed. Yeast strains also vary widely in susceptibility to the enzyme and lysis is dependent on physiological state (Eddy & Williamson, 1957; Holter & Ottolenghi, 1960). Moreover, snail extracts are heterogeneous, containing many enzymes, and precautions may be necessary to avoid artifacts from this source. β-Glucanases are available from a number of other sources in much purer form, (Fleet & Phaff, 1974) but so far these have mainly been used as research tools for studying cell wall structure; moreover, efficient protoplast formation in *Sacch. cerevisiae* probably depends on the combined action of several enzymes (Phaff, 1971).

Glucose analogues, 2-deoxyglucose and fluorodeoxyglucoses are effective inhibitors of cell wall synthesis in yeasts (references in Table 5). When growing cells are treated with these compounds cell lysis occurs after several hours, and the points at which lysis is initiated coincide with the regions of synthesis of glucan layers (Johnson, 1968). The response of various species to these analogues differs: 2-deoxyglucose at 250 μg/ml caused lysis in *Schizosacch. pombe, Pichia farinosa* and *Sacch. cerevisiae* (Johnson, 1968), while 2-fluoro-2-deoxyglucose was effective between 50 and 500 μg/ml on *Sacch. cerevisiae* it did not promote lysis in *Schizosacch. pombe* at 500 μg/ml (Biely *et al.*, 1973). These glucose analogues are readily taken up by yeast and phosphorylated (Cramer & Woodward, 1952; Bessell *et al.*, 1972). They therefore have a wide spectrum of effects on cellular physiology other than the inhibition of glucan synthesis. If lysed cultures obtained in the presence of 2-fluoro-2-deoxyglucose are incubated further, resistant mutants rapidly take over the culture (Biely *et al.*, 1973); these are a potential nuisance if these analogues were to find application in cell lysis.

6. Summary

The practical necessity for inhibition or inactivation of yeasts in non-medical situations arises mainly when yeasts form part of a very heterogeneous flora, e.g.

in food preservation, and preparation of pharmaceuticals. Yeasts are only a problem when they can compete favourably with other organisms in the microflora. Particular products which are susceptible to spoilage by yeasts have been discussed in considerable detail elsewhere (Ingram, 1958; Walker & Ayres, 1970). From these, and considering the response of yeasts to various inactivation methods available, it is possible to identify the general conditions which predispose to yeast spoilage and which should be kept in mind when altering existing procedures for processing or distributing a product or producing a new one. These conditions include: (i) where heat treatment is minimal or not applied, or the product is subject to extended storage after opening; (ii) when the pH of the product is low in the absence of inhibitors such as benzoates, sorbate or acetate; (iii) when the major method of preservation involves lowering the a_w by sugars, or in some cases, brines; (iv) when sublethal doses of radiation are given; (v) at low redox potential, particularly in combination with low pH.

In addition there are situations in which yeasts are more resistant to processes normally leading to their inactivation. The nature of the suspending medium has long been known to be an important factor affecting resistance; examples include the effect of lipids in increasing resistance to heat (Ingram, 1958), and reduction in the lethal effects of radiation in the presence of free radical sinks, or by lowering a_w. For most physical and chemical treatments resistance can arise by mutation, and can be either dominant or recessive. Both are a potential problem since even in diploid populations mutations which are recessive can be made homozygous by gene conversion or mitotic recombination events. Finally, the resistance of a particular species of yeast can vary widely as a function of the physiological state of cells within the culture. Growing cells can differ from starved cells, budded cells from unbudded, and vegetative cells from spores.

7. References

ANAND, J. C. & BROWN, A. D. (1968). Growth rate patterns of the so-called osmophilic and non-osmophilic yeasts in solutions of polyethylene glycol. *Journal of General Microbiology* **52**, 205-212.

AVNER, P. R., COEN, D., DUJON, B. & SLONIMSKI, P. P. (1973). Mitochondrial genetics. IV. Allelism and mapping studies of oligomycin resistant mutants in *S. cerevisiae*. *Molecular and General Genetics* **125**, 9-52.

AZUKAS, J. J., COSTILOW, R. N. & SADOFF, H. L. (1961). Inhibition of alcoholic fermentation by sorbic acid. *Journal of Bacteriology* **81**, 189-194.

BATTANER, E. & VAZQUEZ, D. (1971). Inhibition of protein synthesis by ribosomes of the 80S type. *Biochimica et biophysica acta* **254**, 316-330.

BATTLEY, E. H. & BARTLETT, E. J. (1966). A convenient pH-gradient method for the determination of the maximum and minimum pH for microbial growth. *Antonie van Leeuwenhoek* **32**, 245-255.

BEAM, C. A., MORTIMER, R. K., WOLFE, R. G. & TOBIAS, C. A. (1954). The relation of radioresistance to budding in *Saccharomyces cerevisiae*. *Archives of Biochemistry and Biophysics* **49**, 110-122.

BECK, C. & VON MEYENBURG, H. K. (1968). Enzyme pattern and aerobic growth of *Saccharomyces cerevisiae* under varying degrees of glucose limitation. *Journal of Bacteriology* **96**, 479-486.
BESSELL, E. M., FOSTER, A. B. & WESTWOOD, J. H. (1972). The use of deoxyfluoro-D-glucopyranoses and related compounds in the study of yeast hexokinase specificity. *Biochemical Journal* **128**, 199-204.
BEVAN, E. A. & MAKOWER, M. (1963). The physiological basis of killer character in yeast (abstract only). *Proceedings of the 11th International Congress of Genetics* **1**, 203.
BEVAN, E. A., HERRING, H. J. & MITCHELL, D. J. (1973). Preliminary characterisation of two species of dsRNA in yeast and their relationship to the 'killer' character. *Nature, London* **245**, 81-86.
BHARGAVA, M. M. & HALVORSON, H. O. (1971). Isolation of nuclei from yeast. *Journal of Cell Biology* **49**, 423-429.
BIELY, P., KOVARIK, J. & BAUER, S. (1973). Lysis of *Saccharomyces cerevisiae* with 2-deoxy-2-fluoro-D-glucose, an inhibitor of the cell wall glucan synthesis. *Journal of Bacteriology,* **115**, 1108-1120.
BRIDGES, A. E., OLIVO, J. P. & CHANDLER, V. L. (1956). Relative resistances of microorganisms to cathode rays. II. Yeasts and moulds. *Applied Microbiology* **4**, 147-149.
BROWN, A. D. (1974). Microbial water relations: features of the intracellular composition of sugar-tolerant yeasts. *Journal of Bacteriology* **118**, 769-777.
BROWN, A. D. (1975). Microbial water relations. Effects of solute concentration on the respiratory activity of sugar-tolerant and non-tolerant yeasts. *Journal of General Microbiology* **86**, 241-249.
BROWN, M. R. W. & MELLING, J. (1971). Inhibition and destruction of microorganisms by heat. In *Inhibition and Destruction of the Microbial Cell*, ed. Hugo, W. B. London & New York: Academic Press.
BRUCE, A. K. & PARKER, J. H. (1974). Radiation response of energy-deficient yeast mutants: the relationship of radioresistance to respiratory and catalase activities as measured by oxygen polarography. *Canadian Journal of Microbiology* **20**, 587-593.
BURNS, V. W. (1956). X-ray induced division delay of individual yeast cells. *Radiation Research* **4**, 394-412.
BUSSEY, H., SHERMAN, D. & SOMERS, J. M. (1973). Action of yeast killer factor: a resistant mutant with sensitive spheroplasts. *Journal of Bacteriology* **113**, 1193-1197.
CONWAY, E. J. & DOWNEY, M. (1950). An outer metabolic region of the yeast cell. *Biochemical Journal* **47**, 347-355.
CONWAY, E. J. & DUGGAN, F. (1958). A cation carrier in the yeast cell wall. *Biochemical Journal* **69**, 265-274.
COOPER, D., BANTHORPE, D. V. & WILKIE, D. (1967). Modified ribosomes conferring resistance to cycloheximide in mutants of *Saccharomyces cerevisiae*. *Journal of Molecular Biology* **26**, 347-350.
CRAMER, F. B. & WOODWARD, G. E. (1952). 2-deoxy-D-glucose as an antagonist of glucose in yeast fermentation. *Journal of the Franklin Institute* **253**, 354-360.
CRAMER, J. H., BHARGAVA, M. M. & HALVORSON, H. O. (1972). Isolation and characterization of γ DNA of *Saccharomyces cerevisiae*. *Journal of Molecular Biology* **71**, 11-20.
CULBERTSON, M. R. & HENRY, S. A. (1975). Inositol-requiring mutants of *Saccharomyces cerevisiae*. *Genetics, Princeton* **80**, 23-40.
D'ARCY, P. F. (1971). Inhibition and destruction of moulds and yeasts. In *Inhibition and Destruction of the Microbial Cell*, ed. Hugo, W. B. London & New York: Academic Press.
DAWES, I. W. (1975). Study of cell development using derepressed mutations. *Nature, London* **255**, 707-708.
DAWES, I. W. & HARDIE, I. H. (1974). Selective killing of vegetative cells in sporulated yeast cultures by exposure to diethyl ether. *Molecular and General Genetics* **131**, 281-289.

DONKERSLOOT, J. A. & MATELES, R. I. (1968). Enrichment of auxotrophic mutants of *Aspergillus flavus* by tritium suicide. *Journal of Bacteriology* 96, 1551-1555.
DRAKE, S. D., EVANS, J. B. & NIVEN, C. F. (1958). Microbial flora of packaged frankfurters and their radiation resistance. *Food Research* 23, 291-296.
DRAKE, S. D., EVANS, J. B. & NIVEN, C. F. (1959). The identity of yeasts in the surface flora of packaged frankfurters. *Food Research* 24, 243-246.
DUFFUS, J. (1969). Isolation of nuclei from *Schizosaccharomyces pombe*. *Biochimica et biophysica acta* 195, 230-233.
EATON, N. R. (1962). New press for disruption of microorganisms. *Journal of Bacteriology* 83, 1359-1360.
EDDY, A. A. & WILLIAMSON, D. H. (1957). A method of isolating protoplasts from yeast. *Nature, London* 179, 1262-1253.
EDDY, A. A. & WILLIAMSON, D. H. (1959). Formation of aberrant cell walls and of spores by the growing yeast protoplast. *Nature, London* 183, 1101-1108.
EMEIS, V. & GUTZ, H. (1958). Eine einfache Technik für Massen isolation von Hefersporen. *Zeitschrift für Naturforschung* 13b, 647-650.
ESPOSITO, R. E. (1968). Genetic recombination in synchronized cultures of *Saccharomyces cerevisiae*. *Genetics, Princeton* 59, 191-210.
FARRELL, J. & ROSE, A. H. (1965). Low temperature microbiology. *Advances in Aplied Microbiology* 7, 335-378.
FARRELL, J. & ROSE, A. H. (1967). Temperature effects on microorganisms. In *Thermobiology*, ed. Rose, A. H. London: Academic Press.
FLEET, G. H. & PHAFF, H. J. (1974). Lysis of yeast cell walls: glucanases from *Bacillus circulans* W-12. *Journal of Bacteriology* 119, 207-219.
FOWELL, R. R. (1969). Sporulation and hybridization of yeasts. In *The Yeasts*, Vol. I, eds Rose, A. H. & Harrison, J. S. New York & London: Academic Press.
FRY, R. M. (1966). Freezing and drying of bacteria. In *Cryobiology*, ed. Meryman, H. T. New York & London: Academic Press.
GAME, J. C. & COX, B. S. (1971). Allelism tests of mutants affecting sensitivity to radiation in yeast, and proposed nomenclature. *Mutation Research* 12, 328-333.
GAME, J. C. & COX, B. S. (1974). Repair systems in *Saccharomyces*. *Mutation Research* 26, 257-264.
GAME, J. C. & MORTIMER, R. K. (1974). A genetic study of X-ray sensitive mutants in yeast. *Mutation Research* 24, 281-292.
GARVER, J. C. & EPSTEIN, R. L. (1959). Method for rupturing large quantities of microorganisms. *Applied Microbiology* 7, 318-319.
GENTLES, J. C. & LA TOUCHE, C. J. (1969). Yeasts as human and animal pathogens. In *The Yeasts*, Vol. I, eds Rose, A. H. & Harrison, J. S. New York & London: Academic Press.
GIBSON, B. (1973). The effect of high sugar concentrations on the heat resistance of vegetative microorganisms. *Journal of Applied Bacteriology* 36, 365-376.
GOLDBLITH, S. A. (1967). Basic principles of microwaves and recent developments. *Advances in Food Research* 15, 277-301.
GOLDBLITH, S. A. (1971). The inhibition and destruction of the microbial cell by radiations. In *Inhibition and Destruction of the Microbial Cell*, ed. Hugo, W. B. New York & London: Academic Press.
GOTTLIEB, D. & NICOLAS, G. (1967). Mode of action of lomofungin. *Aplied Microbiology* 18, 35-40.
GRANT, P., SÁNCHEZ, L. & JIMÉNEZ, A. (1974). Cryptopleurine resistance: genetic locus for a 40S ribosomal component in *Saccharomyces cerevisiae*. *Journal of Bacteriology* 120, 1308-1314.
GRENSON, M., MOUSSET, M., WIAME, J. M. & BECHT, J. (1966). Multiplicity of the amino acid permeases in *Saccharomyces cerevisiae*, *Biochimica et biophysica acta* 127, 325-338.
HAGEN, P-O. (1971). The effect of low temperature on microorganisms: conditions under

which cold becomes lethal. In *Inhibition and Destruction of the Microbial Cell*, ed. Hugo, W. B. London & New York: Academic Press.
HAMILTON-MILLER, J. M. T. (1973). Chemistry and biology of the polyene macrolide antibiotics. *Bacteriological Reviews* 37, 166-196.
HARTWELL, L. (1967). Macromolecule synthesis in temperature-sensitive mutants of yeast. *Journal of Bacteriology* 93, 1662-1670.
HENRY, S. A. (1973). Death resulting from fatty acid starvation in yeast. *Journal of Bacteriology*, 116, 1293-1303.
HENRY, S. A., DONAHUE, T. F. & CULBERTSON, M. R. (1975). Selection of spontaneous mutants by inositol starvation in yeast. *Molecular and General Genetics* (in press).
HOFFMAN, C., SCHWEITZER, T. R. & DALBY, G. (1938). Fungistatic properties of fatty acids and possible biochemical significance. *Food Research* 4, 539-545.
HOLTER, H. OTTOLENGHI, P. (1960). Observations on yeast protoplasts. *Comptes rendus des travaux du Laboratoire de Carlsberg, Série physiologique* 31, 409-422.
HUGHES, D. E. (1951). A press for disrupting bacteria and other microorganisms. *British Journal of Experimental Pathology* 32, 97-109.
HUGHES, D. E., WIMPENNY, J. W. T. & LLOYD, D. (1971). The disintegration of micro-organisms. In *Methods in Microbiology* 5B, eds Norris, J. R. & Ribbons, D. W. London & New York: Academic Press.
HUGO, W. B. (1971). *Inhibition and Destruction of the Microbial Cell.* London & New York: Academic Press.
INGRAM, M. (1957). Microorganisms resisting high concentrations of sugars or salts. *Symposium of the Society for General Microbiology* 7, 90-133.
INGRAM, M. (1958). Yeasts in food spoilage. In *The Chemistry and Biology of Yeasts,* ed. Cooke, A. H. New York & London: Academic Press.
JAMES, A. P. & WERNER, M. M. (1965). The radiobiology of yeast. *Radiation Botany* 5, 359-382.
JAY, J. M. (1970). *Modern Food Microbiology.* New York: Van Nostrand Rheinhold.
JIMÉNEZ, A., LITTLEWOOD, B. & DAVIES, J. E. (1972). Inhibition of protein synthesis in yeast. In *Molecular Mechanism of Antibiotic Action on Protein Biosynthesis and Membranes,* eds Munoz, E. Garcia-Ferrandez, F. & Vazquez, D. Amsterdam: Elsevier.
JOHNSON, B. F. (1968). Lysis of yeast cell walls induced by 2-deoxyglucose at their sites of glucan synthesis. *Journal of Bacteriology* 95, 1169-1172.
KOH, W. Y., MOREHOUSE, C. T. & CHANDLER, V. L. (1956). Relative resistances of microorganisms to cathode rays. I. Nonsporeforming bacteria. *Applied Microbiology* 4, 143-146.
KUENZI, M. & FIECHTER, A. (1969). Changes in carbohydrate composition and trehalase activity during the budding cycle of *Saccharomyces cerevisiae. Archiv für Mikrobiologie* 64, 396407.
KURAISHI, H., TAKAMURA, Y., MIZUNAGA, T. & UEMURA, T. (1971). Factors affecting the death of biotin deficient yeast cells. *Journal of General and Applied Microbiology* 17, 29-42.
KUSHNER, D. J. (1971). Influence of solutes and ions on microorganisms. In *Inhibition and Destruction of the Microbial Cell,* ed. Hugo, W. B. New York & London: Academic Press.
LAMPEN, J. O. (1966). Interference by polyenic antifungal antibiotics (especially nystatin and filipin) with specific membrane functions. *Symposium of the Society for General Microbiology* 16, 111-130.
LASKOWSKI, W. (1960). Inaktivier ungsversuche mit homozygoten Hefestämmen vershiedenen Ploidiegrades. Il Aufbau homozygoter Stämme und Dosiseffecktkurven für ionisierende Strahlen, UV und organische Peroxyde. *Zeitschrift für Naturforschung* 15b, 495-506.

LESTER, H. E. & GROSS, S. R. (1959). Efficient method for selection of auxotrophic mutants of *Neurospora*. *Science, New York* **129**, 572.

LEUPOLD, U. (1957). Physiologisch-genetische Studien an adenin-abhängigen Mutaten von *Schizosaccharomyces pombe*. Ein Beitrag zum Problem der Pseudoallelie. *Schweizerische Zeitschrift für allgemeine Pathologie und Bakteriologie* (Now *Pathologia et microbiologia*) **20**, 535-544.

LITTLEWOOD, B. E. (1972). A method for obtaining antibiotic-sensitive mutants in *Saccharomyces cerevisiae*. *Genetics* **71**, 305-308.

LITTLEWOOD, B. E. & DAVIES, J. E. (1973). Enrichment for temperature-sensitive and auxotrophic mutants in *Saccharomyces cerevisiae*. *Mutation Research* **17**, 315-322.

LUBIN, M. (1959). Selection of auxotrophic bacterial mutants by tritium-labeled thymidine. *Science, New York* **129**, 838-839.

LUND, A. (1951). Some beer-spoilage yeasts and their heat resistance. *Journal of the Institute of Brewing* **57**, 36-41.

LUND, A. (1958). Ecology of yeasts. In *The Chemistry and Biology of Yeasts*, ed. Cooke, A. H. New York & London: Academic Press.

MAESEN, T. J. M. & LAKO, E. (1952). The influence of acetate on the fermentation of bakers' yeast. *Biochimica et biophysica acta* **9**, 106-107.

MAZUR, P. & SCHMIDT, J. J. (1968). Interactions of cooling velocity, temperature, and warming velocity on the survival of frozen and thawed yeast. *Cryobiology* **5**, 1-17.

McCORMACK, G. (1950). 'Pink yeast' isolated from oysters growing at temperatures below freezing. *Commercial Fisheries Review* **12**, 28.

MEGNET, R. (1965). Effect of 2-deoxyglucose on *Schizosaccharomyces pombe*. *Journal of Bacteriology* **90**, 1032-1035.

MELNICK, D., LUCKMANN, F. H. & GOODING, C. M. (1954). Sorbic acid as a fungistatic agent for foods. VI. Metabolic degradation of sorbic acid in cheese by molds and the mechanism of mold inhibition. *Food Research* **19**, 44-58.

MITCHELL, J. H. & ENRIGHT, J. J. (1957). Effect of low moisture levels on the thermostability of active dry yeast. *Food Technology, Champaign* **11**, 359-362.

MORTIMER, R. K. (1958). Radiobiological and genetic studies of a polyploid series (haploid to hexaploid) of *Saccharomyces cerevisiae*. *Radiation Research* **9**, 312-326.

MORTIMER, R. K. & HAWTHORNE, D. C. (1969). Yeast genetics. In *The Yeasts* Vol. I, eds Rose, A. H. & Harrison, J. S. New York & London: Academic Press.

MOSSEL, D. A. A. & SAND, F. E. M. J. (1968). Occurrence and prevention of microbial deterioration of confectionary products. *Conserva, Den Haag* **17**, 23-32.

MOUSTACCHI, E. (1965). Induction by physical and chemical agents of mutations for radioresistance in *Saccharomyces cerevisiae*. *Mutation Research* **2**, 403-412.

NEAL, A. L., WEINSTOCK, J. O. & LAMPEN, J. O. (1965). Mechanisms of fatty acid toxicity for yeast. *Journal of Bacteriology* **90**, 12631.

NELSON, F. E. (1972). Plating medium pH as a factor in apparent survival of sublethally stressed yeasts. *Applied Microbiology* **24**, 236-239.

NOVOTNÝ, P. (1964). A simple rotary disintegrator for microorganisms and animal tissues. *Nature, London* **202**, 363-366.

ONISHI, H. (1963). Osmophilic Yeasts. *Advances in Food Research* **12**, 53-92.

OVERCAST, W. W. & WEAKLEY, D. J. (1969). An aureomycin-rose bengal agar for enumeration of yeast and mold in cottage cheese. *Journal of Milk and Food Technology* **32**, 442-445.

PALLERONI, N. J. & DE PRITZ, M. J. R. (1960). Influence of sorbic acid on acetate oxidation by *Saccharomyces cerevisiae* var. *ellipsoideus*. *Nature, London* **185**, 688-689.

PARRY, E. M. (1969). The effects of UV-light post-treatments on the survival characteristics of 21 UV-sensitive mutants of *Saccharomyces cerevisiae*. *Mutation Research* **8**, 545-556.

PATEL, P. V. & JOHNSTON, J. R. (1968). Dominant mutation for nystatin resistance in yeast. *Applied Microbiology* **16**, 164-165.

PHAFF, H. J. (1971). Structure and biosynthesis of the yeast cell envelope. In *The Yeasts*, Vol II, eds Rose, A. H. & Harrison, J. S. London & New York: Academic Press.
PLISCHKE, M. E., VON BORSTEL, R. C., MORTIMER, R. K. & COHN, W. E. (1975). Genetic markers and associated gene products in *Saccharomyces cerevisiae*. In *Handbook of Biochemistry and Molecular Biology*, 3rd edn, ed. Fasman, G. D. Cleveland: Chemical Rubber Co. Press.
PONTECORVO, G., ROPER, J. A., HEMMONS, L. M., MACDONALD, K. D. & BUFTON, A. W. J. (1953). The genetics of *Aspergillus nidulans*. *Advances in Genetics* **5**, 141-238.
PRINCE, H. N. (1959). Effect of pH on the antifungal activity of undecylenic acid and its calcium salt. *Journal of Bacteriology* **78**, 78891.
PUT, H. M. C. & SAND, F. E. M. J. (1975). Some notes on the heat resistance of ascosporogenous yeasts. *Journal of Applied Bacteriology* **39** (3), iii.
RECCA, J. & MRAK, E. M. (1952). Yeasts occurring in citrus products. *Food Technology, Champaign* **6**, 450-454.
ŘEHÁČEK, J., BERAN, K. & BIČÍK, V. (1969). Disintegration of microorganisms and preparation of yeast cell walls in a new type of disintegrator. *Applied Microbiology* **17**, 462-466.
RESNICK, M. A. (1969). A photoreactivationless mutant of *Saccharomyces cerevisiae*. *Genetics* **62**, 51931.
RESNICK, M. A. & SETLOW, J. K. (1972). Repair of pyrimidine dimer damage induced in yeast by ultraviolet light. *Journal of Bacteriology* **109**, 979-986.
RESNICK, M., TIPPETS, R. D. & MORTIMER, R. K. (1967). Separation of spores from diploid cells of yeast by stable flow free-boundary electrophoresis. *Science, New York* **158**, 80304.
RIDGWAY, C. J. & DOUGLAS, H. C. (1958). Unbalanced growth of yeast due to inositol deficiency. *Journal of Bacteriology* **76**, 163-166.
ROSE, D. (1970). Some factors influencing the survival of freeze-dried yeast cultures. *Journal of Applied Bacteriology* **33**, 228-232.
ROSE, D. (1972). Fermentation of cane molasses by yeasts after preservation under conditions of high sugar concentration. *Journal of Applied Bacteriology* **55**, 499-503.
ROUSSEAU, P. & HALVORSON, H. O. (1969). Preparation and storage of single spores of *Saccharomyces cerevisiae*. *Journal of Bacteriology* **100**, 1426-1427.
ROZIJN, T. H. & TONINO, G. J. M. (1964). Studies on the yeast nucleus: the isolation of nuclei. *Biochimica et biophysica acta* **91**, 105-112.
SCHINDLER, D., GRANT, P. & DAVIES, J. E. (1974). Trichodermin resistance, a mutation affecting eukaryotic ribosomes. *Nature, London* **248**, 535-36.
SCOTT, W. J. (1957). Water relations of food spoilage microorganisms. *Advances in Food Research* **7**, 83-127.
SHERMAN, F. (1959). The effects of elevated temperatures on yeast. II. Induction of respiratory-deficient mutants. *Journal of Cellular and Comparative Physiology* **54**, 372.
SHERMAN, F. & ROMAN, H. (1963). Evidence for two types of allelic recombination in yeast. *Genetics, Princeton* **48**, 255-261.
SHIMADA, S., KURAISHI, H. & AIDA, K. (1972). Unbalanced growth death of yeast due to pantothenate deficiency. *Journal of General and Applied Microbiology* **18**, 383-397.
SILVERMAN, G. J. & SINSKEY, T. J. (1968). The destruction of microorganisms by ionizing irradiation. In *Disinfection, Sterilization and Preservation*, eds Lawrence, C. A. & Block, S. S. Philadelphia: Lea & Febiger.
SKOGERSON, L., MCLAUGHLIN, C. & WAKATAMA, E. (1973). Modification of ribosomes in cryptopleurine-resistant mutants of yeast. *Journal of Bacteriology* **116**, 818-822.
SLATER, M. L. (1973). Effect of reversible inhibition of deoxyribonucleic acid synthesis

on the yeast cell cycle. *Journal of Bacteriology* **113**, 263-270.
SNOW, R. (1966). An enrichment method for auxotrophic yeast mutants using the antibiotic 'Nystatin'. *Nature, London* **211**, 20607.
SOMERS, J. & BEVAN, E. A. (1969). The inheritance of the killer character in yeast. *Genetical Research* **13**, 71-83.
STOKES, J. L. (1971). Influence of temperature on the growth and metabolism of yeasts. In *The Yeasts*, Vol. II, eds Rose, A. H. & Harrison, J. S. New York & London: Academic Press.
SUOMALAINEN, H. & OURA, E. (1971). Yeast nutrition and solute uptake. In *The Yeasts*, Vol II, eds Rose, A. H. & Harrison, J. S. New York & London: Academic Press.
SWANN, M. M. (1962). Gene replication, ultraviolet sensitivity and the cell cycle. *Nature, London* **193**, 1222-1227.
SYKES, G. (1965). *Disinfection and Sterilization*, 2nd edn. London: E. & F. N. Spon Ltd.
TANNER, F. W. & WILLIAMSON, B. W. (1928). The effect of freezing on yeasts. *Proceedings of the Society for Experimental Biology and Medicine, New York* **25**, 377-381.
TAURO, P., HALVORSON, H. O. & EPSTEIN, R. L. (1968). Time of gene expression in relation to centromere distance during the cell cycle of *Saccharomyces cerevisiae*. *Proceedings of the National Academy of Sciences, USA* **59**, 277-284.
UNRAU, P., WHEATCROFT, R. & COX, B. S. (1971). The excision of pyrimidine dimers from DNA of ultraviolet irradiated yeast. *Molecular and General Genetics* **113**, 359362.
VAN UDEN, N. (1971). Kinetics and energetics of yeast growth. In *The Yeasts*, Vol. II, eds Rose, A. H. & Harrison, J. S. New York & London: Academic Press.
WALKER, H. W. & AYERS, J. C. (1970). Yeasts as spoilage organisms. In *The Yeasts*, Vol. III, eds Rose, A. H. & Harrison, J. S. New York & London: Academic Press.
WALTER, P., LARDY, H. A. & JOHNSON, D. (1967). Antibiotics as tools for metabolic studies. X. Inhibition of phosphoryl transfer reaction in mitochondria by peliomycin, ossamycin and venturicidin. *Journal of Biological Chemistry* **242**, 5014-5018.
WICKERHAM, L. J. & BURTON, K. A. (1954). A simple technique for obtaining mating types in heterothallic diploid yeasts, with special reference to their uses in the genus Hansenula. *Journal of Bacteriology* **67**, 303-308.
WILLIAMSON, D. H. (1970). The effect of environmental and genetic factors on the replication of mitochondrial DNA in yeast. *Symposium of the Society for Experimental Biology* **24**, 247-276.
WILLIAMSON, D. H., MOUSTACCHI, E. & FENNELL, D. (1971). A procedure for rapidly extracting and estimating the nuclear and cytoplasmic DNA contents of yeast cells. *Biochimica et biophysica acta* **238**, 369-374.
WINTERSBERGER, U. & WINTERSBERGER, E. (1970). Studies on deoxyribonucleic acid polymerases from yeast. 2. Partial purification and characterization of mitochondrial DNA polymerase from wild type and respiration-deficient yeast cells. *European Journal of Biochemistry* **13**, 20-27.
WITTER, L. M., BERRY, J. M. & FOLINAZZO, J. F. (1958). The viability of *Escherichia coli* and spoilage yeast in carbonated beverages. *Food Research* **23**, 133-142.
WOLF, K., DUJON, B. & SLONIMSKI, P. P. (1973). Mitochondrial genetics. V. Multifactorial crosses involving a mutation conferring paromomycin-resistance in *Saccharomyces cerevisiae*. *Molecular and General Genetics* **115**, 530.
WOOD, W. A. (1966). Carbohydrate metabolism. In *Methods in Enzymology*, Vol. IX, eds Colowick, S. P. & Kaplan, H. O. New York & London: Academic Press.
WOODS, D. R. & BEVAN, E. A. (1968). Studies on the nature of the killer factor produced by *Saccharomyces cerevisiae*. *Journal of General Microbiology* **51**, 115-126.
ZAKHAROV, I. A. & INGE-VECHTOMOV, S. G. (1964). Ascospore isolation of yeast for genetic analysis without a micromanipulator. *Issledovaniya po Genetike* (English translation) **2**, 134-139.

The Survival of Bacteria in Toiletries

S. A. MALCOLM

*Unilever Research, Isleworth Laboratory, Unilever Limited,
455 London Road, Isleworth, Middlesex, England*

CONTENTS

1. Introduction 305
2. Contamination 306
   - (a) Definition 306
   - (b) Influence of product type 306
   - (c) Incidence rates of contamination 307
3. Origins of contamination 308
4. Consequences of contamination 310
   - (a) The health hazard 310
   - (b) Product spoilage 311
5. Microbiological standards for toiletries 311
6. Conclusion 314
7. References 314

1. Introduction

IN A MONOGRAPH entitled 'The hygienic manufacture and preservation of toiletries and cosmetics' published by the Society of Cosmetic Chemists of Great Britain (*Anon.*, 1970), cosmetics and toiletries are defined as substances or preparations (i) intended to cleanse, beautify or modify the appearance of a person by external application to the skin, nails, hair, eyes or the oral cavity (but not intended to be swallowed); or (ii) which exert a non-systemic action on or modification of local physiological functions so as to prevent, reduce or correct minor undesirable surface conditions, blemishes or defects of the skin, nails, hair, eyes or the oral cavity (but are not intended to be swallowed). In the remainder of this paper toiletries and cosmetics are taken to be synonymous and both will be referred to as toiletries.

The number of different toiletry products available is very large as is the range of raw materials used in their manufacture. Therefore toiletries vary widely in the types of environment which they provide for micro-organisms, ranging from the hostile conditions of an anti-perspirant (an oily suspension of aluminium chlorhydrate in an aerosol can) to the hospitable conditions of a skin-cream (a mixture of often nutritious ingredients in an aqueous environment). Consequently generalized statements about the microbiology of toiletries can be misleading. Unfortunately, this does not mean that they are not made!

The fact that bacteria can survive in toiletries has been known for many years but it is only comparatively recently that the contaminated toiletry has become a matter for the serious concern of manufacturers. However, despite the large and ever increasing amount of literature on the subject there has been, as yet, no clear definition of the problem. It is hoped that this paper will go at least a small way towards remedying this situation.

2. Contamination

We exist in an environment in which almost everything we eat, drink, touch and see bears viable bacteria, i.e. is contaminated. Yet nearly all of these things are quite acceptable to us because the bacteria which are present will harm neither us nor the article carrying them; contamination is undesirable only if it results in spoilage of the contaminated material or if it can act as a source of infection. Thus a toiletry containing bacteria which are harmless to both the product and the consumer should be considered as acceptable. This is not always the case and the presence of bacteria in a toiletry is often considered automatically, both in the literature and elsewhere, as undesirable.

(a) *Definition*

If the word 'contaminated' is to be applied to any toiletry product containing one bacterium (or more) then to avoid confusion it is necessary to make a qualification. This can best be done by considering the types of contamination which may occur, namely static and dynamic contamination. Static contamination describes the situation in which a product contains non-metabolizing bacteria whilst dynamic contamination exists where the organisms are actively metabolizing. In the former situation, the numbers of bacteria in the product will remain constant (assuming that death is not occurring) whilst in the latter situation the numbers may increase.

(b) *Influence of product type*

Since it is the environment which determines the ability of bacteria to metabolize it follows that the type of contamination, static or dynamic, will be controlled by the nature of the toiletry product. For example, in anhydrous systems such as oils, talcs, face powders etc. bacteria are unable to metabolize and consequently only static contamination is possible. Similarly in aqueous products containing materials which are inhibitory to bacteria, e.g. alcohols, glycols, germicides, preservatives and high concentrations of humectants, only static contamination may occur. However, in aqueous products which do not contain inhibitory materials both static and dynamic contamination are possible.

(c) *Incidence rates of contamination*

In our experience static contamination of toiletries is most likely to be due to the presence of Gram positive spore-forming bacteria and other Gram positive rods and cocci and one might expect that, considering the ubiquitous nature of bacteria and the lack of asepsis in the production of toiletries, all toiletry products would demonstrate static contamination. However, this does not appear to be the case although the reported rates of incidence of contamination in toiletries vary widely. The most recently published survey is that which is currently being performed by manufacturers of toiletries and cosmetics in the U.S.A. This survey was initiated by and is under the supervision of the Microbial Content Subcommittee of the Cosmetics, Toiletries and Fragrance Association (CTFA) of the U.S.A. An interim report (*Anon.*, 1975) states that of 2680 products so far examined only 2.2% have been found to be contaminated. Similarly low incidence rates have been recorded by Wilson *et al.* (1971) and Wolven & Levenstein (1972) who found 3.5% of 58 products and 3.6% of 228 products, respectively, to be contaminated. Other surveys have indicated much higher contamination rates. Thus, Heiss (1967) found that 43% of 60 skin-creams were contaminated whilst Baird, R. M. (pers. comm., 1975) has shown a contamination rate of *c.* 29% and Dunnigan & Evans (1970), performing a survey initiated by the FDA, found that *c.* 20% of 169 creams and lotions were contaminated.

The very large discrepancies in the findings of the different workers may be ascribed, at least in part, to the fact that they have used different definitions of contamination and different methods of evaluating the microbial status of the products they were examining. Thus it seems that whilst Heiss, Dunnigan & Evans and Baird have recorded a product as contaminated if a positive result was obtained with sterility-type or enrichment tests others have used only plate counts to check for contamination (Wilson *et al.*, Wolven & Levenstein and CTFA). In these latter cases the recorded incidence will depend on the worker's attitude to small numbers of colonies on the plates. He may choose to disregard them as unimportant or to consider them as evidence of contamination. It is reasonable to assume that a higher incidence of contamination will be recorded if a sterility-type test is used rather than a plate count. Indeed Lühdorf & Tybring (1969) showed that 23.8% of pharmaceutical liquids were contaminated when examined by a plate count but that the incidence rate was 46.0% when samples were tested by membrane filtration. This was largely due to the failure of the plate count method to reveal the occurrence of a large number of products containing between one and 10 micro-organisms/g. Lühdorf & Tybring used these two methods because membrane filtration was not suitable for testing emulsions and suspensions, so the results obtained are not strictly comparable, having been obtained in different ways for different products. It would, however, be surprising if a large proportion of pharmaceutical solutions contained between one and 10 bacteria/g whilst emulsions and suspensions did not.

From the above it would seem that if contamination is defined simply as the presence of viable micro-organisms (i.e. including both static and dynamic contamination) then the contamination rate of toiletries is $c.$ 20-40%. Unfortunately, no author has drawn a distinction between static and dynamic contamination nor do most surveys indicate numbers of micro-organisms recovered, or if they do then the type recovered is not specified (the survey by Dunnigan & Evans, 1970, is an exception). Thus it is difficult to draw conclusions about the rate of incidence of dynamic contamination. It is well accepted, however, that Gram negative bacteria are the only types which will multiply in toiletries; thus if a product is found to contain a large number of Gram negative bacteria of a single species it is good evidence for dynamic contamination of that product. Jarvis et al. (1974) stated that the predominant micro-organisms isolated from high count products were Gram negative rods. These authors do not define 'high count' but if we assume it to mean $> 10^4$ bacteria/g then they observed a dynamic contamination rate of 2.6%. Baird (1975) found that 10 (i.e. 4.7%) of 212 products which she examined contained large numbers of Gram negative bacteria. A much lower rate of dynamic contamination seems to have been observed in the CTFA survey (Anon., 1975), only four (i.e. 0.15%) of 2680 products being found to contain Gram negative bacteria.

As has been stated earlier susceptibility to contamination will be a function of the product formulation and therefore when comparing reported incidences of contamination it is necessary to consider the types of product included in the survey. A survey made of toiletries which are largely aqueous in nature will certainly give different results from a survey of products many of which are anhydrous; therefore any estimate of the current microbiological status of toiletries should be quoted separately for each product type.

3. Origins of Contamination

All toiletry products will become contaminated during manufacture and packing. The sources of contaminants will be varied but most important of these are the raw materials. Pedersen & Ulrich (1968) examined 226 batches of 84 different pharmaceutical raw materials and found that 51% of the raw materials were contaminated in one or more batches tested and that 14% of all batches contained 1000 or more micro-organisms/g. Westwood & Pin-Lim (1971) examined 410 batches of 22 different pharmaceutical raw materials and found that all were contaminated in one or more batches and that 19% of all batches tested contained more than 1000 bacteria/g. Although the author is not aware of a similarly exhaustive survey of raw materials for toiletries it is reasonable to believe that similarly high incidence rates of contamination might be expected to occur, particularly in materials of natural origin.

Water is also a raw material and certainly the most widely used in toiletries. Unless sterilized it will always contain bacteria and may, if stored before use (Schuster & Modde, 1968) or passed through a deionizer (Eisman *et al.*, 1949), contain millions of bacteria/ml. Other sources of contamination will be the manufacturing plant, airborne bacteria, operatives and packaging.

Contamination from all of the above sources can be reduced by using raw materials of good microbiological quality and by efficient factory hygiene, but it will never by practicable to eliminate all sources of contamination. However, from the published results of others and from the author's own experience it is clear that many products are free from viable micro-organisms. It follows therefore that many of the contaminants must be killed by product processing and by the product ingredients.

Heat sufficient to kill most vegetative bacteria is frequently used in the processing of toiletries, and the wide use of anionic surface active agents in toiletries may also account for the death of vegetative cells, especially of the Gram positive bacteria (Baker *et al.*, 1941). Gram positive spores would be expected to survive the heating phase of processing but may suffer 'lethal germination' (Rode & Foster, 1960) in the presence of surface active agents. The resistant spore is induced to germinate by the surface active agent which then kills the non-resistant vegetative cell. A wide range of other ingredients will also be lethal to bacteria. For example, colourants, perfumes, alcohols, glycols, aluminium salts, zinc salts etc. may all be capable of eliminating bacteria added unintentionally during manufacture. The physical properties of the product such as reduced water activity and high or low pH can also be lethal to bacteria. (The author has, however, encountered products at pH 3.5 and pH 10 which have been heavily contaminated with Gram negative bacteria.)

Except where raw materials of a natural origin such as talc are used, static contamination usually occurs at a level of < 100 micro-organisms/g and typically the organisms present represent a variety of different species. However, when the bacteria in a dynamically contaminated product have reached their maximum population the numbers are usually very high, i.e. 10^4 bacteria/g or more and are often in pure culture. These bacteria are almost invariably Gram negative.

Thus the bacteria present in a statically contaminated product are those of the original contaminants which have survived the rigours of processing or have been added during packing, whilst those in the dynamically contaminated product have resulted from the multiplication of perhaps only a relatively small number of bacteria added during manufacture (Malcolm & Woodroffe, 1975). This has implications for factory hygiene. Any improvement in the microbiological quality of raw materials and in factory hygiene is likely to be reflected by a reduction in the number of static contaminants in a product but it may not necessarily have any effect on the number of dynamic contaminants unless the improvements have been sufficient to eliminate all the organisms capable of multiplication in the product.

4. Consequences of Contamination

Baker (1959) has described bacteria as "that unwanted cosmetic ingredient", and many who work in the toiletry industry would agree with him. However, it is important to consider under what circumstances and for what reasons bacteria are undesirable. The first consideration must be the health of the consumer and whether a contaminated product can give rise to infection. The second consideration is product stability.

(a) *The health hazard*

In theory both static and dynamic contamination may pose a threat to the health of the consumer and indeed the only two published cases where a connection between an infection and the use of a contaminated toiletry product have been demonstrated provide examples of each type of contamination. In 1946, cases of tetanus in the newborn were observed in New Zealand following the use of talcum powder containing spores of *Clostridium tetani* (Tremewan, 1946) whilst Morse *et al.* (1967) found that a strain of *Klebsiella pneumoniae* involved in an outbreak of septicaemia in patients with intravenous catheters, was present in large numbers in a hand-cream dispenser. Parker (1972) has claimed to be aware of several outbreaks of neonatal otitis in British hospitals in which the causative organism was found in antiseptic skin-care lotions. Strictly speaking, this preparation is not a toiletry, but it would have a similar formulation and usage pattern.

Parker (1972) has also observed that most infections attributable to medicaments are caused by Gram negative bacteria. This observation is supported by the literature. Of 13 reports of infections or colonizations associated with the use of non-sterile preparations (pharmaceuticals, detergents and disinfectants) in hospitals, all involve Gram negative bacteria (Kunz & Ouchterlony, 1955; Pyrah *et al.*, 1955; Noble & Savin, 1966; Phillips, 1966; Plotkin & McKitrick, 1966; Lang *et al.*, 1967; Mertz *et al.*, 1967; Victorin, 1967; Shooter *et al.*, 1969; Bassett *et al.*, 1970, Cooke *et al.*, 1970; Bassett *et al.*, 1973; Schaffner *et al.*, 1973) and all but two (Kunz & Ouchterlony, 1955; Lang *et al.*, 1967) appear to be instances where multiplication to high numbers has occurred within the preparation, i.e. dynamic contamination. Thus if contaminated toiletries were to be a hazard to the health of the consumer then it seems that it would be the dynamically contaminated product which would pose the greatest threat.

There are few recorded instances of the isolation of pathogens, other than *Pseudomonas aeruginosa*, from toiletries. We are aware of only one report of the recovery of *Staphylococcus aureus* from a toiletry (Heiss, 1967) although other workers have reported the isolation of other Gram positive cocci (Wolven & Levenstein, 1969; Dunnigan & Evans, 1970; Myers & Pasutto, 1973). There have, however, been at least two reports of the recovery of *Staph. aureus* from pharmaceutical preparations (*Anon.*, 1971; Lambin *et al.*, 1972).

Heiss (1967) and Myers & Pasutto (1973) have reported the presence of *Streptococcus* spp. in toiletry creams but only on a single occasion each.

Except where grossly contaminated raw materials are used static contamination will be at a relatively low level and comparable with that of the environment. Consequently it is hard to believe that these products can be regarded as a significant threat to the health of the consumer. Even if pathogens were present they would be present in such small numbers as to be unlikely to give rise to infection under normal conditions of use.

Dynamically contaminated products, however, may contain millions of bacteria/g and will therefore present a challenge to the body surfaces of the consumer which is much greater than will occur as a result of contact with most other materials in the environment. It is also known that the opportunist pathogen *Ps. aeruginosa* will thrive in toiletries (Tenenbaum, 1967) and so presumably may other pathogenic members of this genus. Therefore the presence of bacteria in a product in which they are capable of multiplication must be regarded as undesirable.

(b) *Product spoilage*

Product spoilage where the presence of contaminants results in changes in the product which are perceivable by the consumer is clearly undesirable. These changes can only occur in the presence of actively metabolizing organisms, therefore only the dynamically contaminated product can become spoiled. Obvious spoilage is, however, an infrequent occurrence and the majority of dynamically contaminated products even though contaminated with large numbers of bacteria will be indistinguishable from uncontaminated products.

5. Microbiological Standards for Toiletries

The manufacturer will impose microbiological standards to protect the good name of the company since a spoiled product will damage the product image as would a product which initiates infection. There are as yet no legally imposed microbiological standards for toiletries but if legislation is passed in the future it will be concerned not with product spoilage but with the potential hazard of toiletries to the health of the consumer.

There are at the moment two schools of thought on which is the most appropriate standard to impose on toiletries in order to protect the consumer. The first recommends that the standard should define the maximum number of organisms permitted in a given weight or volume of product, whilst the second recommends that the standard should demand the absence of certain named pathogens from a given weight or volume.

The numerical limit is criticized on the grounds that total viable counts on toiletries have poor reproducibility and that it would permit the sale of products containing pathogenic micro-organisms.

A standard which would restrict the types of organisms present has been criticized because it is difficult or impossible to construct a meaningful list of pathogens and because the microbiological techniques required would make it difficult for many manufacturers to be able to check to ensure that their products met the standard. Since poor reproducibility of results is a measure of the difficulties of sampling and separation of bacteria from toiletry products it follows that the criticism of poor reproducibility can also be levelled at tests for the absence of named organisms.

A useful compromise between the different forms of standard might be the absence of a limited number of named organisms from a limited number of product types together with a numerical limit for all other products. An example would be the absence of *Ps. aeruginosa* from baby products and products for use in and around the eye and absence of *Cl. tetani* from talc, together with a limit of not more than 1000 micro-organisms/g for other products.

All the currently proposed numerical limits for toiletries have been selected on an empirical basis since at the moment there is insufficient evidence to decide what is the maximum number of micro-organisms which may be safely applied to the body surfaces. This number will vary with the type of organism as well as with the type of product and the limit should also, ideally, be determined by taking into consideration the quantity of product to be applied. It seems obvious that no practical standard, even if the information were available, could encompass all these parameters. Therefore what is required of the standard is that it distinguishes between those products having a microbial flora quantitatively similar to that of the environment and those which contain numbers of organisms in excess of normal environmental contamination. The most commonly proposed numerical limit for toiletries is 1000 organisms/g. This limit is of value in that it is sufficiently high to prevent the rejection of most statically contaminated products whilst at the same time it is considerably lower than the number of organisms achieved in most dynamically contaminated products.

As important as the value of the numerical limit is the time after manufacture at which the product is tested for compliance with the limit. In most products the period immediately following manufacture will be a time of change; some micro-organisms will be dying and others will be multiplying. The time taken to reach an equilibrium situation will vary from product to product. In products which are inhibitory to micro-organisms the equilibrium might be reached within minutes or even seconds of packing into the final container. With products in which multiplication can occur it may be two to three weeks before the stationary phase population is achieved (this of course is not a true equilibrium situation since the stationary phase will pass into a decline phase and eventual death of all organisms, but this process may take anything from a few weeks to more than two years).

Clearly any estimate of either the types or numbers of organism present which is made before equilibrium is reached will provide little information about the quality of the product as it will be when purchased by the consumer. It is debatable whether the manufacturer should be expected to be responsible for the microbiological status of a product once the consumer starts to use it, but assuming normal handling and storage it must be the manufacturer's responsibility until the consumer purchases it. Therefore if a microbiological standard is imposed on toiletries it is likely that it would be expected to be applicable until that time. Any agency whose function is to check products to ensure that they comply with prescribed standards will select products from the market place for testing and is therefore likely to have a good estimate of the status of the product when it is eventually purchased. The manufacturer, however, must perform his examination shortly after the product is produced and will have to rely on experience to interpret the significance of the results.

In products where only static contamination is possible and multiplication cannot occur it will not be difficult to interpret the results in relation to a numerical limit, since any change in the numbers of organisms present after manufacture will be small. But if the standard is one which demands the absence of named organisms then the manufacturer may, for example, have to decide whether the low number of prescribed organisms found shortly after manufacture will still be viable when the product comes onto the market. He may in fact have to reject products which would, in time, have come within the standard.

With products capable of dynamic contamination the problem is even more difficult, since it will be a matter of judgment as to whether or not, for example, a low count recorded a few days after manufacture indicates a satisfactory or unsatisfactory product or whether or not a test showing the absence of *Ps. aeruginosa* from one gram of product can be repeated two weeks or two months later with the same negative result.

Whether or not legal standards are imposed on toiletries it is clear that the toiletry industry must ensure that its products are microbiologically safe. Decisions about safety will have to be based on considerations of the types and numbers of bacteria present, the type of product and past experience.

6. Conclusion

The concern now being shown by manufacturers of toiletries for the microbiological quality of their products must be welcomed. However, it is important to realize that despite the fact that toiletries are used every day by almost everybody there is little evidence to suggest that they are responsible for initiating infections. This is due to the good microbiological quality of the majority of toiletries and perhaps also because contaminated toiletries are unlikely to act as a source of infection. However, since it will always be difficult to obtain evidence to show that contaminated toiletries cannot or are unlikely to give rise to disease it is important that manufacturers should maintain their vigilance.

7. References

ANON. (1970). The hygienic manufacture and preservation of toiletries and cosmetics. *The Journal of the Society of Cosmetic Chemists* **21**, 719-800.

ANON. (1971). (P.H.L.S. Working Party). Microbial contamination of medicines administered to hospital patients. *The Pharmaceutical Journal* **207**, 96-99.

ANON. (1975). *CTFA Cosmetic Journal* **7**, 3.

BAKER, J. H. (1959). That unwanted cosmetic ingredient – bacteria. *The Journal of the Society of Cosmetic Chemists* **10**, 133-143.

BAKER, Z., HARRISON, R. W. & MILLER, B. F. (1941). Action of synthetic detergents on the metabolism of bacteria. *The Journal of Experimental Medicine* **73**, 249-271.

BASSETT, D. C. J., STOKES, K. J. & THOMAS, W. R. G. (1970). Wound infection with *Pseudomonas multivorans* – a water-borne contaminant of disinfectant solutions. *The Lancet* **i**, 1188-1191.

BASSETT, D. C. J., DICKSON, J. A. S. & HUNT, G. H. (1973). Infection of Holter valve by pseudomonas-contaminated chlorhexidine. *The Lancet* **i**, 1263-1264.

COOKE, E. M., SHOOTER, R. A., O'FARRELL, S. M. & MARTIN, D. R. (1970). Faecal carriage of *Pseudomonas aeruginosa* by newborn babies. *The Lancet* **ii**, 1045-1046.

DUNNIGAN, A. P. & EVANS, J. R. (1970). Report of a special survey: Microbiological contamination of topical drugs and cosmetics. *TGA Cosmetic Journal* **2**(4), 39-41.

EISMAN, P. C., KULL, F. C. & MAYER, R. L. (1949). The bacteriological aspects of deionized water. *The Journal of the American Pharmaceutical Association, Scientific Edition* **38**, 88-91.

HEISS, F. (1967). Keimgehalt von Körperpflegemitteln. *Fette, Seifen, Anstrichmittel* **69**, 365-369.

JARVIS, B., REYNOLDS, A. J., RHODES, A. C. & ARMSTRONG, M. (1974). A survey of microbiological contamination in cosmetics and toiletries in the U.K. (1971). *The Journal of the Society of Cosmetic Chemists* **25**, 563-575.

KUNZ, L. J. & OUCHTERLONY, O. T. G. (1955). Salmonellosis originating in hospital—a newly recognized source of infection. *The New England Journal of Medicine* **253**, 761-763.

LANG, D. J., KUNZ, L. J., MARTIN, A. R., SCHROEDER, S. A. & THOMSON, L. A. (1967). Carmine as a source of nosocomial salmonellosis. *The New England Journal of Medicine* **276**, 829-832.

LAMBIN, S., DESVIGNES, A., KIGER, J. L. & AZRIA, M. (1972). Étude de la contamination microbienne des sirops pharmaceutiques. *Annales pharmaceutiques françaises* **30**, 161-168.
LÜHDORF, H. & TYBRING, L. (1969). Microbial content in non-sterile pharmaceuticals. VI. Oral liquid preparations. *Dansk Tidsskrift for Farmaci* **43**, 229-237.
MALCOLM, S. A. & WOODROFFE, R. C. S. (1975). The relationship between water-borne bacteria and shampoo spoilage. *The Journal of the Society of Cosmetic Chemists* **26**, 277-288.
MERTZ, J. J., SCHARER, L. & McCLEMENT, J. H. (1967). A hospital outbreak of *Klebsiella pneumoniae* from inhalation therapy with contaminated aerosol solutions. *American Review of Respiratory Diseases* **95**, 454-460.
MORSE, L. J., WILLIAMS, H. L., GRENN, F. P., ELDRIGE, E. E. & ROTTA, J. R. (1967). Septicaemia due to *Klebsiella pneumoniae* originating from a hand-cream dispenser. *The New England Journal of Medicine* **277**, 472-473.
MYERS, G. E. & PASUTTO, F. M. (1973). Microbial contamination of cosmetics and toiletries. *The Canadian Journal of Pharmaceutical Sciences* **8**, 19-23.
NOBLE, W. C. & SAVIN, J. A. (1966). Steroid cream contaminated with *Pseudomonas aeruginosa*. *The Lancet* **i**, 347-349.
PARKER, M. T. (1972). The clinical significance of the presence of micro-organisms in pharmaceutical and cosmetic preparations. *The Journal of the Society of Cosmetic Chemists* **23**, 415-426.
PEDERSEN, E. A. & ULRICH, K. (1968). Microbial content in non-sterile pharmaceuticals. III. Raw materials. *Dansk Tidsskrift for Farmaci* **42**, 71-83.
PHILLIPS, I. (1966). Postoperative respiratory-tract infection with *Pseudomonas aeruginosa* due to contaminated lignocaine jelly. *The Lancet* **i**, 903-904.
PLOTKIN, S. A. & McKITRICK, J. C. (1966). Nosocomial meningitis of the newborn caused by a flavobacterium. *The Journal of the American Medical Association* **198**, 662-664.
PYRAH, L. N., GOLDIE, W., PARSONS, F. M. & ROPER, F. P. (1955). Control of *Pseudomonas pyocyanea* infection in a urological ward. *The Lancet* **ii**, 314-317.
RODE, L. J. & FOSTER, J. W. (1960). The action of surfactants on bacterial spores. *Archiv für Mikrobiologie* **36**, 67-94.
SCHAFFNER, W., REISIG, G. & VERRALL, R. A. (1973). Outbreak of *Ps. cepacia* infections due to contaminated anaesthetics. *The Lancet* **i**, 1050-1051.
SCHUSTER, G. & MODDE, H. (1968). Erfahrungen aus der Praxis uber die Kontaminierung kosmetischer Praparate bei der Herstellung. *Seifen-Ole-Fette-Wachse* **94**, 709-712.
SHOOTER, R. A., COOKE, E. M., GAYA, H., KUMAR, P., PATEL, N., PARKER, M. T., THOM, B. T. & FRANCE, D. R. (1969). Food and medicaments as possible sources of hospital strains of *Pseudomonas aeruginosa*. *The Lancet* **i**, 1227-1229.
TENENBAUM, S. (1971). Significance of pseudomonads in cosmetic products. *American Perfumer and Cosmetics* **86**, 33-37.
TREMEWAN, H. C. (1946). Tetanus neonatorum in New Zealand. *The New Zealand Medical Journal* **45**, 312-313.
VICTORIN, L. (1967). An epidemic of otitis in newborns due to infection with *Pseudomonas aeruginosa*. *Acta Paediatrica Scandinavica* **56**, 344-348.
WESTWOOD, N. & PIN-LIM, B. (1971). Microbial contamination of some pharmaceutical raw materials. *The Pharmaceutical Journal* **207**, 99-102.
WILSON, L. A., KUEHNE, J. W., HALL, S. W. & AHEARN, D. G. (1971). Microbial contamination in ocular cosmetics. *The American Journal of Opthamology* **71**, 1298-1302.
WOLVEN, A. & LEVENSTEIN, I. (1969). Cosmetics – contaminated or not. *TGA Cosmetics Journal* **1**(1), 34-37.
WOLVEN, A. & LEVENSTEIN, I. (1972). Microbiological examination of cosmetics. *American Cosmetics and Perfumery* **87**, 63-65.

Resuscitation of Injured Bacteria in Foods

M. VAN SCHOTHORST

*Laboratory for Zoonoses and Food Microbiology,
National Institute of Public Health, P.O. Box 1,
Bilthoven, The Netherlands*

CONTENTS

1. Introduction 317
2. Factors involved in recovery 318
   (a) The type of stress 318
   (b) The type of organism 318
   (c) Conditions before and during stress 319
   (d) Factors involved during recovery 320
3. Recovery of bacteria in foods 321
4. Resuscitation treatments 323
5. Future outlook 324
6. References 325

1. Introduction

THE BORDERLINE between life and death has been found to be indistinct for many living organisms, and there is no reason why bacteria should be an exception to this rule. Several authors have indeed suggested that there is a gradual decline in viability towards death (Eijkman, 1908; Arpai, 1962; Hansen & Rieman, 1963). In ageing cultures bacteria may show various changes in physiology such as extended lag phase, changes in nutritional requirements and increased sensitivity to NaCl, bile salts, dyes etc. before they die (Griffiths & Haight, 1973; Hurst *et al.*, 1974; Jackson, 1974; Fung & Vandenbosch, 1975). These changes are reversible up to a point and therefore terms like 'injury' or 'sublethal impairment' have been introduced. Following recovery from injury the bacteria are able to multiply and show the same properties, including pathogenicity (Sorrels *et al.*, 1970; Tiwari & Maxcy, 1972) or enterotoxinogenicity (Collins-Thompson *et al.*, 1973; Fung & Vandenbosch, 1975) as before injury. When recovery is not possible, the bacteria are considered to be dead.

In practical analytical food microbiology the phenomenon of injury may present problems since many foods are processed in order to kill a proportion of the bacteria present. As a consequence of processing, some surviving cells exhibit the characteristics of injury, causing variations in plate counts, even when non-selective agar is used (Sherman, 1916; Wright, 1917; Nelson, 1943). The increased dependence of 'metabolically' injured cells on certain nutrients (Straka & Stokes, 1959), and the loss of tolerance towards some substances used in selective media, even at concentrations normally used and to which uninjured bacteria are not sensitive (Clark & Ordal, 1969; Maxcy, 1970; Erwin & Haight, 1973), may lead to a false sense of security when the presence of pathogenic or indicator organisms has to be determined (Gunderson & Rose, 1948; Hartsell, 1951; Labots, 1959). In this review several aspects of so-called resuscitation treatments (Allen *et al.*, 1952) which are intended to avoid the difficulties mentioned above, will be discussed.

2. Factors Involved in Recovery

It is not within the scope of this paper to discuss in detail all the factors involved in recovery, since some may bear no relation to practical problems that face the food microbiologist. The earlier literature will not be reviewed since this has been done by Harris (1963); moreover this paper will be limited to recovery of non-spore forming bacteria, since the recovery of spores has been reviewed by Roberts (1970). Only those aspects of recovery experiments that may influence the resuscitation of bacteria in food microbiological work will be mentioned.

(a) *The type of stress*

The influence of different types of stress, i.e. heating, freezing, irradiation, drying or disinfection, on the survival of bacteria has been studied extensively (see other papers of this Symposium), but only a few experiments in which the consequences of different types of stress on the conditions necessary for recovery were studied, have been described (Heinmetz *et al.*, 1954; Corry *et al.*, 1969; Maxcy, 1973). From these studies on the influence of differences in stress on recovery requirements, no conclusions can be drawn.

(b) *The type of organism*

Dabbah & Moats (1969) used several psychotrophic organisms in heat injury and recovery experiments. Some cells of *Pseudomonas fragi* and *Achromobacter* species survived the treatment (30 min at 55°) in trypticase-soy broth (TSB) and grew normally on TS agar afterwards, while another *Pseudomonas* sp. needed a recovery time of 48 h. Tomlins *et al.* (1971) studied the effect of thermal injury on the TCA cycle enzymes of *Staphylococcus aureus* and *Salmonella typhimurium*. They concluded that for *Staph. aureus* there was a specifically increased nutritional requirement for recovery, but not necessary as complex as that for optimal growth, but for the recovery of *Salm. typhimurium* the nutritional requirements were minimal. Roth *et al.* (1973*a*) found that one strain of *Escherichia coli* recovered from heat injury in lactose broth, while another strain did not, whereas in similar conditions in TSB both strains were resuscitated. In studies on metabolic freeze injury of *E. coli* and *Aerobacter aerogenes*, McLeod *et al.* (1966) found that the latter bacteria responded better to an amino acid mixture in the minimal recovery medium than did the former, while *E. coli* showed a marked increase in count when cystine was added to the minimal medium. From these studies one can conclude that organisms of different type differ in their resuscitation requirements.

(c) *Conditions before and during stress*

Many factors which have an influence on the injury and death of cells will probably also have an influence on their recovery, but it is difficult to find reference in the literature to studies on this subject. Most experiments dealing with different conditions before and during stress were designed to study the damage but not the effects on recovery requirements.

The initial number of bacteria present before stressing seems to have an influence on the recovery time as was found by Dabbah & Moats (1969), but the effect of the physiological age of the bacteria is less clear. Hurst *et al.* (1974) studied this effect of different physiological ages of heat injured *Staph. aureus* and found no influence on the repair of salt tolerance. Fung & Vandenbosch (1975) also found that the physiological age of *Staph. aureus* before freeze-drying had no effect on the recovery time.

One of the most important factors to be taken into account is the complexity of the medium in which the organisms are grown before stress is applied. Many authors (Straka & Stokes, 1959; Nakamura & Dawson, 1962; Moss & Speck, 1966; McLeod *et al.*, 1966; Ray & Speck, 1972*b*) found lower recoveries on minimal media than on rich media when the organisms were grown before stressing in nutritionally rich media such as nutrient broth or TSB. Sometimes this effect was also found when the bacteria were grown in minimal media before being stressed (Sinskey & Silverman, 1970). Hurst *et al.* (1973) found better recovery of heat injured *Staph. aureus* when the organism was grown in a simple defined medium containing peptones than on a rich one. Gomez *et al.* (1973) related the growth requirements after stressing with the conditions before stressing. When heat treated *Salm. typhimurium* was grown in TSY broth the classical metabolic injury was observed: TSY agar yielding higher recovery of stressed bacteria than mineral salts agar (M^9). On the other hand when the culture was grown in M^9 before exposure to heat, reduction of viable counts on TSY agar relative to M^9 was observed (the so-called minimal medium recovery phenomenon).

(d) *Factors involved during recovery*

Other factors involved in recovery include components of the recovery medium, pH, redox potential and temperature. Medium constituents which limit recovery, such as salt, dyes and bile salts, will not be discussed here since they are used mainly to determine the extent of injury in recovery studies (Busta & Jezeski, 1963; Ray & Speck, 1973a; Smolka et al., 1974). The composition of the medium, however, needs special attention. When bacteria are able to grow on rich media but not on minimal salts medium, they are said to be metabolically injured. Recovery from metabolic injury by exposure to media containing a wide variety of metabolites, vitamins, amino acids, peptones and other nutrients has been studied extensively (Heinmetz et al., 1954; Baird-Parker & Davenport, 1965; McLeod et al., 1966; Russell & Harries, 1968; Sinskey & Silverman, 1970; Ray & Speck, 1972a; Ray et al., 1972). In general, the presence of amino acids, an energy source, magnesium and phosphates in the recovery medium seems to be necessary to repair the RNA and to restore the damaged cell membranes, but not all results are consistent in this respect. On the other hand some peptones should not be added to M^9 mineral salts medium to avoid DNA breakage during the 'minimal medium recovery' (Gomez & Sinskey, 1973).

The optimal temperature for recovery has been studied by various workers for different injured organisms, such as *E. coli* injured by chlorine (Sato et al., 1972), freeze-dried *E. coli* (Sinskey & Silverman, 1970), heat-injured *Staph. aureus* (Iandolo & Ordal, 1966) and freeze-injured *Salm. anatum* (Ray et al., 1971). In most instances temperatures between 30 and 37° have been found to be adequate. Stiles et al. (1973) found no resuscitation of heat-injured *E. coli* in TSB at 4.4° and slower recovery at 20° than at 37°, while Ray & Speck (1972b) found optimal recovery of freeze-injured *E. coli* at temperatures from 25 to 35°, but advise incubation at 25°.

The optimal pH for recovery has been studied by Iandolo & Ordal (1966) for heat-injured *Staph. aureus* and by Janssen & Busta (1973) for freeze-injured *Salm. anatum*. Values of pH between 6.0 and 7.4 were found to be adequate, but Ray & Speck (1972b) found for freeze-injured *E. coli*, optimum repair at pH 8.0-9.0 in the presence of 0.75% KH_2PO_4. Smolka et al. (1974) studied the influence of the pH at different concentrations of NaCl in nutrient agar on the recovery of heat-injured *Staph. aureus*. They found no differences in the range of pH 5.5-8.0 when the salt concentration was $< 4\%$.

Harries & Russell (1966) drew attention to the fact that heat-injured *E. coli* recovered better under anaerobic conditions than in the presence of oxygen. Also, Gomez & Sinskey (1974) found that the phenomenon of minimal medium recovery did not occur under anaerobic conditions. Most of the studies on the recovery of bacteria do not mention the use of anaerobic techniques and therefore it can be assumed that these studies have been carried out in the presence of oxygen.

The diluent used in counting procedures may also influence recovery. Ray & Speck (1973b) found that 1% peptone solution was superior to solutions of $MgSO_4$, KH_2PO_4 or 1% non-fat dry milk. Stiles et al. (1973) found no differences in viable counts of heat-injured E. coli in PCA and VRBA during storage at 0 or 20° for 3 h when 0.1% peptone solution was used. Weiler & Hartsell (1969) found 0.1% tryptone to be superior to phosphate buffer solution or deionized water in the recovery of freeze-injured E. coli.

3. Recovery of Bacteria in Foods

Many factors influence the recovery of injured bacteria in foods and the resuscitation treatments required. For instance, in a heat-processed food the principal stress which has caused the injury (i.e. heat) is known and it would be better to use a resuscitation treatment known to be satisfactory for heat-damaged cells. To attempt the recovery of cells from a food processed differently (e.g. by drying) might necessitate the use of a different resuscitation treatment. Furthermore, the situation is complicated by the fact that food constituents can affect the processes of repair of micro-organisms in the food, and hence the choice of resuscitation medium. Most experiments carried out to study the recovery of injured cells from food are therefore more or less practical trials in which methods are compared to find the best one for the purpose (Clark & Ordal, 1969; Corry et al., 1969; Licciardello et al., 1970; van Schothorst & van Leusden, 1972; Tiwari & Maxcy, 1972; Roth et al., 1973a, b; Warseck et al., 1973). Without discussing these results a general conclusion can be made, namely, that many selective media currently in use are not optimal for the examination of foods which have been heated, dried, frozen or treated in some other way to inhibit growth of bacteria. Before reviewing the recovery treatments which are described to remedy this unfortunate situation the factor of time needed for the isolation or enumeration of bacteria present in foods has to be discussed. This factor is of great importance in food microbiology because the result of an examination is often required as soon as possible.

The results are confusing, especially with regard to the time needed for recovery of the bacteria. Some authors recommend 1-2 h for certain purposes (Mossel & Ratto, 1970; Ray & Speck, 1973c), while others have demonstrated that under certain conditions even 6 h is not sufficient (Dabbah & Moats, 1969; Sinskey & Silverman, 1970; Tomlins et al., 1971; van Schothorst & van Leusden, 1972). The differences in time needed for resuscitation may be due not only to differences in stress and other important factors in the experimental design, but also to differences in the components in the selective media used to recover the organisms. For instance, it is well known that bile salts, which are widely used in media for the isolation of Gram negative bacteria, consist of a non-standardized mixture of various components with different toxicities to injured cells (Northolt, 1972; Ray & Speck, 1973b).

To illustrate this point an experiment on the isolation of salmonellae from artificially contaminated milk powder can be used (van Schothorst & van Leusden, 1972). Decimal dilutions of milk powder were incubated in buffered peptone water (BPW) and after various times the cultures were transferred to tetrathionate-brilliant green broth (TBBG) with taurocholate, and to tetrathionate-brilliant green broth with ox bile. After 3 h the recovery in BPW was complete as determined with the taurocholate-containing medium compared with 4 h with the medium containing ox bile. Duitschaever & Jordan (1974) working with heat-injured *Streptococcus faecium* have also demonstrated the influence of the selectivity of the medium on the time necessary for repair.

Another important parameter which influences the recovery time, is the degree of injury (van Schothorst & van Leusden, 1975a). During prolonged storage of milk powder which had been inoculated artificially with salmonellae it was found that the number of dead cells increased and the number of uninjured cells decreased (Fig. 1). Using different resuscitation times several degrees in severity of injury could arbitrarily be established. The cells injured to the first degree could be determined because their number was greater than the number of uninjured cells. The same of course is true for the bacteria injured to the third degree in comparison with those injured to the second degree. However, if the population of salmonellae in the milk powder had consisted of a large

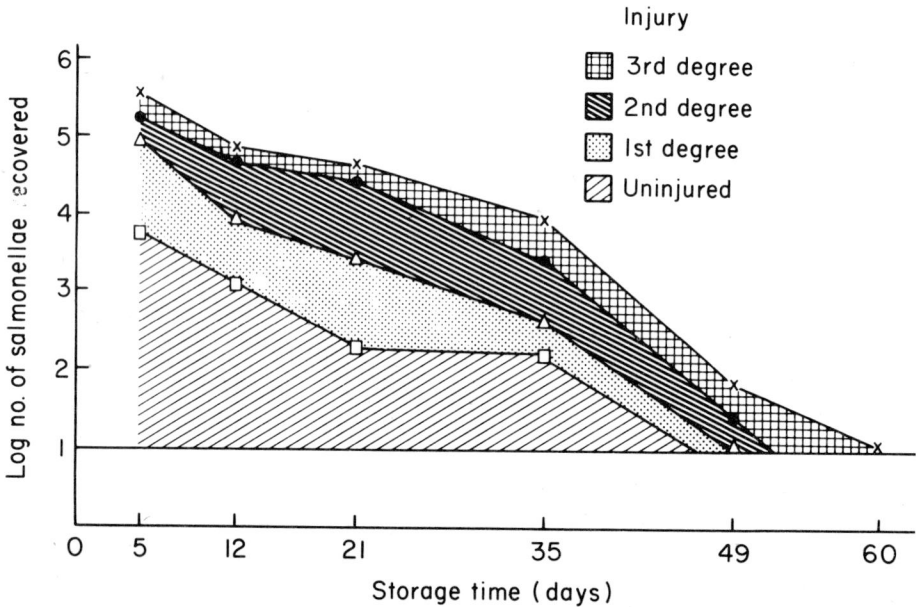

Fig. 1. Effect of pre-enrichment times on recovery of salmonellae from milk powder (a_w, 0.30-0.40). Pre-enrichment times: □, 0 h; △, 2 h; ●, 6 h; ×, 18 h.

proportion of cells which were injured to the first degree (i.e., which could be recovered during 2 h of resuscitation) and only a minor proportion of cells which were more severely injured, the number of bacteria found after a resuscitation treatment of 2 h would have been essentially the same as would have been found after a longer period of resuscitation. In studying one of the experiments (Warseck *et al.*, 1973) which led to the recommendation of a 1-2 h resuscitation time for the enumeration of freeze-injured *E. coli*, it becomes apparent that large populations ($c.$ 10^8) were always used, and that during freezing and thawing the whole population survived with a relatively large number ($c.$ 10^7) of uninjured bacteria. It is likely that only slightly damaged cells were present in such populations, which could explain the short resuscitation times needed. Studies of Collins-Thompson *et al.* (1973) seem to support this line of reasoning since more cells of *Staph. aureus* were damaged during heating in potassium buffer than in sodium buffer and more time was needed for complete recovery of the 'potassium cells' than of the 'sodium cells'.

The results of the studies mentioned above on the isolation of salmonellae from milk powder can also be used to demonstrate the need for longer periods of time for the recovery of severely damaged cells (Fig. 1). During the first stages of storage salmonellae could be isolated even without resuscitation. After 49 days of storage, pre-enrichment in a non-selective medium, during which resuscitation could take place, was necessary to isolate the salmonellae. After 55 days the pre-enrichment time had to be extended to $>$ 6 h in order to isolate salmonellae. The consequences of these findings for the establishment of resuscitation treatments will be discussed later.

4. Resuscitation Treatments

Such procedures should ideally be applicable to all kinds of foods, micro-organisms, injuries and isolation or enumeration techniques, since many factors which may influence the recovery are unknown. No universal resuscitation treatment has been described up-to-date. However, in presence–absence (PA) tests or MPN determinations the establishment of a resuscitation treatment will not be difficult, even when long resuscitation times are needed, when a non-selective pre-enrichment for 18 h is used. This period of treatment will in most cases allow sufficient time for resuscitation; moreover, multiplication of cells after resuscitation does not lead to unreliable results. The complexity of the medium in PA tests will be of minor importance since the food under examination will determine the 'richness' of the resuscitation medium selected. In MPN procedures the nutrients in the pre-enrichment medium may be of importance in the higher dilutions and the use of TSB has been recommended (Clark & Ordal, 1969; Dabbah & Moats, 1969; Mossel & Ratto, 1970; Warseck *et al.*, 1973), but in several situations the nutrients in buffered peptone water seem also to be sufficient (Ray & Speck, 1973b; van Schothorst & van Leusden, 1972, and unpublished results). Buffering the pre-enrichment medium may be of importance since a drop in pH, as has been shown to occur when lactose broth is used (van Schothorst & van Leusden, 1975b), may induce another injury. Moreover when phosphate buffers are used resuscitation may be enhanced (Iandolo & Ordal, 1966; Ray et al., 1972; Ray & Speck, 1972a). When the time for PA tests or MPN determinations is limited, transfer of the total amount of pre-enrichment medium into the enrichment medium (Alford & Knight, 1969) may be considered. In an enumeration procedure by plating, resuscitation can also be carried out in a liquid medium, but in this case multiplication has to be avoided. This can be done by choosing a time that does not exceed the lag time of uninjured cells under such conditions. The reliability of the results of such a treatment is dependent among other factors on the population under study and the selectivity of the agar used.

In order to avoid problems which may arise when such short incubation times are chosen, other procedures have been proposed: (i) plating on non-selective media and replication on selective media (Mossel et al., 1965); (ii) an overlay technique in which the selective agar is carefully poured over the non-selective one (Hartman et al., 1975; Speck et al., 1975); (iii) a membrane filter technique in which the membrane, after incubation on a non-selective medium is transferred to the selective one (Rose & Litsky, 1965). At the moment all methods seem to have merits for certain purposes but no final recommendation on any of these procedures for general use can be given.

In some instances no resuscitation treatment in the isolation or enumeration of specific organisms is necessary, namely, when the selective agar is not inhibitory to injured cells. This has been found with the enumeration of staphylococci on Baird-Parker's agar (Sinskey et al., 1964; Collins-Thompson et al., 1974; Gray et al., 1974).

5. Future Outlook

From the foregoing sections it can be concluded that many investigations have been carried out to clarify several aspects of injury and recovery of bacteria in general. However, much information is lacking with regard to the establishment of resuscitation treatments in food microbiology.

There is a great need for such treatments since by using resuscitation a source of variation in the microbiological analysis of foods may be eliminated. The availability of selective media for micro-organisms other than staphylococci, media which do not inhibit the multiplication of injured bacteria, would be of great importance because in this way some problems in the control of foods for consumer protection or trade may be avoided.

6. References

ALFORD, J. A. & KNIGHT, N. L. (1969). Applicability of aeration and delayed addition of selenite to the isolation of salmonellae. *Applied Microbiology* **18**, 1060-1064.

ALLEN, L. A., PASLEY, S. M. & PIERCE, M. S. F. (1952). Conditions affecting the growth of *Bacterium coli* on bile salts media. Enumeration of this organism in polluted waters. *Journal of General Microbiology* **7**, 257-267.

ARPAI, J. (1962). Nonlethal freezing injury to metabolism and motility of *Pseudomonas fluorescens* and *Escherichia coli*. *Applied Microbiology* **10**, 297-301.

BAIRD-PARKER, A. C. & DAVENPORT, E. (1965). The effect of recovery medium on the isolation of *Staphylococcus aureus* after heat treatment and after the storage of frozen or dried cells. *Journal of Applied Bacteriology* **28**, 390-402.

BUSTA, F. F. & JEZESKI, J. J. (1963). Effect of sodium chloride concentration in an agar medium on growth of heat-shocked *Staphylococcus aureus*. *Applied Microbiology* **11**, 404-407.

CLARK, C. W. & ORDAL, Z. J. (1969). Thermal injury and recovery of *Salmonella typhimurium* and its effect on enumeration procedures. *Applied Microbiology* **18**, 332-336.

CORRY, J. E. L., KITCHELL, A. G. & ROBERTS, T. A. (1969). Interaction in the recovery of *Salmonella typhimurium* damaged by heat or gamma radiation. *Journal of Applied Bacteriology* **32**, 415-428.

COLLINS-THOMPSON, D. L., HURST, A. & KRUSE, H. (1973). Synthesis of enterotoxin B by *Staphylococcus aureus* strain S6 after recovery from heat injury. *Canadian Journal of Microbiology* **19**, 1463-1468.

COLLINS-THOMPSON, D. L., HURST, A. & ARIS, B. (1974). Comparison of selective media for the enumeration of sublethally heated food-poisoning strains of *Staphylococcus aureus*. *Canadian Journal of Microbiology* **20**, 1072-1075.

DABBAH, E. & MOATS, W. A. (1969). Factors affecting resistance to heat and recovery of heat-injured bacteria. *Journal of Dairy Science* **52**, 608-614.

DUITSCHAEVER, C. L. & JORDAN, D. C. (1974). Development of resistance to heat and sodium chloride in *Streptococcus faecium* recovering from thermal injury. *Journal of Milk and Food Technology* **37**, 382-386.

ERWIN, D. G. & HAIGHT, R. D. (1973). Lethal and inhibitory effects of sodium chloride on thermally stressed *Staphylococcus aureus*. *Journal of Bacteriology* **116**, 337-340.

EIJKMAN, C. (1908). Die Ueberlebungskurve bei Abtötung von Bakterien durch Hitze. *Biochemische Zeitschrift* **11**, 12-20.

FUNG, D. Y. C. & VANDENBOSCH, L. L. (1975). Repair, growth, and enterotoxigenesis of *Staphylococcus aureus* S-6 injured by freeze-drying. *Journal of Milk and Food Technology* **38**, 212-218.

GOMEZ, R. F. & SINSKEY, A. J. (1973). Deoxyribonucleic acid breaks in heated *Salmonella typhimurium* LT-2 after exposure to nutritionally complex media. *Journal of Bacteriology* **115**, 522-528.

GOMEZ, R. F. & SINSKEY, A. J. (1974). Effect of aeration on minimal medium recovery of heated *Salmonella typhimurium*. *Journal of Bacteriology* **122**, 106-109.

GOMEZ, R. F., SINSKEY, T. J., DAVIES, R. & LABUZU, T. P. (1973). Minimal medium recovery of heated *Salmonella typhimurium* LT2. *Journal of General Microbiology* **74**, 267-274.

GRAY, R. J. H., GASKE, M. A. & ORDAL, Z. J. (1974). Enumeration of thermally stressed *Staphylococcus aureus* MF31. *Journal of Food Science* **39**, 844-846.

GRIFFITHS, R. F. & HAIGHT, R. D. (1973). Reversible heat injury in the marine psychrophilic bacterium *Vibrio marinus* MP-1. *Canadian Journal of Microbiology* **19**, 557-561.

GUNDERSON, M. F. & ROSE, K. D. (1948). Survival of bacteria in a precooked, fresh-frozen food. *Food Research* **13**, 254-263.

HANSEN, N. H. & RIEMAN, H. (1963). Factors affecting the heat resistance of non-sporulating organisms. *Journal of Applied Bacteriology* **26**, 315-333.

HARRIES, D. & RUSSELL, A. D. (1966). Revival of heat-damaged *Escherichia coli*. *Experienta* **22**, 803.

HARRIS, N. D. (1963). The influence of the recovery medium and the incubation temperature on the survival of damaged bacteria. *Journal of Applied Bacteriology* **26**, 387-397.

HARTMAN, P. A., HARTMAN, P. S. & LANZ, W. W. (1975). Violet red bile 2 agar for stressed coliforms. *Applied Microbiology* **29**, 537-539.

HARTSELL, S. E. (1951). The longevity and behaviour of pathogenic bacteria in frozen foods: The influence of plating media. *American Journal of Public Health* **41**, 1072-1077.

HEINMETZ, F., TAYLOR, W. W. & LEHMAN, J. J. (1954). The use of metabolites in the restoration of the viability of heat and chemically inactivated *Escherichia coli*. *Journal of Bacteriology* **67**, 5-12.

HURST, A., HUGHES, A. BEARE-ROGERS, J. L. & COLLINS-THOMPSON, D. L. (1973). Physiological studies on the recovery of salt tolerance by *Staphylococcus aureus* after sublethal heating. *Journal of Bacteriology* **116**, 901-907.

HURST, A., HUGHES, A. & COLLINS-THOMPSON, D. L. (1974). The effect of sublethal heating on *Staphylococcus aureus* at different physiological ages. *Canadian Journal of Microbiology* **20**, 765-768.

IANDOLO, J. J. & ORDAL, Z. J. (1966). Repair of thermal injury of *Staphylococcus aureus*. *Journal of Bacteriology* **91**, 134-142.

JACKSON, H. (1974). Loss of viability and metabolic injury of *Staphylococcus aureus* resulting from storage at 5°C. *Journal of Applied Bacteriology* **37**, 59-64.

JANSSEN, D. W. & BUSTA, F. F. (1973). Repair of injury in *Salmonella anatum* cells after freezing and thawing in milk. *Cryobiology* **10**, 386-392.

LABOTS, H. (1959). The behaviour of sublethally heated *Escherichia coli* in milk and other media. *XI International Dairy Congress* **3**, 1355-1360.

LICCIARDELLO, J. J., NICKERSON, J. T. R. & GOLDBLITH, S. A. (1970). Recovery of salmonellae from irradiated and unirradiated foods. *Journal of Food Science* **35**, 620-624.

McLEOD, R. A., SMITH, L. D. H. & GELINAS, R. (1966). Metabolic injury to bacteria. 1. Effect of freezing and storage on the requirements of *Aerobacter aerogenes* and *Escherichia coli* for growth. *Canadian Journal of Microbiology* **12**, 61-72.

MAXCY, R. B. (1970). Non-lethal injury and limitations of recovery of coliform organisms on selective media. *Journal of Milk and Food Technology* **33**, 445-448.

MAXCY, R. B. (1973). Condition of coliform organisms influencing recovery of subcultures on selective media. *Journal of Milk and Food Technology* **36**, 414-416.

MOSS, C. W. & SPECK, M. L. (1966). Release of biologically active peptides from *Escherichia coli* at subzero temperatures. *Journal of Bacteriology* **91**, 1105-1111.

MOSSEL, D. A. A., JONGERIUS, E. & KOOPMAN, M. J. (1965). Sur la nécessité d'une revivification préalable pour le dénombrement des Enterobacteroaceae dans les aliments déshydrates, irradiés ou non. *Annales de l'Institut Pasteur, Lille* **16**, 119-125.

MOSSEL, D. A. A. & RATTO, M. A. (1970). Rapid detection of sublethally impaired cells of Enterobacteriaceae in dried foods. *Applied Microbiology* **20**, 273-275.

NAKAMURA, M. & DAWSON, D. A. (1962). Role of suspending and recovery media in the survival of frozen *Shigella sonnei*. *Applied Microbiology* **10**, 40-43.

NELSON, F. E. (1943). Factors which influence the growth of heat-treated bacteria. *Journal of Bacteriology* **45**, 395-403.

NORTHOLT, M. D. (1972). Composition of oxbile and of commercial preparations. *Antonie van Leeuwenhoek* **38**, 632.

RAY, B. & SPECK, M. L. (1972*a*). Metabolic process during the repair of freeze-injury in *Escherichia coli*. *Applied Microbiology* **24**, 585-590.

RAY, B. & SPECK, M. L. (1972*b*). Repair of injury induced by freezing *Escherichia coli* as influenced by recovery medium. *Applied Microbiology* **24**, 258-263.

RAY, B. & SPECK, M. L. (1973a). Freeze-injury in bacteria. *CRC Critical Review in Clinical Laboratory Science* **4**, 161-213.
RAY, B. & SPECK, M. L. (1973b). Discrepancies in the enumeration of *Escherichia coli*. *Applied Microbiology* **25**, 494-498.
RAY, B. & SPECK, M. L. (1973c). Enumeration of *Escherichia coli* in frozen samples after recovery from injury. *Applied Microbiology* **25**, 499-503.
RAY, B., JEZESKI, J. J. & BUSTA, F. F. (1971). Effect of rehydration on recovery, repair and growth of injury freeze-dried Salmonella anatum. *Applied Microbiology* **22**, 184-189.
RAY, B., JANSSEN, D. W. & BUSTA, F. F. (1972). Characterization of the repair of injury induced by freezing *Salmonella anatum*. *Applied Microbiology* **23**, 803-809.
ROBERTS, T. A. (1970). Recovering spores damaged by heat, ionizing radiations or ethylene oxide. *Journal of Applied Bacteriology* **33**, 74-94.
ROSE, R. E. & LITSKY, W. (1965). Enrichment procedures for use with the membrane filter for the isolation and enumeration of fecal streptococci in water. *Applied Microbiology* **13**, 106-108.
ROTH, L. A., STILES, M. E. & CLEGG, L. F. L. (1973a). Reliability of selective media for the enumeration and estimation of *Escherichia coli*. *Canadian Institute of Food Science and Technology Journal* **6**, 230-234.
ROTH, L. A., STILES, M. E. & CLEGG, L. F. L. (1973b). Reliability of enrichment and selective media for the estimation of *Salmonella typhimurium*. *Canadian Institute of Food Science and Technology Journal* **6**, 235-238.
RUSSELL, A. D. & HARRIES, D. (1968). Factors influencing the survival and revival of heat-treated *Escherichia coli*. *Applied Microbiology* **16**, 335-339.
SATO, T., IZAKI, K. & TAKAHASHI, H. (1972). Recovery of cells of *Escherichia coli* from injury induced by sodium chloride. *Journal of General and Applied Microbiology* **18**, 307-317.
SHERMAN, J. M. (1916). The advantages of a carbohydrate medium in the routine bacterial examination of milk. *Journal of Bacteriology* **1**, 481-488.
SINSKEY, T. J. & SILVERMAN, G. J. (1970). Characterization of injury incurred by *Escherichia coli* upon freeze-drying. *Journal of Bacteriology* **101**, 429-437.
SINSKEY, T. J., McINTOSH, A. H., PABLO, I. S., SILVERMAN, G. T. & GOLDBLITH, S. A. (1964). Considerations in the recovery of microorganisms from freeze-dried foods. *Health Laboratory Science* **1**, 297-306.
SMOLKA, L. R., NELSON, F. E. & KELLEY, L. M. (1974). Interaction of pH and NaCl on enumeration of heat-stressed *Staphylococcus aureus*. *Applied Microbiology* **27**, 443-447.
SORRELS, K. M., SPECK, M. L. & WARREN, J. A. (1970). Pathogenicity of *Salmonella gallinarum* after metabolic injury by freezing. *Applied Microbiology* **19**, 39-43.
SPECK, M. L., RAY, B. & READ, R. B. Jr. (1975). Repair and enumeration of injured coliforms by a plating procedure. *Applied Microbiology* **29**, 549-550.
STILES, M. E., ROTH, L. A. & CLEGG, L. F. L. (1973). Heat injury and resuscitation of *Escherichia coli*. *Canadian Institute of Food Science and Technology* **6**, 226-229.
STRAKA, R. P. & STOKES, J. L. (1959). Metabolic injury to bacteria at low temperatures. *Journal of Bacteriology* **78**, 181-185.
TIWARI, N. P. & MAXCY, R. B. (1972). Post-irradiation evaluation of pathogens and indicator bacteria. *Journal of Food Science* **37**, 485-487.
TOMLINS, R. I., PIERSON, M. D. & ORDAL, Z. J. (1971). Effect of thermal injury on the TCA cycle enzymes of *Staphylococcus aureus* MF 31 and *Salmonella typhimurium* 7136. *Canadian Journal of Microbiology* **17**, 759-765.
TOMLINS, R. I., VAALER, G. L. & ORDAL, Z. J. (1972). Lipid biosynthesis during the recovery of *Salmonella typhimurium* from the thermal injury. *Canadian Journal of Microbiology* **18**, 1015-1021.
VAN SCHOTHORST, M. & VAN LEUSDEN, F. M. (1972). Studies on the isolation of injured salmonellae from foods. *Zentralblatt für Bakteriologie, Parasitenkunde, Infektionskrankheiten und Hygiene, Abteilung* I. *Originale A* **221**, 19-29.

VAN SCHOTHORST, M. & VAN LEUSDEN, F. M. (1975a). Further studies on the isolation of injured salmonellae from foods. *Zentralblatt für Bakteriologie, Parasitenkunde, Infektionskrankheiten und Hygiene, Abteilung* 1. *Originale A* **230**, 186-191.
VAN SCHOTHORST, M. & VAN LEUSDEN, F. M. (1975b). Comparison of several methods for the isolation of salmonellae from egg products. *Canadian Journal of Microbiology* **21**, 1041-1045.
WARSECK, M., RAY, B. & SPECK, M. L. (1973). Repair and enumeration of injured coliforms in frozen foods. *Applied Microbiology* **26**, 919-924.
WEILER, W. A. & HARTSELL, S. E. (1969). Diluent composition and the recovery of *Escherichia coli. Applied Microbiology* **18**, 956-957.
WRIGHT, J. H. (1917). The importance of uniform culture media in the bacteriological examination of disinfectants. *Journal of Bacteriology* **11**, 315-346.

Inactivation of Bacteria by Freeze-drying

I. J. BOUSFIELD AND A. R. MACKENZIE

*National Collection of Industrial Bacteria
Torry Research Station,
Aberdeen, Scotland*

CONTENTS

1. Introduction 329
2. The freeze-drying process 329
3. Lethal effects of freeze-drying 330
   - (a) Freezing 331
   - (b) Drying 332
4. Protection of bacteria during and after freeze-drying 333
5. Storage 337
6. Resuscitation 338
7. Variation between organisms 339
8. References 340

1. Introduction

MUCH OF THE WORK described in this Symposium is directed at determining the optimum conditions for inactivating or killing microbes. In this review microbial inactivation will be considered from the opposite point of view; the reduction rather than the enhancement of deleterious effects and the preservation rather than the destruction of, specifically, bacteria. The proliferation of culture collections throughout the world during the last 20 years or so is ample evidence of the increasing interest in the long-term preservation of reference cultures. Freeze-drying is probably the most widely used method of preserving large collections of bacterial cultures, although storage of seed stocks in liquid nitrogen is becoming increasingly popular. However, there are several reasons why preservation by freeze-drying is at present the most convenient method for service culture collections (see Perry *et al.*, 1975) and therefore in this contribution only the effects of freeze-drying on bacteria will be considered.

2. The Freeze-drying Process

The theoretical aspects of freeze-drying have been discussed in detail by Meryman (1966*a*) and practical considerations of the design of methods and equipment may be found in the reviews by Rowe (1970, 1971). For those interested in the early development of the freeze-drying of bacteria the article by Fry (1954) still provides an excellent starting point.

Briefly, freeze-drying is dehydration by the sublimative removal of water from a frozen material, which for the purposes of this review is a frozen bacterial suspension. Cells may be prefrozen at atmospheric pressure or 'snap frozen' by the rapid removal of water vapour in a vacuum chamber. If evaporative freezing is used suspensions must either be degassed or centrifuged to prevent frothing. The frozen suspensions are held under vacuum, when the ice will sublime. Heat is applied to the sample to maintain continued sublimation and the resulting water vapour is trapped either on a cold surface or with a chemical desiccant. Bacterial suspensions are usually dried to a residual moisture content of c. 1% after which they are stored under vacuum.

Most British culture collections use modifications of the centrifugal freeze-drying method of Greaves (1944, 1946). Full details of the method used by the National Collection of Type Cultures (NCTC) and by the National Collection of Industrial Bacteria (NCIB) are given in the articles by Lapage *et al.* (1970) and Bousfield & MacKenzie (1972). Methods for freeze-drying a large number of different organisms have been listed recently by Lapage & Redway (1974).

3. Lethal Effects of Freeze-drying

There is no doubt that freeze-drying is injurious to most bacteria and cells undergoing the process are obviously subjected to great stress. The main stages in the preservation of bacteria by freeze-drying are freezing, drying, storage and resuscitation; loss of viability may occur as a result of any of these. Fry & Greaves (1951) showed that the highest death rate occurs during the first 2 h of freeze-drying, when water removal is at its most rapid. Curves plotted from their results are shown in Fig. 1. With the very sensitive *Azomonas insignis* we have found that the sharpest drop in viability occurs in the first few minutes, when rapid cooling and freezing take place.

It is tempting to assume that the cells surviving freeze-drying are in some way inherently more resistant than are those which die. However, Fry & Greaves (1951) showed that when cultures of *Klebsiella pneumoniae* D201 H were subjected to three successive cycles of drying and rehydration, the percentage of survivors after each cycle was more or less constant. Steel & Ross (1963) reporting on their experience of freeze-drying in the NCTC said that they also found no evidence of resistant populations being selected during drying. What then are the factors which affect the viability of bacterial populations during and after freeze-drying? Death may result from various physical and chemical effects induced by the removal of water and these are discussed below. For convenience of presentation freezing and drying will be treated separately, but it must be

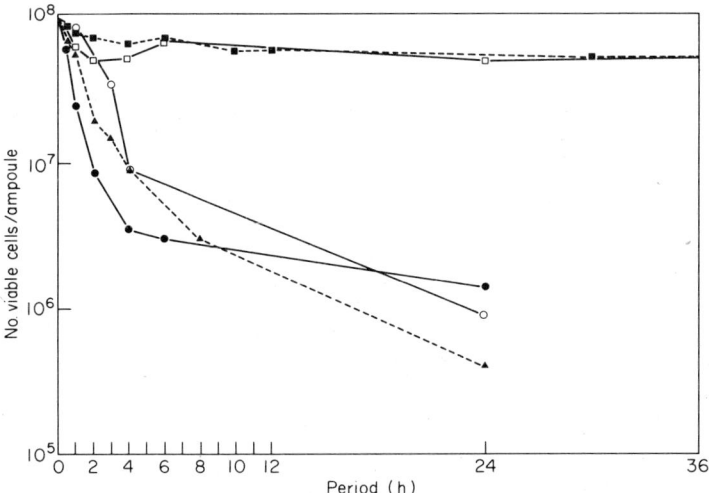

Fig. 1. Death during freeze-drying of *Kl. pneumoniae* D 201 H in various suspending media; 7.5% glucose in nutrient broth (□ ■), 5% gelatin (○), 5% albumin (▲), 10% haemoglobin (●). (Curves plotted from the results of Fry & Greaves, 1951.)

remembered that both of these processes are dehydrative and many destructive effects may be common to both.

(a) *Freezing*

Much has been published on the effects of freezing on micro-organisms and no attempt will be made to review this literature. The extensive articles by Meryman (1966b) and Mazur (1966) deal fully with the subject and a useful and very readable summary was published a few years ago by Davies (1968a, b). The effect of low temperatures on micro-organisms is also discussed by Ingram elsewhere in this Symposium

Briefly, freezing injury has been ascribed by various authorities to some or all of the following: intracellular ice formation (Mazur, 1963), electrolyte concentration and cell permeability changes (Lovelock, 1953), precipitation of proteins or the formation of abnormal chemical bonds by dehydration (Lovelock, 1957; Levitt, 1962), pH changes (Lea & Hawke, 1952), changes in the concentration of dissolved gases (Lovelock, 1957), removal of 'structural' water (Meryman, 1966b), reduction of cell volume by dehydration (Meryman, 1970) and redistribution of intracellular water due to different vapour pressures inside and outside the cell (Litvan, 1972).

Meryman (1966b) has said that freezing injury and protection against it can be summarized in terms of two controllable variables; rate of freezing and protective additives. (The latter have been discussed in reviews by Doebbler, 1966 and Meryman, 1971.) The extent of the destructive effects mentioned above depends entirely on these two factors. (Greaves & Davies, 1965; Greaves et al., 1967). Work by Greaves et al. (1967) has shown quite clearly that for the frost sensitive organism *Pseudomonas* sp. 10H at least, the number of survivors obtained after freezing and thawing varies considerably with the rate of freezing and the composition of the suspending medium. Davies (1968b) pointed out that to obtain optimal survival after freeze-drying it is essential to determine the rate of freezing which gives maximum survival of a particular culture in a particular suspending medium. Although desirable, this is not feasible for a large culture collection where several different kinds of bacteria may be freeze-dried simultaneously and where time is not available for such studies. Also, the centrifugal freeze-dryers used by most British culture collections lack facilities for controlling the rate of freezing. Consequently the tendency is to accept possibly suboptimal survival levels by using a 'general purpose' suspending medium and accepting whatever freezing rate that a particular freeze-dryer imposes.

(b) *Drying*

Following any initial loss of viability during freezing, there is a further decline in survival levels during the subsequent drying process. Prolonged drying may cause injury by the removal of 'bound' water at low residual moisture levels, although how this happens is not known. Meryman (1966a) has suggested that the removal of surface water from reactive cell proteins may be harmful. Scott (1960) put forward the hypothesis that reactions between cell proteins and carbonyl compounds are a major cause of the death of dried cells (see Section 4). Injury also occurs if the drying temperature following initial freezing is too high and it has been shown several times that lower drying temperatures give better survival levels (see e.g. Davies, 1968b). This improvement has been attributed to the slowing down of chemical reactions and the elimination of liquid phases in suspensions containing eutectic solutions (Meryman, 1966a). However, there is an apparent paradox in that drying from the liquid phase without any initial freezing often results in higher survival levels than those obtained by freeze-drying similar suspensions. Indeed, we routinely use Annear's (1958) L-drying method for organisms such as *A. insignis* which survive freeze-drying very poorly. L-drying differs from freeze-drying in that the ampoules are held in a water bath at $20°$ to prevent their contents freezing during evacuation. Drying proceeds by rapid evaporation of water from the liquid suspension. A detailed description of the L-drying method has been given by Lapage et al. (1970). Figure 2 shows a comparison of the survival levels of *A. insignis* after freeze-drying and after L-drying.

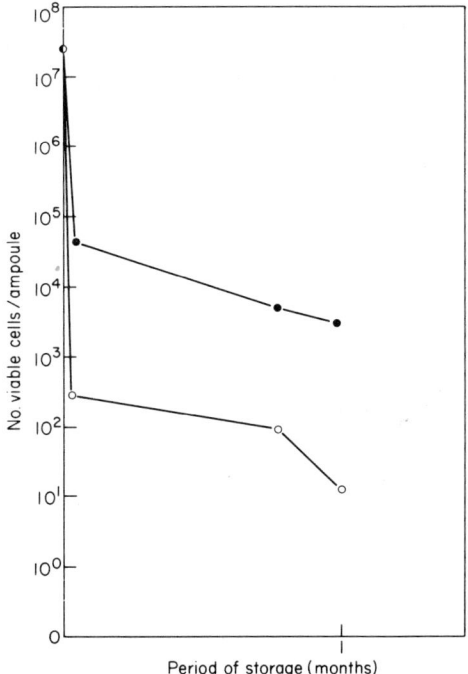

Fig. 2. Death of *Az. insignis* NCIB 9127 suspended in *'Mist desiccans'* during and after freeze-drying (○) and L-drying (●).

4. Protection of Bacteria During and After Freeze-drying

Apart from the nature of the organisms themselves, perhaps the most important single influence on the survival of bacteria during and after freeze-drying is the composition of the suspending medium. It is generally accepted that freeze-drying 'naked' bacteria, i.e. cells suspended in distilled water, usually does not lead to high survival levels. In addition to any osmotic damage which might be caused by suspending the cells in a hypotonic medium, there is no protection against the destructive effects of freezing and drying. However, Lion & Bergmann (1961a, b) claimed that high survival levels can be obtained in the absence of a protective medium, provided the cells do not come into contact with oxygen. They reported survival levels of 50% for *Escherichia coli* freeze-dried from distilled water: the 50% loss in viability was ascribed mainly to the effects of oxygen. While it has long been accepted that oxygen is deleterious to dried bacteria, especially during storage (see Section 5) most workers still believe that freeze-drying itself is a major cause of death for unprotected cells, and that some kind of protective suspending medium is necessary. Much of the earlier work was done on a hit-or-miss basis with broth, serum, blood or milk providing most of the hits. Fry (1954, 1966) and others have reviewed the early work thoroughly and we shall not dwell on it here, other than to mention a few of the more significant contributions.

Otten (1930, 1932) was the first to comment on the role of a 'protective colloid' in freeze-drying, after he had found that meat extract or a thick suspension of killed bacteria exerted a significant protective effect in comparison with water or physiological saline. Elser *et al.* (1935) found that serum gave better results than did meat broth for certain neisseriae, and Leifson (1936) observed that the addition of blood to pork infusion was superior to the meat extract alone for the drying of *V. cholerae*. The idea of a protective colloid soon gained ground — although much later Scott (1958) showed that the protective effect of certain substances was not due to their colloidal nature — and Naylor & Smith (1946) and Stamp (1947) carried out systematic investigations to devise suitable suspending media. Stamp used 10% nutrient gelatin containing 0.5% ascorbic acid as an anti-oxidant and obtained good results with *Serratia marcescens* and other enterobacteria, but poor survival rates with more sensitive organisms such as *V. cholerae*. After many experiments with gelatin, Naylor & Smith (1946) finally settled on 2% dextrin as the protective colloid with 0.5% each of ascorbic acid, thiourea and ammonium chloride added. Although giving good survival levels with *Serr. marcescens,* later work showed that this medium was not effective for some other organisms (see e.g. Hornibrook, 1950; Splitstoesser & Foster, 1957; Fry, 1966).

Milk was used as a suspending medium as early as 1914 when Rogers dried lactic acid bacteria in skim milk. Since then, milk has been used quite commonly, both with and without various additives. Hornibrook (1949) successfully dried *V. cholerae* in skim milk and later (1950) he discovered that a solution made up of the dialysable constituents of milk was equally, if not more effective. Damjanovic & Radulovic (1967) successfully freeze-dried *Lactobacillus bifidus* in a mixture of skim milk, sucrose and gelatin and in 1972 Jakovljevic reported freeze-drying the very sensitive *V. fetus* in whole milk, obtaining survival after storage for eight years. Nikolov (1973) used sour milk for the freeze-drying of lactic acid bacteria and Sinha *et al.* (1974) found that skim milk fortified with ascorbic acid and thiourea was a more effective suspending medium for *Streptococcus lactis* than was skim milk alone. Double strength skim milk has been used for many years by the American Type Culture Collection as the suspending medium for freeze-drying a wide variety of bacteria (Alexander, 1973).

Fry (1951) and Fry & Greaves (1951) reported the results of several years' work on the survival of bacteria during and after drying, in which they paid close attention to the effect of the suspending medium. Using the D201H strain of

Kl. pneumoniae, they studied the effect of various substances including nutrient broth, nutrient gelatin, serum, haemoglobin, albumin and gum acacia. They found that survival rates varied immediately after drying and were very poor in broth-free media after storage for a few months (Fig. 1). Fry & Greaves concluded that the nature of the protective colloid was of great importance for survival; an observation directly contradicting the statement made by Proom & Hemmons (1949) that it was of little importance. They also made the significant observations that 'over-drying' resulted in low viabilities and that the residual moisture content of the product was important.

The report by Leshchinskaya (1944) that BCG could be dried successfully in 50% aqueous glucose led Fry & Greaves to add glucose to nutrient broth. Survival levels were found to be much higher than with nutrient broth alone, the optimum glucose concentration being between 5 and 10%. However, viability levels of *Neisseria gonorrheae* in glucose broth were poor, but the addition of 75% of serum gave much higher survival rates. These observations resulted in the formulation of the now famous *'Mist. desiccans'* comprising 1 part of nutrient broth and 3 parts of serum with 7.5% of glucose added. *'Mist. desiccans'* gives adequate survival levels for a wide variety of bacteria during both drying and storage and has been used routinely for many years in the NCIB for the preservation of almost all cultures. Fry & Greaves experimented with other carbohydrates but found none superior to glucose. Their results led to what Fry (1966) has since called the 'carbohydrate era' during which various carbohydrates and related substances were added to different protective colloids in attempts to improve survival after freeze-drying. Morichi (1970) tested 17 carbohydrates and polyalcohols in aqueous solution for their protective effect on several bacteria, finding that disaccharides were the best and pentoses the worst. Recently Redway & Lapage (1974) tried 21 carbohydrates and related compounds in serum, concluding that *meso*-inositol, non-reducing disaccharides and certain polyalcohols gave the best results.

Work in the nineteen-fifties on the preservation of BCG vaccine and the retention of its potency at tropical storage temperatures showed that addition of sodium glutamate to the suspending medium markedly improved the storage potential of the dried product (see e.g. Miller & Goodner, 1953; Cho & Obayashi, 1956; Cho *et al.*, 1956). Greaves (1960) demonstrated that *Kl. pneumoniae* D201H dried in 5% sodium glutamate survived boiling for longer than when dried in either 5% peptone or *'Mist. desiccans'*. Further work by the Japanese seemed to confirm the effectiveness of glutamate as a protective agent (Obayashi & Cho, 1957; Obayashi *et al.*, 1957; Obayashi, 1960) although Ungar *et al.* (1962) were unable to find any advantage in its use for the Copenhagen strain of BCG, preferring a mixture of dextran, glucose and Triton 1 WR 1339.

Most of the suspending media used for freeze-drying have been chosen empirically, although systematic investigations into the effects of a variety of compounds were carried out by Morichi (1970), and even now the way in which they act is not altogether clear. More work has been done on the mechanisms of protection against freezing than against drying as such. Detailed discussions of cryoprotective mechanisms may be found in the reviews by Mazur (1966), Davies (1968b) and Meryman (1971).

Scott's (1960) hypothesis that reactions between carbonyl compounds and amino groups of cell proteins are a major cause of death during the drying and storage of bacteria has more recently been supported by Marshall & Scott (1970) and Marshall *et al.* (1974). This hypothesis immediately explains the effectiveness of glutamate as a protective agent, since this compound will neutralize carbonyl groups. As Fry (1966) has pointed out, the theory would also explain the enhanced survival obtained when nutrient broth or peptone is added to the drying medium, since amino acids present in these would react with and neutralize carbonyl groups. Paradoxically, glucose is used as a protective agent when according to Scott's theory it should be deleterious. Fry (1966) said that when he dried *V. cholerae* in a peptone solution containing 5% glucose or sucrose, there was little difference between the two in their effects on survival both immediately after drying and after several years of storage. However, glucose is a weak carbonyl compound and because Fry was using peptone in his suspending medium, presumably any damaging effect of the glucose would be neutralized. Redway & Lapage (1974) on the other hand did not include broth or supplementary amino acids in any of their serum-carbohydrate suspending media and they found that glucose and other carbonyl compounds, although having some effect, were not very effective as protective agents. Although these results might be construed as further evidence to support Scott's hypothesis, Redway and Lapage preferred to explain the efficiency of *meso*-inositol, non-reducing disaccharides and certain polyalcohols in terms of their water-retention capacity. They also suggested that *meso*-inositol might protect the DNA of the cells being dried. From our own very limited experience of *meso*-inositol we have found that for *A. insignis* immediately after drying, *meso*-inositol in serum is inferior as a protective agent to both 'Mist. desiccans' and glucose in serum. However, we have obtained enhanced survival levels by replacing the glucose in 'Mist desiccans' with 5% of *meso*-inositol. We do not yet have any information on survival levels after storage.

The protective effect of carbohydrates was ascribed by Fry & Greaves (1951) to their water-binding capabilities, an explanation which has since become widely accepted. Morichi (1970) considered that protective effects might be due to physicochemical reactions such as hydrogen bonding with various cellular constituents, saying that

"protective solutes may stabilize the conformation of cellular constituents in place of bound water in the process of dehydration".

Orndorff & MacKenzie (1973) on the other hand believed the role of additives to be fairly non-specific and based on general physicochemical properties rather than on the presence or absence of particular functional groups. They considered that the maintenance of an amorphous matrix throughout the freeze-drying process was the most important factor influencing survival.

5. Storage

Since the *raison d'être* of culture collections is the storage of viable cultures for long periods, a low death rate during storage is more important than high immediate postdrying viability levels. Apart from the nature of the organism itself, the main factors influencing survival during storage are the suspending medium, the storage temperature, the atmosphere of storage, and the residual moisture content. All of these factors are interrelated.

In the manufacture of many pharmaceutical and biological preparations, the maxim seems to be 'the drier the better' but this is not so for living organisms. 'Overdrying' of bacteria results in loss of viability (Fisher, 1950; Fry & Greaves, 1951) and it is widely held that the enhanced survival obtained when a carbohydrate is added to the suspending medium is due to the water-retaining capacity of the carbohydrate. Such a 'water-buffer' obviates the need for carefully controlled drying times. There have been numerous studies on the effects of residual moisture on survival to determine the optimum water level for dried bacteria, which seems to be *c.* 1%. In most of these studies various suspending media were used, but a few years ago, Nei *et al.* (1966) argued that the use of protective agents made it impossible to determine the true effects of residual moisture. For this reason, they devised a technique for freeze-drying *E. coli* cells suspended in distilled water to predetermined moisture levels. They found that the relationship of residual water content to survival depended on the atmosphere and temperature of storage. For cells stored *in vacuo* or under nitrogen, higher survival levels were obtained at 5% water content than at 15 or 20%. For cells stored in air, the reverse was the case, an effect attributed by Nei *et al.* to a decreased rate of oxidation of cellular constitutents at the higher moisture levels. Scott (1958) investigated the effect of water activity on survival levels, finding that the optimum varied according to the suspending medium and to the atmosphere of storage. Later Marshall & Scott (1970) said that the storage temperature was more important than the effect of residual moisture and in 1973 Marshall, Coote & Scott found that residual moisture levels influenced the effectiveness of various storage atmospheres. Later (1974) they came to the conclusion that within certain limits, the water activity was not critical for survival.

The effect of the storage atmosphere on survival seems to depend on the extent to which traces of oxygen are present. The lethal effect of oxygen on dried bacteria has long been known (Rogers, 1914) and has been the subject of investigations by various workers (see e.g. Lion & Bergmann, 1961a, b; Benedict et al., 1961; Dimmick et al., 1961; Lion, 1963; Heckly et al., 1963; Swartz, 1970, 1971; Cox & Heckly, 1973; Cox, 1973; Israeli et al., 1975). Several of these workers believed that free radical formation was the cause of death on exposure to oxygen (Lion et al., 1961; Dimmick et al., 1961) but this theory has been disputed by Cox & Heckly (1973) who could not correlate loss of viability with the production of oxygen-induced free radicals. Recently, Israeli et al. (1975) have suggested that freeze-drying damages the DNA-initiation mechanism and that exposure to oxygen makes this damage irreversible.

Poor survival levels are obtained generally by storing freeze-dried cells in air, but it is interesting to note that Obayashi et al. (1961) obtained better survival of *L. bifidus* stored in air than under vacuum. However, Damjanovic & Radulovic (1969) came to the opposite conclusion. They also found that storage in nitrogen and carbon dioxide were as effective as storage under vacuum. Gheorghiu & Sturdza (1969) said that storage of BCG vaccine in argon was as effective as storage *in vacuo*. Marshall et al. (1973) showed that survival levels of *Ps. fluorescens* and *Salmonella newport* after storage in various gases depended upon the composition of the suspending medium and the residual moisture content. When the bacteria were dried in papain digest broth, death in air was rapid at zero and fairly high ($0.40\ a_w$) water activities. Storage in nitrogen or under vacuum gave better results than did storage in carbon dioxide or argon. When a sucrose-glutamate suspending medium was used, differences between the various atmospheres were small and the residual moisture content was less important.

In most culture collections freeze-dried cultures are stored in sealed, evacuated ampoules. This is probably a more reliable method than using ampoules containing gases which may not be free from traces of oxygen. Also, Greiff et al. (1975) have shown recently that glass ampoules containing gases at atmospheric pressure may in fact become porous during 'sealing' and for long-term high temperature storage they advise the use of a neoprene 'over-seal'.

It is generally recognized that survival levels decrease with increasing storage temperature and there is little to discuss here other than to say that many workers (e.g. Redway & Lapage, 1974) now use the 'accelerated storage test' (Greiff & Rightsel, 1964; Damjanovic & Radulovic, 1968) whereby the long-term storage potential of a particular dried organism is predicted from its known survival following short-term storage at elevated temperatures.

6. Resuscitation

The terms viability and survival levels have been used throughout this review, but it must be remembered that these are both relative terms, depending on the conditions under which dried bacteria are resuscitated. Freeze-drying causes varying degrees of sublethal cellular injury and vastly different counts may be obtained for the same organism freeeze-dried under the same conditions but resuscitated on different media. Since the introduction of the concept of temporary metabolic injury (Straka & Stokes, 1959) it has become increasingly recognized that, in general, the use of nutritionally deficient media for resuscitation may result in metabolically injured cells being unable to grow and hence misleading estimates of survival levels may be obtained.

Apart from failure to recover injured cells during resuscitation, there is evidence that rehydration itself may result in loss of viability. Fry (1966) has discussed the rather conflicting reports which have appeared on this subject and in general it seems that the rate and temperature of rehydration and the osmolality of the recovery medium may influence resuscitation (Leach & Scott, 1959; Morichi *et al.*, 1967; Choate & Alexander, 1967; Ray *et al.*, 1971). At the NCIB, liquid growth media at room temperature are used routinely for the rehydration of most cultures. Rehydration occurs at an uncontrolled rate and while these conditions may be far from optimal, the prime concern is to obtain a viable culture; quantitative recovery is normally not important.

7. Variation Between Organisms

A major influence on the survival of freeze-dried bacteria which has already been mentioned in passing is the nature of the organism itself. There is some evidence that the age of a particular culture when it is freeze-dried can affect its subsequent survival. According to Fry (1966) most organisms are least resistant to drying during the logarithmic growth phase but as he points out, in the case of BCG at least, there is conflicting evidence. Amarger *et al.* (1972) claimed recently that logarithmic cultures of *Rhizobium meliloti* gave higher counts immediately after freeze-drying, but that older cultures were better able to survive storage at $30°$ in the freeze-dried state.

Although the age of the culture has some influence on survival, much more significant is the great variation between different organisms. All of the factors influencing survival which we have discussed depend to some extent on the nature of the organism being freeze-dried and every newly described species is an unknown quantity in terms of its survival potential in the freeze-dried state. The literature abounds in information on the immediate postdrying recovery and shelf life of different bacteria in different suspending media and the collation of such information will not be attempted here. A very useful Table giving data on the survival during storage of *c.* 200 species may be found in the recent monograph by Lapage & Redway (1974).

It is often said that Gram positive bacteria survive freeze-drying better than do Gram negative bacteria. Certainly there are Gram negative bacteria which have never been freeze-dried successfully (e.g. extreme halophiles) and there are some which are very sensitive to freeze-drying (e.g. spirochaetes, vibrios, neisseriae). However, such generalizations should be viewed with caution. Higher postdrying recoveries may be generally obtained with Gram positive organisms and the death rate during storage may be lower, but from the culture collection point of view, what matters is that some cells remain viable so that subcultures may be made from the freeze-dried material. Many of the results to be found in the literature on freeze-drying express survival levels as percentages of the original population, but these figures may be misleading at first sight. In our opinion, the expressing of results in terms of numbers of decimal reductions (see e.g. Redway & Lapage, 1974) has more meaning. After all, if 99.9% of a population of 10^9 cells are killed during freeze-drying, there still remain 10^6 viable cells, which is quite a lot.

8. References

ALEXANDER, M. T. (1973). A review of the freezing and freeze-drying procedures at the ATCC (abstract). *Cryobiology* **10**, 468.
AMARGER, N., JACQUEMETTON, M. & BLOND, G. (1972). Influence de l'âge de la culture sur la survie de *Rhizobium meilioti* à la lyophilisation et à la conservation après lyophilisation. *Archiv für Mikrobiologie* **81**, 361-366.
ANNEAR, D. I. (1958). Observations on drying bacteria from the frozen state and from the liquid state. *Australian Journal of Experimental Biology and Medical Science* **36**, 211-221.
BENEDICT, R. G., SHARPE, E. S., CORMAN, J., MEYERS, G. B., BAER, E. F., HALL, H. H. & JACKSON, R. W. (1961). Preservation of microorganisms by freeze-drying. II. The destructive action of oxygen. Additional stabilizers for *Serratia marcescens*. Experiments with other microorganisms. *Applied Microbiology* **9**, 256-262.
BOUSFIELD, I. J. & MacKENZIE, A. R. (1972). The routine control of contamination in a culture collection. In *Safety in Microbiology,* eds Shapton, D. A. & Board, R. G. London & New York: Academic Press.
CHO, C. & OBAYASHI, Y. (1956). Effect of adjuvant on preservability of dried BCG vaccine at 37°C. *Bulletin of the World Health Organization* **14**, 657-669.
CHO, C., OBAYASHI, Y., IWASAKI, T. & KAWASAKI, J. (1956). Effect of storage at 37°C on immunizing power of dried BCG vaccine. *Bulletin of the World Health Organization* **14**, 671-680.
CHOATE, R. V. & ALEXANDER, M. T. (1967). The effect of the rehydration temperature and rehydration medium on the viability of freeze-dried *Spirillum atlanticum*. *Cryobiology* **3**, 419-422.
COX, C. S. (1973). Oxygen-induced free radicals and viable decay in freeze-dried bacteria. *Annexe 1973-5 to the Bulletin of the International Institute of Refrigeration, Commission Cl.,* pp. 55-60.
COX, C. S. & HECKLY, R. J. (1973). Efects of oxygen upon freeze-dried and freeze-thawed bacteria: viability and free radical studies. *Canadian Journal of Microbiology* **19**, 189-194.
DAMJANOVIC, V. & RADULOVIC, D. (1967). Survival of *Lactobacillus bifidus* after freeze-drying. *Cryobiology* **4**, 30-32.
DAMJANOVIC, V. & RADULOVIC, D. (1968). Predicting the stability of freeze-dried *Lactobacillus bifidus* by the accelerated storage test. *Cryobiology* **5**, 101-104.
DAMJANOVIC, V. & RADULOVIC, D. (1969). The effect of different gases on the stabilities of freeze-dried suspensions of *Lactobacillus bifidus*. *Annexe 1969-9 to the*

Bulletin of the International Institute for Refrigeration, Commission X, pp. 205-210.
DAVIES, J. D. (1968a). Freeze-drying biological materials—Part 1. *Process Biochemistry* 3, 11-14 & 21.
DAVIES, J. D. (1968b). Freeze-drying biological materials—Part 2. *Process Biochemistry* 3, 48-52.
DIMMICK, R. L., HECKLY, R. J. & HOLLIS, D. P. (1961). Free radical formation during storage of freeze-dried *Serratia marcescens*. *Nature, London* 192, 776-777.
DOEBBLER, G. F. (1966). Cryoprotective compounds. Review and discussion of structure and function. *Cryobiology* 3, 2-11.
ELSER, W. J., THOMAS, R. A. & STEFFEN, G. I. (1935). The desiccation of sera and other biological products (including microorganisms) in the frozen state with preservation of the original qualities of products so treated. *Journal of Immunology* 28, 433-473.
FISHER, P. J. (1950). Viability of dried cultures. A note on the immediate death rate. *Journal of General Microbiology* 4, 455-456.
FRY, R. M. (1951). The influence of the suspending fluid on the survival of bacteria after drying. In *Freezing and Drying*, ed. Harris, R. J. C. London: Institute of Biology.
FRY, R. M. (1954). The preservation of bacteria. In *Biological Applications of Freezing and Drying*, ed. Harris, R. J. C. London & New York: Academic Press.
FRY, R. M. (1966). Freezing and drying of bacteria. In *Cryobiology*, ed. Meryman, H. T. London & New York: Academic Press.
FRY, R. M. & GREAVES, R. I. N. (1951). The survival of bacteria during and after drying. *Journal of Hygiene* 49, 220-246.
GHEORGHIU, M. & STURDZA, S. A. (1969). L'argon tampon protecteur dans le conditionnement du BCG lyophilisé. *Annexe 1969-9 to the Bulletin of the International Institute for Refrigeration, Commission X*, pp. 211-216.
GREAVES, R. I. N. (1944). Centrifugal vacuum freezing. Its application to the drying of biological materials from the frozen state. *Nature, London* 153, 485-487.
GREAVES, R. I. N. (1946). The preservation of proteins by drying. *MRC Special Report Series* No. 258. London: HMSO.
GREAVES, R. I. N. (1960). Some factors which influence the stability of freeze-dried cultures. In *Recent Research in Freezing and Drying*, eds Parkes, A. S. & Smith, A. U. Oxford: Blackwell Scientific Publications.
GREAVES, R. I. N. & DAVIES, J. D. (1965). Separate effects of freezing, thawing and drying living cells. *Annals of the New York Academy of Sciences* 125, 548-558.
GREAVES, R. I. N., DAVIES, J. D. & STEELE, P. R. M. (1967). The freeze-drying of frost-sensitive organisms. *Cryobiology* 3, 283-287.
GREIFF, D. & RIGHTSEL, W. (1964). An accelerated storage test for predicting the stability of suspensions of measles virus dried by sublimation *in vacuo*. *Journal of Immunology* 94, 395-400.
GREIFF, D., MELTON, H. & ROWE, T. W. G. (1975). On the sealing of gas-filled glass ampoules. *Cryobiology* 12, 1-14.
HECKLY, R. J., DIMMICK, R. L. & WINDLE, J. J. (1963). Free radical formation and survival of lyophilized microorganisms. *Journal of Bacteriology* 85, 961-965.
HORNIBROOK, J. W. (1949). A simple, inexpensive apparatus for the desiccation of bacteria and other substances. *Journal of Laboratory and Clinical Medicine* 34, 1315-1320.
HORNIBROOK, J. W. (1950). A useful menstruum for drying organisms and viruses. *Journal of Laboratory and Clinical Medicine* 35, 788-792.
ISRAELI, E., KOHN, A. & GITELMAN, J. (1975). The molecular nature of damage by oxygen to freeze-dried *Escherichia coli*. *Cryobiology* 12, 15-25.
JAKOVLJEVIC, D. (1972). Long term storage of *Vibrio fetus* cultures by freeze-drying. *Australian Veterinary Journal* 48, 21.
LAPAGE, S. P. & REDWAY, K. F. (1974). Preservation of bacteria with notes on other microorganisms. *Public Health Laboratory Service Monograph Series*, No. 7. London: H.M.S.O.

LAPAGE, S. P., SHELTON, J. E., MITCHELL, T. G. & MacKENZIE, A. R. (1970). Culture collections and the preservation of bacteria. In *Methods in Microbiology* Vol. 3A, eds Norris, J. R. & Ribbons, D. W. London & New York: Academic Press.

LEA, C. H. & HAWKE, J. C. (1952). Lipovitellin. 2. The influence of water on the stability of lipovitellin and the effects of freezing and of drying. *Biochemical Journal* **52**, 105-114.

LEACH, R. H. & SCOTT, W. J. (1959). The influence of rehydration on the viability of dried microorganisms. *Journal of General Microbiology* **21**, 295-307.

LEIFSON, E. (1936). The preservation of bacteria by drying *in vacuo*. *American Journal of Hygiene* **23**, 231-236.

LESHCHINSKAYA, E. N. (1944). The immunizing value of the BCG dry glucose vaccine. *Problemy Tuberkulosis* **6**, 55-59. (English translation; *American Review of Soviet Medicine,* Feb. 1946, pp. 210-215.)

LEVITT, J. (1962). A sulfhydryl-disulfide hypothesis of frost injury and resistance in plants. *Journal of Theoretical Biology* **3**, 355-391.

LION, M. B. (1963). Quantitative aspects of the protection of freeze-dried *Escherichia coli* against the toxic effect of oxygen. *Journal of General Microbiology* **32**, 321-329.

LION, M. B. & BERGMANN, E. D. (1961a). The effect of oxygen on freeze-dried *Escherichia coli*. *Journal of General Microbiology* **24**, 191-200.

LION, M. B. & BERGMANN, E. D. (1961b). Substances which protect lyophilized *Escherichia coli* against the lethal effect of oxygen. *Journal of General Microbiology* **25**, 291-296.

LION, M. B., KIRBY-SMITH, J. S. & RANDOLPH, M. L. (1961). Electronspin resonance signals from lyophilized bacterial cells exposed to oxygen. *Nature, London* **192**, 34-36.

LITVAN, G. G. (1972). Mechanism of cryoinjury in biological systems. *Cryobiology* **9**, 182-191.

LOVELOCK, J. E. (1953). The haemolysis of human red blood cells by freezing and thawing. *Biochima et biophysica acta* **10**, 414-426.

LOVELOCK, J. E. (1957). The denaturation of lipid protein complexes as a cause of damage by freezing. *Proceedings of the Royal Society, Series B,* **147**, 427-433.

MARSHALL, B. J., COOTE, G. G. & SCOTT, W. J. (1973). Effects of various gases on the survival of dried bacteria during storage. *Applied Microbiology* **26**, 206-210.

MARSHALL, B. J., COOTE, G. G. & SCOTT, W. J. (1974). A study of factors affecting the survival of dried bacteria during storage. *Commonwealth Scientific and Industrial Research Organization, Australia. Division of Food Research Technical Paper* No. 39, 1-29.

MARSHALL, B. J. & SCOTT, W. J. (1970). The effects of some solutes on preservation of dried bacteria during storage *in vacuo*. In *Culture Collections of Microorganisms,* eds Iizuka, H. & Hasegawa, T. Baltimore & Manchester: University Park Press.

MAZUR, P. (1963). Kinetics of water loss from cells at subzero temperatures and the likelihood of intracellular freezing. *Journal of General Physiology* **47**, 347-369.

MAZUR, P. (1966). Physical and chemical basis of injury in single-celled microorganisms subjected to freezing and thawing. In *Cryobiology,* ed. Meryman, H. T. London & New York: Academic Press.

MERYMAN, H. T. (1966a). Freeze-drying. In *Cryobiology,* ed. Meryman, H. T. London and New York: Academic Press.

MERYMAN, H. T. (1966b). Review of biological freezing. In *Cryobiology,* ed. Meryman, H. T. London & New York: Academic Press.

MERYMAN, H. T. (1970). The exceeding of a minimum tolerable cell volume in hypertonic suspension as a cause of freezing injury. In *The Frozen Cell,* eds Wolstenholme, G. E. W. & O'Connor, M. London: Churchill.

MERYMAN, H. T. (1971). Cryoprotective agents. *Cryobiology* **8**, 173-183.

MILLER, R. & GOODNER, K. (1953). Studies on the stability of lyophilized BCG vaccine. *Yale Journal of Biology and Medicine* **25**, 262-283.

MORICHI, T. (1970). Nature and action of protective solutes in freeze-drying of bacteria. In *Culture Collections of Microorganisms,* eds Iizuka, H. & Hasegawa, T. Baltimore and Manchester: University Park Press.
MORICHI, T., IRIE, R., YANO, N. & KEMBO, H. (1967). Death of freeze-dried *Lactobacillus bulgaricus* during dehydration. *Agricultural and Biological Chemistry* 31, 137-141.
NAYLOR, H. B. & SMITH, P. A. (1946). Factors affecting the viability of *Serratia marcescens* during dehydration and storage. *Journal of Bacteriology* 52, 565-573.
NEI, T., SOUZU, H. & ARAKI, T. (1966). Effect of residual moisture content on the survival of freeze-dried bacteria during storage under various conditions. *Cryobiology* 2, 276-279.
NIKOLOV, N. M. (1973). Properties of lyophilized cultures of lactic acid bacteria in sour milk. *Mikrobiologiya* 42, 1049-1051 (932-934 Engl.).
OBAYASHI, Y. (1960). The preservation of BCG. In *Recent Research in Freezing and Drying,* eds Parkes, A. S. & Smith, A. U. Oxford: Blackwell Scientific Publications.
OBAYASHI, Y. & CHO, C. (1957). Further studies on the adjuvant for dried BCG vaccine. *Bulletin of the World Health Organization* 17, 255-274.
OBAYASHI, Y., KAWASAKI, J., YOSHIOKA, T., SHIMAO, T. & NOGUCHI, T. (1957). Effect of storage at 37° on the allergenic potency of dried BCG vaccine. *Bulletin of the World Health Organization* 17, 275-287.
OBAYASHI, Y., OTA, S. & ARAI, S. (1961). Some factors affecting the preservability of freeze-dried bacteria. *Journal of Hygiene* 59, 77-91.
ORNDORFF, G. R. & MacKENZIE, A. P. (1973). The function of the suspending medium during the freeze-drying preservation of *Escherichia coli. Cryobiology* 10, 475-487.
OTTEN, L. (1930). Die Trockenkonservierung von pathogenen Bakterien. *Zentralblatt für Bakteriologie, Parasitenkunde, Infektionskrankheit und Hygiene, Abteilung* I. *Originale* 116, 199-210.
OTTEN, L. (1932). The preservation of viability and virulence in dried pathogen bacteria. *Transactions of the 8th Congress of the Far East Association for Tropical Medicine* 1930, p. 89.
PERRY, L. B., BOUSFIELD, I. J. & SHEWAN, J. M. (1975). Notes on the preservation and checking of vitamin assay bacteria. In *Some Methods for Microbiological Assay* eds Board, R. G. & Lovelock, D. W. London & New York: Academic Press.
PROOM, H. & HEMMONS, L. M. (1949). The drying and preservation of bacterial cultures. *Journal of General Microbiology* 3, 7-18.
RAY, B., JEZESKI, J. J. & BUSTA, F. F. (1971). Effect of rehydration on recovery, repair and growth of injured freeze-dried *Salmonella anatum. Applied Microbiology* 22, 184-189.
REDWAY, K. F. & LAPAGE, S. P. (1974). Effect of carbohydrates and related compounds on the long term preservation of freeze-dried bacteria. *Cryobiology* 11, 73-79.
ROGERS, L. A. (1914). The preparation of dried cultures. *Journal of Infectious Diseases* 14, 100-123.
ROWE, T. W. G. (1970). Freeze-drying of biological materials: some physical and engineering aspects. In *Current Trends in Cryobiology,* ed. Smith, A. U. New York: Plenum Press.
ROWE, T. W. G. (1971). Machinery and methods in freeze-drying. *Cryobiology* 8, 153-172.
SCOTT, W. J. (1958). The effect of residual water on the survival of dried bacteria during storage. *Journal of General Microbiology* 19, 624-633.
SCOTT, W. J. (1960). A mechanism causing death during storage of dried microorganisms. In *Recent Research in Freezing and Drying,* eds Parkes, A. S. & Smith, A. U. Oxford: Blackwell Scientific Publications.
SINHA, R. N., DUDANI, A. T. & RANGANATHAN, B. (1974). Effect of individual ingredients of fortified skim milk as suspending media on survival of freeze-dried cells of *Streptococcus lactis. Cryobiology* 11, 368-370.
SPLITSTOESSER, D. F. & FOSTER, E. M. (1957). The influence of drying conditions on

the survival of *Serratia marcescens*. *Applied Microbiology* **5**, 333-339.
STAMP, LORD (1947). The preservation of bacteria by drying. *Journal of General Microbiology* **1**, 251-265.
STEEL, K. J. & ROSS, H. E. (1963). Survival of freeze-dried bacterial cultures. *Journal of Applied Bacteriology* **26**, 370-375.
STRAKA, R. P. & STOKES, J. L. (1959). Metabolic injury to bacteria at low temperatures. *Journal of Bacteriology* **78**, 181-185.
SWARTZ, H. M. (1970). Effect of oxygen on freezing damage. I. Effect on survival of *Escherichia coli* B/r and *Escherichia coli* B_{s-1}. *Cryobiology* **6**, 546-551.
SWARTZ, H. M. (1971). Effect of oxygen on freezing damage. II. Physical-chemical effects. *Cryobiology* **8**, 255-264.
UNGAR, J., MUGGLETON, P. W., DUDLEY, J. A. R. & GRIFFITHS, M. I. (1962). Preparation and properties of a freeze-dried BCG vaccine of increased stability. *British Medical Journal,* Oct. 27, **ii,** 1086-1089.

Practical and Legislative Aspects of the Chemical Preservation of Food

B. JARVIS AND CAROLE S. BURKE

British Food Manufacturing Industries Research Association, Randalls Road, Leatherhead, Surrey, England

CONTENTS

1. Introduction 345
2. Food preservative legislation in Great Britain 346
3. Legislation on preservatives in other countries 348
4. Why do we need chemical preservation of foods? 354
5. Practical considerations affecting the choice of chemical preservative 357
6. Future trends in chemical preservation of foods 362
7. Summary 364
8. Acknowledgments 365
9. References to cited food legislation for England and Wales 365
10. References 367

1. Introduction

SINCE EARLIEST TIMES man has found it necessary to preserve foods for use in times of shortage and processes such as drying, salting and fermentation have been used for many centuries. With the transition which has occurred over the past 200 years from a self-sufficient rural population to a dependent urban population, food preservation has become of increasing economic importance. The nineteenth century industrial revolution, which led to a massive increase in the sizes of urban populations, stimulated the development of large-scale food manufacture and distribution.

During the latter part of the eighteenth and the first half of the nineteenth centuries adulteration of foods was widespread. The objective of such adulteration was not to preserve foods but to add bulk or colour, or to otherwise disguise poor quality materials (Accum, 1820; Amos, 1960). Concern about the adulteration of foods in Great Britain led to the establishment in 1850 of the Lancet Analytical Sanitary Commission and in 1855 a Parliamentary Committee was established to investigate the problem. The outcome was the introduction of the first ever 'pure food' legislation (Adulteration of Food and Drink Act, 1860). Although this was administratively a poor law it led to more specific legislation (The Sale of Food and Drugs Act, 1875) which was concerned *a priori* with ensuring that foods (and drugs) were "of the proper nature, substance and quality". This

Act laid down provision for the appointment and duties of analysts, and stipulated penalties for infringement of the provisions of the Act (e.g. mixing injurious substances with food).

The addition to foods of chemical preservatives was not specifically proscribed by the 1875 Act, although action could be taken against a manufacturer on the basis that the food sold was "not of the nature, substance or quality demanded by the consumer" or that the food was adulterated with "injurious substances". The types of chemical preservative often added to foods during the latter part of the nineteenth and early twentieth centuries suggest that the food manufacturers of that time were perhaps more impressed by the work of Lister on antiseptics than they were by the studies of Pasteur. Preservatives commonly used during that period included alum, boric acid and borax, carbolic acid, creosote, formaldehyde, salicylic acid, thymol, acetic, benzoic and sulphurous acids. Very properly most of these 'preservatives' are now totally excluded from foods.

2. Food Preservative Legislation in Great Britain

Legislation to control preservatives in butter and margarine was introduced in the 1907 Amendment to the Sale of Food and Drugs Act, 1899. Controls for milk and cream were introduced by Regulations passed in 1912 in accordance with the Public Health (Regulations as to Food) Act, 1907. In spite of many public objections, some food manufacturers continued to add a variety of preservatives to foods; others, however, avoided their use so that the label "free from preservatives" became regarded by the consumer as a mark of quality. Legislation to control chemical preservatives in foods was eventually introduced in 1927-8 (The Public Health (Preservatives, etc., in Food) Regulations 1925, Amend. 1926 and 1927) following the Report of a Government Committee (*Anon.* 1924).

The 1925 Regulations (amended) prohibited the sale of foods containing preservative, other than those foods and preservatives which were specifically scheduled; it was required also that foods containing preservative, and compounds sold for purposes of food preservation, should bear appropriate labels. Preservatives were defined in the Regulations as

> "any substance capable of inhibiting, retarding, or arresting the process of fermentation, acidification, or other decomposition of food or of masking any of the evidence of putrefaction, but does not include common salt (sodium chloride), saltpetre (sodium or potassium nitrate), sugars, lactic acid, acetic acid or vinegar, glycerine, alcohol or potable spirits, herbs, hop extracts, spices and essential oils used for flavouring purposes, or any substance added to food by the process of curing known as smoking".

CHEMICAL PRESERVATION OF FOOD 347

The only preservatives specifically permitted by these regulations were sulphur dioxide and benzoic acid. Nevertheless, prosecutions for addition of prohibited preservatives continued long after the introduction of legislation (Liverseege, 1932).

The 1925 Regulations were amended several times over the next 33 years. The first major amendment (other than those of 1926-27) was passed in 1940 (S.R. & O. 1940/633) and permitted the presence in bacon, ham and pickled meats of up to 200 p/m of sodium nitrite at the time of sale. This amendment was introduced as a result of observations that cured meats invariably contained nitrite which had been produced by bacterial reduction of the nitrate (saltpetre) used in the curing process.

A subsequent amendment (S.I. 1953/1610 revised in S.I. 1953/1820) extended the use of sulphur dioxide to imported dehydrated vegetables and permitted citrus fruits imported in biphenyl-treated wrappers to contain residues of this fungistat. For a limited period, it also permitted the addition of up to 2500 p/m of boric acid to margarine, but this provision ceased to have effect two months after the end of margarine rationing in this country (9th May, 1954).

Further amendments in 1958 (S.I. 1958/1319 and S.I. 1958/2167) permitted various levels of o-phenylphenol residues in specific imported fruits, thereby bringing the U.K. Regulations more closely into line with revised residue tolerances adopted in the U.S.A.

In 1962, completely revised Regulations (Preservatives in Food Regulations, 1962) were introduced. These had the effect of widening the classes of foods to which preservatives could be added and of extending the list of food additives which were specifically excluded from the Regulations. Esters of p-hydroxybenzoic acid (parabens), sorbic acid, propionic acid and tetracyclines were added to the list of permitted preservatives for specified foods; the inclusion of nisin (in cheese, clotted cream and canned foods) and nystatin (on the skin but not the flesh of bananas) were permitted under the exclusions clause (Regulation 3). No restrictions were introduced for nitrate in meats, but the earlier Regulations for nitrite (S.R. & O. 1940/633) were modified to permit the presence at the time of sale of up to 200 p/m of nitrite in cooked pickled meats and up to 500 p/m of nitrite in uncooked pickled meats. However, an anomaly under Regulation 3 permitted nitrite in bacon and ham without restriction on amount. These regulations were subsequently amended (S.I. 1971/882) to restrict nitrite and nitrate levels to 200 p/m and 500 p/m, respectively, in all cooked and uncooked pickled meats and in bacon and ham.

New Regulations were introduced in 1974 (Preservatives in Food Regulations, 1974; S.I. 1974/1119) in order to harmonize the U.K. Regulations with those for the European Economic Community. These made provision for widening the permitted uses of parabens, revocation of permission to use tetracyclines in fish and changes in the Regulations for soft drinks and for wines (e.g. sorbic acid permitted at 200 p/m in wine). The Regulations accepted current changes in the use of preservatives for imported fruits and permit residues of biphenyl (70 p/m), 2-hydroxy-biphenyl (12 p/m) or 2-(thiazol-4-yl)-benzimidazole (TZBZ) (6 p/m) in citrus fruits and 3 p/m TZBZ in bananas. An important innovation in the 1974 Regulations was the introduction of purity criteria for preservatives, based on EEC Directive or FAO specifications.

3. Legislation on Preservatives in Other Countries

Legislative control of food preservatives has generally followed a similar course in other countries, although controls were introduced somewhat earlier in some countries (e.g. 1906, U.S.A.; 1913, Denmark) than in Great Britain. All developed countries now have legislative control of the use of food additives, including preservatives. The current (1975) permitted levels for preservatives in foods in most European countries and in the U.S.A. and Canada are presented in Tables 1-5; a summary of typical levels for some common preservatives in particular foods is shown in Table 6. Primary data for these Tables were obtained from the various national food laws as detailed in the Overseas Food Legislation Manual (Burke, 1970). It should be noted that although pimaricin is permitted for use in cheese rind in several EEC countries (Table 5) it is not an EEC permitted preservative.

It is interesting to compare different national attitudes towards use of specific preservatives in foods. Of the European countries, France permits fewest preservatives; sulphur dioxide is permitted only in certain fruit preparations, wines and mustard; propionates and acetates in bread; hexamine in certain fish products; nisin in processed cheese; pimaricin in cheese rinds; and nitrites in pickled meats. Most other European countries are more catholic in their use of preservatives but notable differences occur, both in the types and levels permitted for particular foods. In fruit pulp, for instance, sulphur dioxide is permitted at levels ranging from 50-2000 p/m (and some countries do not specify a maximum level); benzoates are permitted at levels of 250-1500 p/m; parabens at 800-1000 p/m; and sorbates at 250-2000 p/m. With such differences in national legislation it is perhaps not surprising that present-day attempts to harmonize food legislation within the European Economic Community are fraught with nationalistic problems and that manufacturers exporting foods need to tread very warily in order not to infringe specific national food laws.

Table 1

Maximum permitted levels (p/m) for sorbates (sorbic acid and its salts) in foods in various countries

Food	Belgium	Canada	Denmark	Finland	France	Germany, W	Italy	Netherlands	Norway	Sweden	UK	USA
Baked goods	1000	1000	1000	1000			2000		2000	2000		GMP
Bread*	1000	1000		1000		2000			2000	2000	1000	GMP
Candied fruit	1000	1000	1000	1000			1000		1000	2000		GMP
Cheese, including processed	1000	1000	1000	2000			1000	1000	1000	2000	1000	2000
Fish semi-preserves	3000	3000	1000	1000		2000	1000		2000	2000		GMP
Fruit juice	1000		500	1000		2000		5000	1000	1000		
Fruit pulp		1000		1000		2000		250	1000	2000		GMP
Jam	1000	1000	1000			2000		250	1000	1000		GMP
Liquid egg, whole or yolk	1000		5000			††		250††				
Margarine	1000					1000		12500	1000	2000		
Marzipan		1000	1500	1000		1200		2000	1000	2000	1000	1000
Mayonnaise	1000		1000	1000		1500	1000	1000	1000	2000		GMP
Mustard	250		1000			2500	1000	250	1000	2000		GMP
Pickles		1000	1000	1000		1000			1000	2000		GMP
Sauces, ketchups		1000	1000	1000		1500		1000	1000	2000		GMP
Soft drinks containing fruit juice	100	1000	500	1000		2500		250	1000	1000		GMP
Soft drinks, flavoured, possibly carbonated	100	1000	500	1000		1000‡		75	1000	1000		GMP

GMP   Good manufacturing practice.
*   Certain types only.
†   Surface treatment only.
††   Second grade jam only.
‡   In the base.

Table 2

Maximum permitted levels (p/m) for benzoates (benzoic acid and sodium benzoate) in foods in various countries

Food	Belgium	Canada	Denmark	Finland	France	Germany, W	Italy	Netherlands	Norway	Sweden	UK	USA
Fish semi-preserves	1500	1000	1000	2000		2500	1500	5000	5000	2000		1000
Fruit juice		1000	200	1500		1000		250	1500	1000	800	
Fruit pulp		1000		1500				250	1500	1000		
Jam		1000	500	1500		+*		250††	1500	1000	800	1000
Liquid egg, whole or yolk			500			10000		12500	5000	10000		1000
Margarine	1000			3000				2000		2000		2000
Mayonnaise etc	1000	1000	1000	2000		2500		1000	3000	2000		1000
Mustard	250	1000	1000	2000		1000		250	3000	2000		1000
Pickles		1000	1000	2000		2000			1500	2000		1000
Sauces, ketchups		1000	1000	2000		2500		1000	1500	2000	250	1000
Soft drinks containing fruit juice	100	1000	200	1500		1000†		250	1500	1000	250	1000
Soft drinks, flavoured, possibly carbonated	100	1000	200	1500	160			75	500	1000	800 ‡ 160 ‡‡	1000 GMP

GMP Good manufacturing practice.
*   Surface treatment only.
†   Of the base.
††   Only in second grade jam.
‡   Soft drinks for consumption after dilution.
‡‡   Soft drinks for consumption without dilution.

Table 3

Maximum permitted levels (p/m) for parabens (parahydroxybenzoic acid esters and their salts) in foods in various countries

Food	Belgium	Canada	Denmark	Finland	France	Germany, W	Italy	Netherlands	Norway	Sweden	UK	USA
Fish semi-preserves		1000	300	1000		1000			500	500		
Fruit juice		1000	200	1000					900	1000		
Fruit pulp		1000		1000					900			
Jam		1000		1000			1000		900		800	GMP
Mayonnaise etc			300	1000		1200						
Mustard			300	200		1500	1000					
Pickles		1000	300	1000					900	1000	250	
Sauces, ketchups		1000	300	1000		1500			900	1000	250	
Soft drinks containing fruit juice		1000	200	1000					900	1000		
Soft drinks, flavoured, possibly carbonated		1000	200	1000						1000		GMP

GMP Good manufacturing practice.

Table 4

Maximum permitted levels (p/m) for sulphur dioxide (sulphites and related salts) in foods in various countries

Food	Belgium	Canada	Denmark	Finland	France	Germany, W	Italy	Netherlands	Norway	Sweden	UK	USA
Candied fruit		500	100		GMP	100	100		50	200	100	GMP
Dried fruit	2000	2500	1000	2000	2000	1500	100		2000	1000	2000	GMP
Dried vegetables		2500	1000	800	1000	500			2000	1000	2000	GMP
Fruit juice	50	500		800	100	300	50	100	50		350	
Fruit pulp*	2000	500		800		+	+	75	50		+	GMP
Jam	40	500	100	800	20	50	80	50†	50		100	GMP
Mustard	250	500			500					200		GMP
Pickles		500					50		50	200	100	GMP
Sauces, ketchups		500				20			50	200	100	GMP
Soft drinks containing fruit juice		100	25	100			20	100	35	50	350	††GMP
Soft drinks, flavoured, possibly carbonated	+	100	25	100				75	15	50	70	GMP

GMP Good manufacturing practice.
* Includes pulp for further processing.
† Second grade jam 75 p/m.
†† For consumption after dilution.
‡ For consumption without dilution.

Table 5

Maximum permitted levels (p/m) for various preservatives in foods in various countries

Preservative	Food	Belgium	Canada	Denmark	Finland	France	Germany, W	Italy	Netherlands	Norway	Sweden	UK	USA
Propionates	Baked goods	1000	2000	3000	3000		+*	2000		5000	3000	1000	GMP
	Bread†	1000	2000	3000	3000	5000		2000	3000	5000	3000	3000	3200
	Cheese including processed		2000		3000			+					3000
Nitrates	Pickled meat	500	+	500	500	+	500	250	2000			500	+
Nitrites	Pickled meat	200	200	175	150	150	+	150	500	200	200	200	200
Pimaricin	Cheese rind	+				+		+	+		+	+	
Nisin	Processed cheese	+		200	+	+		+				+	
Ethylene oxide	Spices	50	+						50				50
Hexamine	Fish and caviar	1000		500	5000	1000				1000‡	500	+††	
Acetates Ca²⁺ or Na⁺	Bread	+	3000	+	5000	4300	+	4000	4000	+	+	+	4000

GMP Good manufacturing practice.
*   Permitted without restriction on amount (+).
†   Certain types only.
††   As organicide or fungicide only.
‡   Until December 1976 only.

Table 6

Summary of typical permitted levels (p/m) of various common preservatives used in foods

Food	Benzoates	Parabens	Sorbates	Propionates	Sulphur Dioxide (Range)
Cereal products					
Baked goods			1000	2000	
Bread			1000	3000	
Dairy products					
Cheese including processed			1000	3000	
Liquid egg, whole or yolk	10000		1000		
Fish products					
Fish semi-preserves	2000	1000	2000		
Fruit and vegetable products					
Candied fruit			1000		50- 200
Dried fruit					100-2500
Dried vegetables					500-2500
Fruit juice	750	1000	1000		50- 800
Fruit pulp	1000	1000	1000		50-2000
Jam	1000	1000	1000		40- 800
Mustard	1500	1000	1000		200- 500
Pickles	1000	1000	1000		20- 500
Sauces and ketchups	1000	1000	1000		50- 500
Soft drinks containing fruit juice	1000	1000	1000		20- 350
Soft drinks, flavoured possibly carbonated	1000	1000	1000		15- 100
Oil-containing products					
Margarine	2000		1000		
Marzipan			1000		
Mayonnaise etc.	2000	1000	1000		

4. Why do We Need Chemical Preservation of Foods?

Arguments for and against the use of chemical preservation are many; some are summarized in Table 7. Although it is clear that chemical preservatives have been used in the past to cover up the use of spoiled (or spoiling) foods, the types and levels permitted at present would not adequately prevent rapid decomposition if growth of spoilage organisms were already advanced. The most emotive argument against the use of chemical preservatives (and other food additives) is that of potential toxicity to the consumer. There is no doubt that some of the preservatives in use up to about 1930 were harmful, especially in the amounts used; (e.g. boric acid in cream, *vide* Cullen v. McNair (1908) (Robinson, 1931). Furthermore, some preservatives introduced since then have subsequently been withdrawn on toxicological grounds (e.g. diethyl pyrocarbonate).

Table 7

The case for and against the use of food preservatives

Arguments for the use of preservatives	Maintain keeping quality and stability Maintain nutritional quality Control food poisoning organisms
Arguments against the use of preservatives	Disguise faulty processing or handling and deceive the consumer Are harmful to the consumer Reduce the nutritional quality of foods Are unnecessary if good manufacturing practice adopted

As a means of advising on the safety-in-use of food chemicals, recommendations have been made by the World Health Organization on the maximum acceptable daily intake (ADI) for food additives (Anon., 1974*b*). Those for the most commonly used chemical preservatives are listed in Table 8. These data were obtained by extrapolation to man of the 'no effect' level observed in toxicological evaluation of the additives in laboratory animals. In most cases a safety factor of 100 has been used in the extrapolation, but other safety factors may be used from time to time. Toxicological testing is undertaken not only on the additives themselves but sometimes also on known chemical degradation products. This has led in certain instances (e.g. diethyl pyrocarbonate) to recommendations that there is sufficient toxicological evidence to prohibit the use of a preservative in foods.

In some instances, the recommended ADI's are only temporary (e.g. nitrite). This is of importance in assessing the continued use of nitrites as permitted preservatives for cured meats and fish. Recent attempts in the U.S.A. to invoke the Delaney Amendment to the U.S. Food Law as a means of prohibiting the use of nitrite illustrate the emotive aspects of the case against preservatives. That such an action might reduce a possible long-term carcinogenic hazard from nitrosamines must be balanced by the increased potential hazard to the consumer from botulism (Jarvis & Walters, 1972). In all such matters a finely drawn hazard-benefit analysis must be made by competent authorities working on scientific data and free from political pressures of all types. Although the levels of permitted preservatives are based on the recommended ADI's, concern has been expressed that widening the classes of foods in which specific preservatives are permitted will result in the Potential Daily Intake (PDI) exceeding significantly the recommended ADI. The ratio PDI/ADI can be used as a relative indication

Table 8

Recommended maximum acceptable daily intake of preservatives

Preservative	ADI (mg/kg body weight)
Propionates	no limit
Nisin	33,000 i.u./kg
Sorbates	25
Diacetate	15
Parabens	10
Benzoates	5
Nitrate	5
Formates	3
Sulphites	0.7
Nitrite	0.2
Diethyl pyrocarbonate	not to be used

Anon. (1974 *b*).

of exposure to particular food additives. In practice, the PDI is a hypothetical value based on all foods in which a permitted preservative can occur; it assumes that the preservative is always present at the maximum permitted level; the specific foods are consumed daily during a lifetime; and that preservatives are not decreased during storage, cooking, etc. (*Anon.*, 1975). For some of the more common preservatives the PDI/ADI ratios have been calculated as: benzoates, 0.82-2.8; sulphur dioxide and sulphites, 2.9-17.0; sorbates, 0.42; nitrates, 0.50; nitrites, 1.2. The higher levels quoted for sulphur dioxide and benzoates relate to the excessive consumption of beverages. In practice, the consumption of sulphur dioxide has been calculated to exceed the ADI by 4.7 times in Israel and by 2.8 times in Belgium. In the U.K., the intake of sulphur dioxide could well exceed the ADI if large quantities of sulphite-preserved foods are consumed.

The most difficult argument to refute is that adoption of Good Manufacturing Practice (GMP) renders chemical preservatives unnecessary. It is interesting to note that the U.S. Food Law permits certain preservatives to be used only in conjunction with GMP and for this purpose restricts the quantity of preservative used to that quantity which is necessary for the particular situation. In the nineteen-twenties it was concluded (*Anon.*, 1924) that widespread use of preservatives was undesirable, but that total prohibition was not justifiable without technological improvements in e.g. refrigerated transportation and storage. In spite of the more widespread use of refrigeration, preservatives are still needed for some perishable foods (e.g. sausages). Furthermore, it has been suggested that for economic reasons the shelf lives of some refrigerated foods could be extended by the addition of a suitable preservative (e.g. sorbic acid in yoghurt). A balance must again be drawn between technological or economic necessity and general desirability. For some foods the more widespread use of deep-frozen storage obviates the need for chemical preservatives but such processes cannot be used for all foods without loss of important characteristics e.g. texture and flavour.

Foods require protection from spoilage not only during distribution but also during domestic storage prior to consumption. However, the manufacturer has no control over domestic (and only rarely over retail) storage conditions and can only recommend maximum times and conditions for storage (cf. *Anon.*, 1974). In some cases preservatives may be added in part to prevent spoilage of multiple-use packs during domestic storage when the product may have become contaminated by a variety of micro-organisms.

5. Practical Considerations Affecting the Choice of Chemical Preservative

The desirable characteristics for an ideal food preservative have been discussed in some detail by Ingram *et al.*, (1964), but none of the presently permitted preservatives possesses all of these characteristics. The user must therefore strive to optimize the desirable characteristics of those preservatives available to him. The potential effectiveness of a food preservative is dependent not only upon the inherent properties of the preservative itself, but also upon the micro-environmental conditions in which it is to be employed. Those factors which may modify the effectiveness of a preservative are summarized in Table 9.

Unless used specifically to combat a particular group of organisms (e.g. pimaricin as a fungistat on cheese rind) it is preferable that a preservative should have a broad spectrum of antimicrobial activity. It is implicit that a food preservative should inhibit both food spoilage organisms and those organisms responsible for food poisoning. A preservative which prevents overt spoilage but permits uninhibited growth of a pathogen, predisposes towards a condition wherein the consumer might not become aware that food had become unfit for safe consumption.

The general antimicrobial spectra and solubilities of food preservatives in common use are summarized in Table 10. Solubility is of importance in

Table 9

Factors affecting preservative activity

The preservative itself	antimicrobial spectrum solubility partition coefficient dissociation constant reactivity relative toxicity
The microflora of the food	numbers, types and condition of organisms resistance to antimicrobial agents
Nature of the food	water activity and pH values redox potential fat content reactive components nature of ingredients
Processing factors	thermal processes dehydration developed preservatives (e.g. acid, smoke, etc.)
Other factors	storage conditions (e.g. temperature) type of packaging

practical applications but can be improved by the use of the preservative in chemically modified forms (e.g. use of sodium benzoate, which has a greater solubility in water at neutral pH values than benzoic acid). In the case of the parabens esters, antimicrobial activity increases with increasing chain length, but there is a concomitant reduction in solubility. The partition coefficient of a preservative may be important in lipid-rich foods since if lipid solubility is high the relative concentration remaining in the aqueous phase of the food may be too low for effective preservation. For most weak acid preservatives—e.g. acetic, lactic, benzoic, propionic, sorbic and sulphurous (SO_2) acids—the dissociation constants are of prime importance. Antimicrobial activity is linked largely with the undissociated acid and for any specific pH value of the food the effectiveness will depend upon the degree of dissociation of the preservative. Nunheimer & Fabian (1940) demonstrated that acetic acid was a more effective inhibitor of *Staphylococcus aureus* than lactic acid or the other acids tested. Similarly it has been shown that the acetic acid content of pickles cannot be replaced in part by lactic acid without reducing the stability of the product, unless an additional aid to preservation is included (Shipp & Alborough, 1954).

Chemical reactivity and volatility of preservatives undoubtedly affects their practical application. For instance, reactions between sulphur dioxide

Table 10

Antimicrobial spectrum, pH dependence and solubility of commonly permitted preservatives

Preservative	Spectrum††	pH Dependence	Solubility in water g/100 at 25°
Sorbates	S, B, b, *Y, M*	±	139 (potassium salt)
Parabens	S, *B*, b, *Y, M*	−	0.25-0.015*
Diacetate	S, B, b, Y, M	±	100 (sodium salt)
Sulphites	*B, b,* Y, M	±	V.S.
Benzoates	*Y,* M, (B)	+	50 (sodium salt)
Propionates	S, *M*	+	150 (sodium salt)
Nisin	S, B	∓	*†
Nitrite	*S, B, b*	±	66 (sodium salt)

* Solubility depends on length alkyl chain.
† Very soluble in dilute acid; solubility in water improved by admixture with carrier proteins (e.g. skim milk powder).
†† S, spores; B, Gram positive bacteria; b, Gram negative bacteria; Y, yeasts; M, moulds; ( ), some strains; Italics, major effects.

and carbohydrates result in effective loss of antimicrobial activities. Similarly, the thermal instability of preservatives like nisin, especially at neutral or alkaline pH value, restricts their potential applications in food preservation. Finally, the relative toxicity of a 'preservative' to micro-organisms and to animal species (especially man) will introduce constraints on the quantity which can be used (see Section 4).

An important characteristic for an ideal food preservative is that it should not induce development of resistance in micro-organisms against which it is intended to be used. Although many strains of osmophilic yeasts (e.g. *Saccharomyces bailii, Sacch. bailii* var. *bisporus, Sacch. pombe,* etc.) are initially sensitive to benzoate and sorbate (Table 11), they have been reported to develop resistance to these inhibitors very readily (Harman *et al.*, 1974) and in an industrial environment resistant strains may soon become the dominant microflora. In order to overcome such problems it may be necessary to use mixtures of preservatives, provided that these interact synergistically (e.g. sulphur dioxide and benzoic acid). A more important criterion is that preservatives must not induce resistance against substances used clinically in human or animal medicine, a major objection to the use of tetracyclines in food preservation (Jarvis & Morisetti, 1969). Indeed, this is also one of the major objections to the use of tetracyclines as animal feed additives (*Anon.*, 1969).

Table 11

Effect of preservatives on spoilage of a 60° brix orange concentrate (pH 3.65) by a mixed inoculum of Saccharomyces bailii, Sacch. bailii *var.* bisporus, Sacch. pombe *and* Sacch. cerevisiae

Preservative	Initial conc. (p/m)	Time (weeks) to spoilage at		
		4° C	8° C	25° C
None	–	> 26	4 (60%)	2 (100%)
SO_2	350	> 26	> 26	5 (20%)
Benzoate	1000	> 26	> 26	6 (20%)
Parabens	500	> 26	4 (20%)	2 (100%)
Sorbate	2000	> 26	> 26	6 (20%)
Pimaricin	100	> 26	> 26	6 (100%)

Whether or not particular organisms will be inhibited by a given concentration of preservative in a particular environment will be dependent on the factors considered previously (i.e. antimicrobial spectrum, resistance, chemical environment, etc.), the condition of the organisms (i.e. whether damaged by exposure to adverse physical or other chemical conditions) and the relative numbers of organisms present in the system. No preservative should be used to cover up the use of spoiled (or spoiling) foods and in most cases where a hitherto satisfactory preservative system fails, it is because of the use of unsatisfactory ingredients or of unhygienic processing conditions.

Only rarely are chemical preservatives used as the sole means of ensuring shelf stability or freedom from spoilage in a product. More normally they are used to change the ecological equilibrium from one in which organisms grow rapidly to one in which growth is retarded or inhibited totally for a period of time. Hence chemical preservatives are used in conjunction with other chemical or physical methods of food preservation. Although such interactions are well known, it is possibly worthwhile to illustrate this concept with respect to a few important food commodities.

In fruit juices and concentrates, having high sugar concentrations and low pH values, growth of bacteria and of some yeasts and moulds is inhibited by the reduced water activity (a_w) The predominant spoilage flora consists largely of osmophilic yeasts; in low numbers these are inhibited by the use of low storage temperatures combined with incorporation of appropriate quantities of preservatives such as sulphur dioxide, benzoate or sorbate. Alternatively, the need for chemical preservatives can be eliminated completely by deep-freezing.

The preservative system responsible for the safety and stability of pasteurized cured meats comprises very complex, but as yet incompletely understood, interactions (Ingram, 1976). For pasteurized cured meats the interactions include: salt x nitrite x other additives (e.g. ascorbate, polyphosphate) x numbers and types of micro-organisms x pH value and cut of meat x process temperature x storage temperature x storage time. The situation is complicated further by the possibility of inhibitors derived from nitrite (Ashworth et al., 1973; Pivnick & Chang, 1974).

Where cured meats and fish are smoked, the process will result in surface dehydration (with associated reduction in a_w) and will introduce small quantities of chemicals derived from the wood smoke e.g. phenols, formaldehyde, etc. It is unlikely that modern 'cold smoke' solutions will have any substantial antimicrobial activity, although those used at the beginning of the century in Germany and other European countries undoubtedly did. A typical recipe quoted by Hausner (1902) contained (gallons): pyroligneous acid, 10; creosote, 1; juniper oil, 1; water 100. The use of such a concoction could not be considered today.

Food additives used for other purposes may in some cases exert antimicrobial activity or may potentiate the effects of added preservatives. Certain spices have been added to foods for many years to enhance preservation and a recent literature survey has uncovered several thousand reports on antimicrobials occurring naturally in plant materials (Hargreaves, Wood & Jarvis, unpublished). In the majority of cases plant extracts have been tested only for *in vitro* antimicrobial activity and in only a very few cases has the nature of the active principle been determined (e.g. allicin from garlic, and eugenol from cloves), but some claims have been made for useful preservative effects in foods. One such claim of recent origin (Bullerman, 1974) confirms earlier reports (Bachmann, 1916) that cinnamon is inhibitory to a wide range of microfungi and demonstrates preferential inhibition of aflatoxin production in laboratory medium and in cinnamon bread. Hence foods to which cinnamon is added for culinary purposes are unlikely to support significant mould growth or aflatoxin production. If the active principle can be isolated it may find uses in other foods subject to its acceptability on toxicological and organoleptic grounds.

Other food additives may demonstrate antimicrobial activity in some circumstances e.g. polyphosphates (Hargreaves et al., 1972). In recent work in our laboratory, patented claims for antimicrobial activity of polyphosphates have frequently not been substantiated when tested in foods, although most polyphosphates tested have some antimicrobial activity *in vitro* (Hargreaves, Wood & Jarvis, unpublished). One of the more interesting aspects of polyphosphates, however, is that they can frequently potentiate the antimicrobial action of other chemical preservatives (Ozawa *et al.* 1963a,b). Our recent studies (Ashworth *et al.*, 1976) have been restricted to interactions between polyphosphates and chemical preservatives used in the 'British fresh sausage'. None of a wide range of preservatives tested could provide a viable alternative to sulphur dioxide, when used in sausage formulations made without polyphosphate. However, in the presence of certain phosphate preparations, antimicrobial activity of all preservatives tested was enhanced. The best mixture was found to be sorbic acid (1000 p/m), nisin (400 p/m) and polyphosphate (2.5% w/w). With this, both the lag period for initiation of growth and the mean population doubling time was extended for all important groups of micro-organisms found in sausages (e.g. *Microbacterium thermosphactum,* coliforms, lactic acid bacteria, pseudomonads, etc.) Examples of some of the results obtained are illustrated in Table 12. Sausages prepared with this preservation system could be stored longer at 5 or 15° than sausages prepared concurrently with sulphur dioxide as preservative. The sausages were organoleptically acceptable to a taste panel. Studies of preservation systems incorporating polyphosphate are presently being made in other food products.

6. Future Trends in Chemical Preservation of Foods

At the present time it is difficult to envisage total prohibition of chemical preservatives for foods. However, the range of preservatives available to the manufacturer of non-acid perishable foods is severely limited and the uses of some of these preservatives may become more restricted in the future. There is a need, therefore, for new preservatives which are toxicologically and technologically acceptable, or for new concepts in the application of existing preservation systems. Details of new food preservation systems are patented regularly (Pintauro, 1974) although few succeed in overcoming the legislative hurdles of the various national food laws. Suggestions for new preservatives range from naturally-occurring antimicrobials isolated from plant and animal materials, to naturally-occurring 'food antibiotics' (see Marth, 1966) and to synthetic antimicrobials which are unsuitable for clinical applications. It has been suggested previously (Jarvis & Morisetti, 1969) that amongst the many thousands of pharmaceutical industry rejects some compounds of potential value to the food industry may exist. Unfortunately access to the pharmaceutical industry data banks is rarely

Table 12

Interactions between preservatives and polyphosphate in 'British fresh sausage' stored at 5°

Preservative	Initial concn. (p/m)	Polyphosphate†	Total plate count		Coliform count*		*Microbacterium thermophactum* count	
			Lag time (h)	DT (h)	Lag time (h)	DT (h)	Lag time (h)	DT (h)
None	—	absent	<24	10	<24	43	<24	11
None	—	present	<24	12	<24	13	<24	10
Sulphur dioxide	450	absent	<24	19	192	ND	72	18
Sulphur dioxide	450	present	<24	55	>336	ND	72	23
Sorbate	1000	present	72	48	192	ND	72	18
Nisin	400	present	48	25	48	41	72	13
Sorbate-nisin	1000/400	present	72	115	>336	ND	192	75

ND, not determined because of the length of the lag phase.
DT, population doubling time.
* Presumptive coliform count.
† 2500 p/m; average chain length 10-12 phosphate residues.

available, and the recovery and sorting of information would be very expensive.

An approach to chemical preservation which is possibly worthy of further consideration is the modification of traditional processes, such as fermentation of foods, and their subsequent application in a new technology. Such an approach may be seen in the work of Christiansen et al., (1975), who used a lactic acid fermentation of Thuringer sausages to prevent growth of *Clostridium botulinum* in the presence of lowered levels of sodium nitrite. An approach in which new technological methods are used to develop a preservative *in situ* can be seen in the work of Reiter (1975) in which hydrogen peroxide is developed in milk from the action of glucose oxidase or lactoperoxidase in the presence of thiocyanate. However, the legal implications of such processes are difficult to assess in advance of specific applications for their use in foods.

The biblical quotation
"There is death in the pot" (2 Kings iv, 40),
cited by Accum (1820) in the foreword to his book on chemical adulteration of foods, has been the watchword to those who would totally prohibit the addition of any 'chemical' to foods. One must applaud the work of those who pioneered the development of food legislation since the long-term effects have been to make food safer today than at any previous time. Nonethless, one must ensure that the prohibitionists do not succeed in changing public and official attitudes to the extent that preservatives become totally prohibited, lest "death in the pot" should result from avoidable microbial growth in foods.

7. Summary

The chemical preservation of foods has been practised in an unsophisticated manner for many hundreds of years in processes such as the fermentation, drying, salting and smoking of meats, fish, and other food commodities. Chemical preservation may be achieved to some degree either by the development of preservatives within the food itself (e.g. lactic acid fermentation) or by deliberate addition of chemical 'preservatives' to food commodities. The widespread adulteration of foods during the eighteenth and nineteenth centuries led to legislation in many countries to control the addition to foods of 'foreign substances'. However, it was not until the late nineteenth and early twentieth centuries that legislation was introduced to control the addition of the diverse chemical preservatives which were being used to prevent food spoilage or to mask poor quality raw materials or poor hygienic processing conditions. In the U.K. many traditionally accepted food preservatives (e.g. salt, acetic and lactic acids, sugar, glycerol, spices, etc) were specifically excluded from the Regulations whilst others (e.g.

saltpetre) were initially excluded but were covered by subsequent Amendments. The changes in U.K. Regulations which have occurred over the past 50 years are considered briefly and the present-day legislation in European countries summarized for specific preservatives in typical food commodities.

The practicalities of the chemical preservation of foods is governed not only by legislative restrictions but also by the nature of the food commodity and its method of preparation, processing and distribution. Examples of interactions between added preservatives and the intrinsic and extrinsic properties of the food are discussed to illustrate the optimization of preservative action. The use of multiple preservation systems and of 'natural preservatives' are considered in relation to microbial resistance and to possible future trends in consumer attitudes and international legislation.

8. Acknowledgments

The authors are indebted to Mr N. R. Jones and Miss D. Flowerdew for helpful discussions on the U.K. food legislation.

9. References to Cited Food Legislation for England and Wales

Adulteration of Food & Drink Act (1860).
The Sale of Food and Drugs Act (1875).
The Sale of Food and Drugs Act (1899).
The Butter and Margarine Act (1907).
The Public Health (Regulations as to Food) Act (1907).
The Public Health (Milk and Cream) Regulations (1912).
The Public Health (Preservatives, etc., in Food) Regulations (1925). S.R. & O. 1925/775.
Amendments to S.R. & O. 1925/775:
   S.R. & O. 1926/1557
   S.R. & O. 1927/557
   S.R. & O. 1940/633
   S.I. 1953/1610
   S.I. 1953/1820
   S.I. 1958/1319
   S.I. 1958/2167.
The Preservatives in Food Regulations (1962). S.I. 1962/1532.
Amendment to S.I. 1962/1532: S.I. 1971/882.
The Preservatives in Food Regulations (1974). S.I. 1974/1119.

10. References

ACCUM, F. (1820). *A Treatise on Adulterations of Food and Culinary Poisons.* London: Longman, Hurst, Rees, Orme & Brown.
AMOS, A. J. (1960). *Pure Food and Pure Food Legislation.* London: Butterworths.
ANON. (1924). *Final Report of the Departmental Committee on the Use of Preservatives and Colouring Matters in Food.* Ministry of Health. London: H.M.S.O.
ANON. (1969). *Report of the Joint Committee on the Use of Antibiotics in Animal Husbandry and Veterinary Medicine.* London: H.M.S.O.
ANON. (1974a). *Date Marking of Food. Interim Report of the Steering Group on Food Freshness.* Ministry of Agriculture, Fisheries & Food. London: H.M.S.O.
ANON. (1974b). Toxicological evaluation of certain food additives with a review of general principles and of specifications. *World Health Organization Technical Report Series* No. 539. Rome: WHO.
ANON. (1975). Potential intake of food additives. Paper CX/FA 75/5 prepared by Codex Secretariat, Codex Alimentarius Commission. Rome: WHO/FAO.
ASHWORTH, J., HARGREAVES, L. L. & JARVIS, B. (1973). The production of an antimicrobial effect in pork heated with nitrite under simulated commercial pasteurization conditions. *Journal of Food Technology* **8**, 477-484.
ASHWORTH, J., HARGREAVES, L. L., WOOD, J. M. & JARVIS, B. (1974). The replacement of sulphur dioxide as the microbial preservative in British fresh sausages. *Leatherhead Food R. A. Research Reports* No. 207.
BACHMANN, F. M. (1916). The inhibiting action of certain spices on the growth of micro-organisms. *Journal of Industrial and Engineering Chemistry* 8, 620-623.
BULLERMAN, L. B. (1974). Inhibition of aflatoxin production by cinnamon. *Journal of Food Science* **39**, 1163-1165.
BURKE, C. S. (1970). *Overseas Food Legislation Manual* Leatherhead: Leatherhead Food R. A.
CHRISTIANSEN, L. N., TOMKIN, R. B., SHAPARIS, A. B., JOHNSTON, R. W. & KAUTTER, D. A. (1975). Effect of sodium nitrite on *Cl. botulinum* growth in a summer-style sausage. *Journal of Food Science* **40**, 488-490.
HARGREAVES, L. L., WOOD, J. M. & JARVIS, B. (1972). The antimicrobial effect of phosphates with particular reference to food products. *B.F.M.I.R.A. Scientific and Technical Surveys* No. 76.
HARMAN, D. F., HOCKING, A. D., PITT, J. I. & WARTH, A. D. (1974). Action of preservatives on yeasts. *CSIRO Division of Food Research Report,* 1973-4, 9.
HAUSNER, A. (1902). *Manufacture of Preserved Foods and Sweetmeats.* London: Scott, Greenwood & Co.
INGRAM, M. (1976). The microbiological role of nitrite in meat products. In *Microbiology in Agriculture, Fisheries & Food,* eds Skinner, F. A. & Carr, J. G. Society for Applied Bacteriology Symposium Series No. 4. London & New York: Academic Press.
INGRAM, M., BUTTIAUX, R. & MOSSEL, D. A. A. (1964). The choice of anti-microbial food preservatives. In *Microbial Inhibitors in Food,* ed. N. Molin. Stockholm: Almqvist & Wiksell.
JARVIS, B. & MORISETTI, M. D. (1969). The use of antibiotics in food preservation. *International Biodeterioration Bulletin* **5**, 39-61.
JARVIS, B. & WALTERS, C. L. (1972). Nitrites in trouble. *Nature, London* **240,** 171.
LIVERSEEGE, J. F. (1932). *Adulteration and Analysis of Food and Drugs.* Ch. 9. London: Churchill.
MARTH, E. H. (1966). Antibiotics in foods-naturally occurring, developed and added. *Residue Reviews* **12**, 65-161.
NUNHEIMER, T. D. & FABIAN, F. W. (1940). Influence of organic acids, sugars and sodium chloride upon strains of food poisoning staphylococci. *American Journal of Public Health* **30**, 1040-1049.
OZAWA, T., NAGAOKA, S. & ARAGAKI, M. (1963a). Interaction of additives with food preservatives. II. Effect of condensed phosphates on the antiseptic action of the

antibacterial compounds. *Shokuhin Eiseigaku Zasshi* **4**, 287-290 (through *Chemical Abstracts* 1964, **60**, 7362).

OZAWA, T., NAGAOKA, S. & ARAGAKI, M. (1963b). Interaction of additives with food preservatives. III. Synergism of condensed phosphates with tetracycline food preservatives. *Shokuhin Eiseigaku Zasshi* **4**, 332-339 (through *Chemical Abstracts* 1964, **60**, 13788h).

PINTAURO, N. D. (1974). *Food Additives to Extend Shelf Life*. Food Technology Review No. 17. New Jersey, U.S.A.: Noyes Data Corporation.

PIVNICK, H. & CHANG, P-C. (1974). Perigo effect in pork. In *Nitrite in Meat Products*, eds Krol, B. & Tinbergen, B. J. Wageningen: Pudoc.

ROBINSON, R. A. (ed.) (1931). *Bell's Sale of Food and Drugs*, 8th edn. London: Butterworths, p. 87.

SHIPP, H. L. & ALBOROUGH, M. (1954). The preservation of pickles and sauces. I. The preservative action of mixtures of acetic and lactic acids. *B.F.M.I.R.A. Research Reports* No. 59.

Subject Index

Acceptable daily intake (ADI), for food additives, 355, 356
Acetic acid, as an inhibitor of *Staph. aureus*, 358
Acetomonas spp., production of SO_2-binding compounds by, 92
Acinetobacter-Moraxella group, radiation resistance of, 242
Aerobacter, effect of cold shock on, 124, 127
Aerobacter aerogenes, growth temperature for, 114, 115
Aflatoxin production in peanuts, effect of a_w on, 222
Age of cells, effect of on freezing rate, 136
Agglutinins, as cause of inhibition of streptococci in milk, 49, 50
Alcaligenes viscosus, effect on pressure on viability of, 265
Alkylation, 75, 76, 77
Alternaria citri, minimum growth temperatures of, 113
Amino acids, interaction of with ethylene oxide, 74
Anaerobic conditions, effect of on recovery of bacteria, 320
Anaesthetic equipment, sterilization of, 79
Antibiotics, antifungal, 288, 290
   effect of pressure on inhibitory effects of, 271, 272
Antimicrobial principle, active, in sulphur dioxide solutions, 92, 93, 94
Antimicrobial spectrum, of food preservatives, 357
Antiseptics, lysis caused by, 2
Ascospores, 196, 197, 200
Aspergillus glaucus, limiting a_w for, 113
A. niger, effects of temperature on morphology of, 210, 211, 212
   requirement of for biotin, 192
A. sydowi, dimorphism in, 207
a_w, effect of on minimum growth temperature, 120
   minimal, for growth of yeasts, 283, 284
   reduced, ability to tolerate, 134
a_w-Weit-Messer, 224, 225
Azomonas insignis, survival of after freeze-drying, 332, 336

Baby products, microbiological standards for, 312
Bacillus stearothermophilus, 'apparent' inhibition of by lactoferrin, 37, 38, 39
B. subtilis spores, resistance of to pressurization, 273, 274
   resistance of to propylene oxide, 68
Bacteria, iron requirements of, 36
Bacterial membranes, leakage from, 2, 3
Bacteriological rinses, 21, 23
Bacteriostasis, leakage associated with, 3
BCG, freeze-drying of, 335, 338, 339
BCG vaccine, storage of in argon, 338
Beef, heat resistance of bacteria in, 168
Bicarbonate, effect of in milk, 39
Blankets, sterilization of, 79, 81
Blastomyces dermatitidis, dimorphism in, 204, 205
Blood sausage, a_w of, 226, 230
Bologna type sausages, a_w of, 226, 229, 230
   shelf life of, 226, 229, 230
Bremia lactucae, membrane damage in, 199
Brettanomyces, requirement for thiamine of, 96
Brucella abortus 'ring test', 50
Budding yeast-like cells, induction of in filamentous fungi, 202
Byssochlamys, spoilage potential of, 195, 196

Canned meats, a_w of, 230, 232
Carbohydrates, addition of to protective colloids, 335, 336
Carbon dioxide, in soft drinks, 71
   inhibition of growth of slime-producing bacteria by, 71
Carbonyl compounds, reaction of with amino groups of cell proteins, 336
Caustic soda, use of in the dairy industry, 15
Cell damage, effect of on minimum growth temperature, 118, 119
Cell disruption, by violent decompression, 260
Cell wall, bacterial, destruction of, 1
   of yeasts, physical rupture of, 296
Cellular membranes, of filamentous fungi, 199, 200
Chemostat, *Aerobacter* cells grown in, 127
Chitin, in the Y form cell wall, 206
Chlorhexidine, inhibition of membrane enzyme by, 3
Chlorination, of water supplies, 26, 27

Chlorine, available, disinfectants based on, 14, 15, 18, 22
　gaseous, use of in the dairy industry, 15
Chocolate, heat resistance of salmonella in, 168
Chromomycosis, fungi causing, 208
Ciders, SO_2-binding capacity of, 92
Cinnamon, inhibitory action of, 361
Citrate, effect of in milk, 39
Clostridium bifermentans, effect of high temperature and pressure on, 262
Cl. botulinum, sensitivity of to a_w, 221, 222
　type A, growth temperature for, 143
　type B, growth temperature for, 114, 143
　type E, growth of, 112, 113, 114, 116, 143
　use of lactic acid fermentation to prevent growth of, 364
Cl. perfringens, effect of cold shock on, 124, 125, 142, 144
　growth temperature for, 114, 122
　membrane enzyme inhibited by chlorhexidine in, 3
Cl. septicum, growth of at high pressure and high temperature, 263
Cl. tetani spores, in talcum powder, 310, 312
Coagulation, of bacterial cytoplasm, 6, 7, 8
Co-enzymes, interaction of SO_2 with, 96, 97
Cold shock, 121, 123, 124, 125, 126, 127, 141, 142, 144
　recovery from, 127
Colostrum, bactericidal activity of, 33, 34
　bovine, inhibition of *E. coli* by lactoferrin in, 39, 40
　bovine, specific antibodies in, 33, 34, 35
Complement antibody, nature of bactericidal activity of, 40
Compression-decompression treatment, rapid, 259
Concentration, of gaseous disinfectants, relation of to killing rate, 69
Conglutinin, in colostrum, 51
Contaminants, radioresistant, 249, 251
Contamination of toiletries, incidence rates of, 307, 308
　origins of, 308, 309
　types of, 306
Cooling rate, effect of on death of microorganisms, 135, 136
Coprinus fimetarius Fr., growth of at supraoptimal temperatures, 192
Cryoprotectants, 138, 139

Cryptococcus, lysis in, 211
Culture media, supercooling of, 134
Cyanide, inhibition of streptococci in milk by, 42, 43
Cytoplasm, inactivation of, 1, 6, 7
Cytoplasmic membrane, alterations in during thermal injury and recovery, 179, 180, 181
　inactivation of, 1, 2
Culture media, sterilization of, 80

D value, of *Byssochlamys* asci, 196
D values, 154, 155, 157, 163, 166, 167, 168
　for osmophilic yeasts, 284
　for purified spores and vegetative cells of a spoilage yeast, 281
D_{10} values, 242, 246, 248
Death rate, for cells of *E. coli,* effect of pressure on, 267, 271
Debaryomyces genus, tolerance of to low a_w levels, 221
Decompression, violent, cell disruption by, 260
Decontamination procedures, use of ethylene oxide for, 78, 79
Degree of injury, effect of on recovery time of cells, 322
Desiccation, of yeast strains, 284
Detergent-disinfectants, test procedure for, 16, 18, 19
Deuterium oxide, effect of on bacteria at increased hydrostatic pressure, 272
Dextrin, as a protective colloid, 334
Differential plating, estimation of thermal injury by, 169, 171, 172
Diluent, used in counting procedures, effect of on recovery, 321
Dimorphism, 204, 205, 206, 207
2, 4-dinitrophenol, as an uncoupling agent, 4
Disinfectants, coagulation of cytoplasm caused by, 6, 7, 8
　gaseous, mechanism of action of, 72
　gaseous, physical properties of, 62, 63
　membrane active, 5
Disruption of yeast cells, 296, 297
Dissociation constants, for food preservatives, 358
DNA breakages, in cold-shocked cells, 127, 141, 142
　damage and repair, during thermal injury and recovery, 182, 183, 184
　synthesis, effect of pressure on, 259
Drugs, sterilization of, 79
Dry fruits, packaged, sterilization of, 80

SUBJECT INDEX

Drying, effect of on radiation resistance, 247
   injury caused to bacterial cells during, 332
Dynamic contamination, 306, 308, 310, 313

Eaton press, 296
EG & G hygrometer, modified, use of to determine a_w of meats, 225
Egg, heat resistance of *Salmonella oranienburg* in, 167
Eggs, fumigation of, 80
Enterococci, effect of a_w on heat resistance of, 232
Enterotoxin production, by *Staph. aureus*, effect of a_w on, 222
Enzyme thermostability, in filamentous fungi, 197
Enzymes, active against yeast cell walls, 297
   fruit, inhibition of action of by adding SO_2 to grape musts or fruit juices, 90
   interaction of SO_2 with, 96, 97
Equilibrium relative humidity of meats, determination of, 223
Erythrocytes, bovine, agglutination of, 50
   guineapig, 33
   sheep, 33
Escherichia coli, cold shock in, 123, 124, 126
   counts of in suckled and artificially fed guinea-pigs, 53
   destruction of by lactoperoxidase system, 46
   DNA, damage to during thermal injury, 183
   effect of deuterium oxide on, at increased pressure, 272
   effect of ethylene oxide on, 66
   effect of freezing on, 138, 139, 140
   effect of nitric oxide on, 71
   effects of pressure and temperature on growth cycle of, 265, 267, 268
   effects of pressure on pH of cultures of, 268, 269, 270
   freeze-dried, 333, 337
   glycolysis in, 97
   inhibition of by addition of lithium chloride to meat, 230
   inhibition of by lactoferrin in bovine colostrum, 39, 40, 41
   inhibition of by pressure, 258, 259
   inhibition of growth of, 36
   inhibition of strains of by secretion from non-lactating udder, 39, 40
   leakage from, 3
   human serotypes of, specific antibody against found in bovine colostrum, 35
   radiation resistance of various strains of, 242, 244, 247, 248, 251
   rapid compression-decompression treatment of cells of, 260
   recovery of from injury, 319, 320, 321
   SO_2-induced mutants of, 97
   tolerance of to QAC, 19
Ethylene glycol, conversion of ethylene oxide to, 65
Ethylene oxide, antibacterial activity of, 63, 65, 66
   bacterial resistance to, 65
   mechanism of bactericidal action of, 72
   use of, 78, 79, 80
   used to decontaminate articles handled by tuberculous patients, 65, 78, 79
Exponential phase cells, heat reistance of, 163

F value, of canned meats, relation of a_w to, 230
Faecal flora, of suckled and of artificially fed babies, 53
Faecal streptococci, heat resistance of, 157
Fentichlor, inhibition of energy dependent substrate uptake by, 5
   leaking of cytoplasmic constituents caused by, in *E. coli,* 7
Fermented sausages, importance of a_w for processing of, 228, 229
Filament formation, inhibitor of, 259
Filamentous fungi, effects of temperature on growth of, 192, 193, 194
Fish, frozen, spoilage of, 132, 133
Flagella, lack of in pressurized bacteria, 261
Food-poisoning bacteria, a_w sensitivity of, 221, 222
Foods, acceptable SO_2 levels in, 104, 105
   legislation concerning, 345, 346, 347, 364, 365
   pressure treatment of, 272, 273, 274
   recovery of bacteria in, 321, 322, 323
   use of propylene oxide for processing of, 80
Formaldehyde, antibacterial activity of, 69, 70
   mode of action of, 77, 78
   use of for disinfection of blankets, 81
   use of in poultry houses, 81, 82
Formalin, use of in dairy industry, 15
Freeze-dried cells, storage of in air, 338
Freeze-drying, lethal effects of, 330, 335
   of yeasts, 284

SUBJECT INDEX

Freezing, change in mechanism of, 135
  changes during, 128, 129, 142
  effect of on radiation resistance, 246
  effect of on yeast, 284
  rate of, 126, 135, 136, 142
Freezing injury, causes of, 331
Freezing zone, growth rate in, 131, 132
Frozen foods, bacterial flora of, 132, 133
  pH of, 129
  effect of a_w of on growth of microorganisms in, 227
Fruit beverages, fermented and non-fermented, uses of SO_2 in, 90
Fruits, frozen, absence of bacteria on, 132, 133
Fusarium moniliforme, dimorphism in, 207

Gaseous disinfectants, mechanism of action of, 72
  physical properties of, 62, 63
Gelatin, use of as a suspending medium, 334
Glass beads, use of to homogenize yeasts, 296
α-Glucan, in the Y form cell wall, 206
Glucose, addition of to nutrient broth, 335, 336
Glutamate, addition of to suspending media, 335
Glutaraldehyde, bactericidal effect of, 71
  mode of action of, 78
Glycerol, promotion of supercooling by, 135
Glycidaldehyde, antibacterial effect of, 70, 71
Gram negative bacteria, survival of after freeze-drying, 339
  survival of after freezing, 140, 142, 145
Gram positive bacteria, effect of ozone on, 71
  survival of after freeze-drying, 339
  survival of after freezing, 138, 140
  tolerance of to low a_w levels, 221
Grande strains, of yeasts, 281
Graphical interpolation method, 223
Growth cycle, of *E. coli*, effects of pressure and temperature on, 265, 266
Growth rate, in freezing zone, 130, 131, 132
Growth temperature, minimum, 112
Guanine, interaction of with ethylene oxide, 74, 75
Guinea-pigs, suckled and artificially fed, *E. coli* counts in, 53

Hair hygrometers, modification of to measure a_w of meat products, 223, 224
Halobacterium halobium, growth temperature for, 114
Ham, raw, low a_w level required for, 226, 229
Hansenula, heat resistance of, 281
Heat injury, recovery from, 318, 319
Heat-resistant fungi, 195, 196
Heating menstruum, composition of, 163, 164
Hexachlorophene, inhibition of electron transport chain of *B. megaterium* by, 3
Histoplasma capsulatum, dimorphism in, 204, 205
Homogenization, of yeast, 296
Hughes press, 296
Humicola, temperature–lipid relationship in, 198
Hydrogen peroxide, inhibitory activity of lactoperoxidase in the presence of, 43, 44
Hydroxysulphonates, formation of, 91
Hypha, three zones of a, 201, 202

Inactivation dose, of radiation, 241
Incubation temperature, effect of on heat resistance, 162
Inhibitors, for yeasts, 288, 297
Inhibitory oxidation product of SCN^-, 45
Iodine, available, disinfectants based on, 15, 19
Iodophors, use of in the dairy industry, 15, 16, 22, 24
Incubation temperature/cold shock difference, 125
Ions, interaction of with hydrostatic pressure, 270
Iron requirements, of bacteria, 36

Killer cells, in yeasts, 290, 294
Klebsiella aerogenes I, failure to reduce resazurin by, 33
Kl. pneumoniae, freeze-drying of, 335
  in a hand-cream dispenser, 310

L-drying method, 332
Lactic acid, as an inhibitor of *Staph. aureus*, 358
  use of to prevent growth of *Cl. botulinum*, 364
Lactic acid bacteria, inhibition of by lactoperoxidase system, 45

SUBJECT INDEX

Lactobacilli counts, in suckled and artificially fed guinea-pigs, 53
Lactobacilli, effect of a_w on heat resistance of, 232
 growth of in sulphited cider, 101
Lactobacillus bifidus, freeze-dried, 334, 338
L. plantarum, effect of SO_2 on, 101, 102, 103
L. viridescens, growth temperature for, 114, 116
Lactoferrin, distribution of, 37
Lactoperoxidase system, preservation of milk by, 47, 48
Lag period, effect of pressure on duration of, 265
LD_{90} values, 241
Leakage, of cytoplasmic constituents, 7
 of small molecular weight substances from cytoplasmic membrane, 2, 3
Legislation, on use of preservatives in foods, 345, 346, 348
Lesions, ultrastructural, in bacteria and erythrocytes, 40
Lipids, effect of temperature on, in filamentous fungi, 198
 interaction of SO_2 with, 100
Lisbôa Tube Test, modified, 16, 18, 19
Lithium chloride, effect of addition of on a_w of meat, 230
Liver sausage, a_w of, 230
Low temperature steam, disinfectant use of, 81
Luminescence, of *Photobacterium phosphoreum,* effect of hydrostatic pressure on, 269, 270
β-Lysin, bactericidal property of, 48, 49
Lung ventilators, sterilization of with ethylene oxide, 79
Lysis, of *E. coli,* staphylococci and streptococci, 2
Lysozyme content, in milk, 35, 36

Mastitis, development of, 52
Mastitis pathogens, elimination of, 24, 25, 28
Mastitis streptococci, inhibition of in milk, 42
Meat products, a_w of, 225, 226
 shelf life of, 226
 shelf life of related to a_w, pH and temperature, 233
Meats, measurement of a_w of, 223, 224, 225
 significance of a_w for, 225
Media, suspending, for freeze-drying, 337

Medium, growth, effect of on heat resistance of organisms, 157, 162
 recovery, 320
Melamine formaldehyde, 69
Membrane damage, in filamentous fungi, 199, 200, 201
Membrane electrochemical potentials, attenuation of, 4
Membrane enzymes, inhibition of, 2, 3
Membrane filter technique, for recovery of injured bacteria, 324
Membrane filtration, for assessing bacterial contamination in toiletries, 307
Membrane structure, of bacterial cells, 2, 5
Meso-inositol, use of in freeze-drying, 336
Mesophiles, growth temperatures for, 129
Mesophilic bacteria, heat resistance of, 157
Metabolic damage, during thermal injury, 181, 182
Methyl bromide, antibacterial effect of, 70
 mode of action of, 78
Microbiological standards for toiletries, 311, 312, 313
Micrococcus radiodurans, 242, 244, 247, 251
M. radiophilus, 244, 247
Micro-organisms, in fermented fruit juices, effects of SO_2 on viability of, 101
Milk, pressure treatment of, 272, 273
 use of as a suspending medium, 334
Milk powder, artificially contaminated with salmonellae, 322, 323
Minimal medium recovery, 320
Minimum growth temperature for foodborne pathogens, 143, 144
'Mist. desiccans', formulation of, 335, 336
Mitchell chemiosmotic hypothesis, 4, 6
Monilinia fructicola, membrane damage in, 199
Morphology, of filamentous fungi, effects of temperature on, 201, 202
Mortadella, shelf life of, 226, 230
Motility of bacteria, effect of pressure on, 260, 261
Moulds, on frozen foods, 132, 133
 tolerance of to a_w, 221
MPN determinations, 323, 324
Musts, from grapes, addition of SO_2 to, 90
Mutants, radioresistant, 247
Mutations, SO_2-induced, 97, 98
M ⇌ Y dimorphism, in filamentous fungi, 204, 205, 206, 207
Mycobacterium smegmatis, effect of ethylene oxide on, 65, 66
Mycotoxin production, a_w requirements for, 222

Mycotypha species, dimorphism in, 207
Myrothecium verrucaria, effect of heat on spores of, 200

NaCl tolerance, of micro-organisms, calculation of a_w from, 221
Narcotics, influence of on luminescence of *Photobacterium phosphoreum* under increased hydrostatic pressure, 270, 271
Neisser-Wechsberg effect, 35
Neisseria gonorrheae, freeze-drying of, 335
Neonate, resistance of to enteric infections, 53
Neurospora crassa, effect of temperature on growth of, 193, 197, 199, 200, 212
Nitrate in foods, permitted levels, 347
Nitric acid, use of in dairy industry, 16
Nitric oxide, effect of on *E. coli,* 71
Nitrite levels, permitted, 347
Nitrogen, storage of bacteria in, 338
Non-hygroscopic surfaces, sensitivity to ethylene oxide of bacteria dried on, 67
Non-lactating udders, inhibition of strains of *E. coli* by secretion from, 39, 40
Non-selective media, use of in resuscitation procedures, 324
Nucleic acids, interaction of SO_2 with, 97, 98
  reaction of with formaldehyde, 77
Numerical limits, for micro-organisms in toiletries, 312, 313
Nystatin, use of, 288, 289, 290

Olefinic compounds, reaction of SO_2 with, 92
Ophthalmic instruments, sterilization of, 79
Organic matter, decrease of bactericidal effect of formaldehyde by, 69
Orthophenylphenol residues, 347
Osmophilic yeasts, growth of, 283, 284
Osmotic damage to cell membrane, 135, 136, 141, 142
Osmotic pressure, effect of on sensitivity of bacteria to hydrostatic pressure, 270
Oxygen, effect of on dried bacteria, 333, 337, 338
  influence of on effect of ionizing radiation, 244, 245
Ozone, bactericidal effect of, 71

PA tests, 323, 324
Paecilomyces varioti, heat resistance of, 195
Palm wine, pH value of, 94

Paracoccidioides brasiliensis, Y → M conversion in, 204, 205, 206
Paraformaldehyde, use of, 69, 70
Paraformaldehyde-produced formaldehyde gas, 70
Partition coefficient, of food preservatives, 358
Pasteurella septica, bactericidal activity of complement/antibody system against 40
Pathogenic fungi, induction of budding, yeast-like growth phase in, 206, 207
Peanuts, aflatoxin production in, 222
Penicillin, sterilization of, 79
Penicillium, tolerance of genus to low a_w levels, 221
Penicillium lilacinum, dimorphism in, 207
P. urticae, induction of microcycle conidiation in, 212
Petites, induction of, 281
pH, effect of on minimum growth temperature, 114, 115, 117
  effects of hydrostatic pressure at different values, 268, 269, 270
  of frozen foods, 129, 133, 135, 139
  of heating menstruum, 168
  optimal, for recovery, 320
  relation of to antimicrobial effect of acid antiseptics, 93, 94
pH gradient, 4, 5
pH range, for growth of yeasts, 284, 285
Phagocytin, bactericidal activity of, 49
Phagocytosis, 49, 50, 51, 52
Phosphate, effect of on heat resistance of bacteria, 166
Phosphate buffers, effect of on resuscitation, 321, 324
Photobacterium phosphoreum, effect of hydrostatic pressure on luminescence of, 269, 270
Plant extracts, antimicrobial activity of, 361
Plate count for bacteria in toiletries, 307
Pleomorphism, non-filamentous, 259, 260
Polyphosphates, antimicrobial activity of, 361, 362
  effect of on heat resistance of bacteria, 166
Pork, heat resistance of bacteria in, 168
Postcolostral milk, bovine, detection of complement (C) in, 33, 34
Potential Daily Intake (PDI) for common food preservatives, 356
Poultry houses, use of formaldehyde in, 81, 82
Pre-enrichment medium, nutrients in, 324
Premises, disinfection of, 81

SUBJECT INDEX

Preservatives, food, arguments for and against use, 354, 355
  desirable characteristics for, 357, 358
  legal control in use of, 346, 347, 348, 354, 355
Pressure, sensitivity of bacteria and yeasts to, 262
Pressure-induced filamentation, 258, 259
Properdin, in milk, 50
β-Propiolactone, antibacterial activity of, 68
  mode of action of, 76, 77
  possible carcinogenicity of, 80
  uses of, 80
Propylene oxide, antibacterial activity of, 67
  mode of action of, 76
  uses of, 80
Protective agents, against freezing, 137, 138
'Protective colloid', role of in freeze-drying, 334, 335
Protein, interaction of with formaldehyde, 77
Proteins, interaction of with ethylene oxide, 74
  pretreatment of urea with, 95
Proteus vulgaris, effect of pressure on viability of, 265
Prozone effect, in colostrum, 35
Pseudomonas aeruginosa, isolation of from toiletries, 311
Ps. fluorescens, growth temperature for, 115, 116
  heat resistance of, 167
  recovery of from thermal injury, 178, 179
  tolerance of to QAC, 19
Ps. perfectomarinus, inhibition of by pressure, 258
Ps. putrefaciens, waters polluted with, 27
Psychrophiles, growth temperatures for, 131, 134
  heat resistance of, 157
Psychrophilic yeasts, growth temperatures for, 282
Psychrotrophs, ability of to survive frozen storage, 140, 144
  heat resistance of, 157
  multiplication of in stored milk, 47

Quaternary ammonium compounds (QAC), use of in the dairy industry, 15, 19, 20, 27
Quaternary ammonium detergents, damage to bacterial cell membrane by, 2
Quinones, inhibitory effect of, 43

Radiation, effects of on yeasts, 285, 286, 287, 288
Radiation sensitive species, 241, 242
Radiation survival curves, 240, 241
Radioresistant mutants, development of, 247, 248, 249
Radioresistant organisms, 244, 247, 251
Rate of freezing, importance of, 126, 135
Raw ham, low a_w level required for, 226, 229
Reactivity, of food preservatives, 358
Recovery of organisms, optimal temperature for, 320
Recovery time, factors affecting, 321, 322, 323
Rehydration, of freeze-dried bacteria, 339
Relative humidity, effect of on action of glycidaldehyde, 71
  effect of on action of β-propiolactone, 68
  influence of on action of ethylene oxide, 65, 76
  influence of on action of propylene oxide, 67
Resistance, of micro-organisms to food preservatives, development of, 359
  of yeasts to radiation, 286, 288
Resuscitation, of freeze-dried bacteria, 339
Resuscitation requirements, of different organisms, 319
Reversible inactivation, 123
  during frozen storage, 138, 139, 140, 141
Rhizopus stolonifer, membrane damage in, 199
Ribonuclease activity, stimulation of, 176
Ribosomal precursor particles, identified during recovery from thermal injury, 178, 179
Ribosomes, thermal denaturation of, 173, 175, 176, 179
RNA polymerase inhibitor, 204
RNA synthesis, during recovery from thermal injury, 173, 176, 177, 178
  effect of pressure on, 259
rRNA, degradation of during thermal injury, 173, 175, 176, 179
Rupture of cell wall, of yeasts, 296, 297

Saccharomyces, heat resistance of, 281
Saccharomyces cerevisiae, effect of pressure on pH range of growth for, 268, 269
  effect of SO_2 on, 103
  effect of superoptimal temperature on, 282
  glycolysis in, 97

Saccharomyces cerevisiae—cont.
  pH optimum for, 284, 285
  radiation sensitivity of, 287
  resistance of to antimicrobial substances, 288, 289
  suicide mutants of, 291
  survival of after freezing, 138
  treatment of with sodium metabisulphite, 93, 94
  use of solvents to destroy vegetative cells of, 295
Sacch. rouxii, ability of to grow at low a_w, 283, 284
Salmonella anatum, freeze injured, recovery of, 320
  heated, survival of, 166, 168
  repair in after freezing and thawing, 139
Salm. enteritidis, survival of, 123
Salm. oranienburg, heat resistance of, 167
Salm. senftenberg cells, effect of ethylene oxide on, 66
Salm. typhimurium, destruction of lipid species during injury, 180
  differential plating, curve for, 171, 172
  effects of thermal injury on TCA cycle enzymes of, 182
  incubated in phosphate buffer, 175
  radioresistant mutants of, 248
  recovery of from heat injury, 177, 178, 179, 180, 181, 182, 183, 318, 319
Salmonellae, elimination of from foods by radiation, 251
  growth temperatures for, 114
  recovery of, 322, 323
Salt tolerance, return of during recovery from thermal injury, 180, 181
Salts, effect of at low temperatures, 122, 123, 137, 144
Sausages, fermented, importance of a_w of, 228, 229
  polyphosphates in, 362
Sclerotic cells, produced by dematiaceous fungi, 208
Selective media, use of in resuscitation procedures, 324
Selectivity of medium, effect of on recovery, 321
Serpula lachrymans, membrane damage in, 199
Serratia marcescens, cells of grown at ordinary pressures in seawater broth, 258
  effect of pressure on motility of, 260
  freeze-drying of, 334
  radiation resistance of, 242, 245
SINA sensor, 224, 225

Skimmed milk, heat resistance of bacteria in, 167
Slime-producing bacteria, inhibition of growth of by carbon dioxide, 71
Sodium chloride, effect of addition of on a_w of meat, 225
  effect of on heat resistance of salmonellae, 163, 167
Sodium hypochlorite, use of in the dairy industry, 14, 16, 18, 22, 23, 24, 25
Sodium nitrite, as a food additive, 347
Solubility, of food preservatives, 357
Solvents, use of to kill vegetative yeast, 295
Spacecraft, sterilization of, 80
'Spherule' form, fungi with a, 208
Spoilage of toiletries, 311
Spores, effect of hydrostatic pressure on, 273, 274, 275
Sporotrichum schenckii, dimorphism in, 204
  M → Y transformation of, 205
Staphylococcus aureus, components of total electrochemical potential of, 4
  effect of ethylene oxide on, 65, 66
  effect of sublethal heating on, 172
  effects of thermal injury on TCA cycle enzymes of, 182
  leakage from, 3
  minimum growth temperature of, 117, 121
  recovery of from a toiletry, 311
  recovery of from heat injury, 177, 178, 179, 180, 181, 182, 318, 319, 320
  resistance of to cold shock, 123, 142
  resistance of to ethylene oxide, 79
  tolerance of to low a_w levels, 222
  tolerance of to QAC, 19
Staph. epidermidis, inhibition of by lactoferrin, 37, 38
Static contamination, 306, 308, 310, 313
Stationary phase cells, heat resistance of, 162, 163
Storage, factors affecting survival of cultures during, 337, 338
Streptococci, faecal, heat resistance of, 157
  Group A, absence of mastitis caused by, 52
  Group A, destruction of in milk, 42
  Group N, inhibition of in milk, 42
Streptococcus spp., in toiletry creams, 311
Streptococcus cremoris, effect of agglutinins in milk on, 49
  inhibition of strains of, 42
  resistance of, 44
Strep. faecalis, effect of tetrachlorosalicylanilide (TCS) on pH gradient of, 5

SUBJECT INDEX

inhibition of membrane enzyme in, 3
leakage from, 3
Strep. faecalis var. *liquefaciens*, minimum growth temperature for, 116
Strep. faecium, radiation resistance of, 246, 247
Strep. lactis, freeze-drying of, 334
Strep. pyogenes, bactericidal activity against, 42
  Group A, destruction of by lactoperoxidase system, 44, 45
Strep. uberis, susceptibility of to lactoperoxidase system, 52
Subzero growth zone, 129
Sucrose, effect of on heat resistance of bacteria, 164, 165, 166, 167
Sugar-tolerant yeasts, 282, 283
Sugars, effect of on freezing injury, 142, 143
Suicide mutants, of *Sacch. cerevisiae*, 291, 294
Sulphite addition-compounds in fruit beverages, 91, 92
Sulphite auto-oxidation, 100
Sulphite-neomycin agar, effect of cold shock on *Clostridium perfringens* in, 125
Sulphite reductase, isolated from micro-organisms, 99
Sulphur, microbial biochemistry of, 99
Sulphur dioxide, acceptable daily intake of, for man, 104, 105
  acceptable levels of in foods and beverages, 104, 105
  ionization of, in aqueous solutions, 90, 91
  mechanism of action of, 94, 95
  use of in foods, 347
Sulphurous acid, existence of, 91
Supercooling, of culture media, 134
Surface active agents, in toiletries, effect of on vegetative bacteria, 309
Survival curves, radiation, 240, 242
  types of, 154 155
Survival levels, of freeze-dried bacteria, 338, 340

TCS, as an uncoupling agent, 4, 5
Teat canal, as defence against bacteria, 52
Teats, cows, washing of, 24, 25, 26
Teat dips, disinfectant, 25, 26
Temperature, effect of on growth of bacteria at high pressure, 262, 263
  effects of on growth of filamentous fungi, 192, 193, 194

effects of on yeast, 280, 281, 282
optimal for recovery, 320
Temperatures during irradiation, effects of, 245, 246
Test procedure, for detergent–disinfectants, 16, 18
Thawing, rate of, 141
Thermophilic fungi, cardinal temperatures of, 192
Thermostability, of enzymes, in filamentous fungi, 197
Thiamine, requirement of micro-organisms for, 96
Thiol group, action of disinfectants on, 8
  interaction of SO_2 with, 95, 96
Tomatoes, heat resistance of micro-organisms from, 168
Toxicity, of food additives, 354, 359
Trehalase, activity of in filamentous fungi, 200
Trehalose, release of, 200
Trichinae, a_w levels required for inactivation of in raw ham and fermented sausages, 229
Tritium suicide, 292, 293, 294
Tryptophan, breakdown of, 99
Tube test for disinfectants, 16, 19, 20

'Ubiquitin', discovery of, 49, 52
Udder, dry, susceptibility of to bacterial infection, 52, 53
Udders, cows, washing of, 24
  non-lactating, inhibition of strains of *E. coli* by secretion from, 39, 40
Unbalanced growth, of yeasts, 290, 291
Unfrozen phase of substrates, 129
Uracil, converting cytosine to, 97
Urea, pretreatment of proteins with, 95
Urea formaldehyde, 69
Ureteral catheters, sterilization of, 79
UV irradiation, of yeasts, 287

Vegetative cells, methods for killing, 290, 294, 295
Viability, effect of pressure on, 265
Vibrio cholerae, freeze-drying of, 334, 336
V. fetus, freeze-drying of, 334
V. marinus, growth of at high hydrostatic pressure, 270
Vinylglycollic acid, as inhibitor of bacterial transport mechanism, 6
Vitamin binders, as antibacterial factors in milk, 51
Vitamins, interaction of SO_2 with, 97

Water activity (a_w), effect of on growth of filamentous fungi, 193
  effect of on heat resistance of salmonellae, 163, 164, 165
  influence of on minimum growth temperature, 112, 113
Water content, optimum, for dried bacteria, 337, 338
Water supplies, chlorination of, 26, 27
Wines, effect of SO_2 on, 90

Yeast glycolysis, inhibition of by SO_2, 98
Yeasts, ability of to produce SO_2, 100
  effect of temperature on cellular membranes of, 200
  on frozen foods, 132, 133
  SO_2-induced mutations in, 97
  tolerance of to a_w, 219, 220
  uptake of SO_2 by, 94, 95

z value, 154, 157, 167